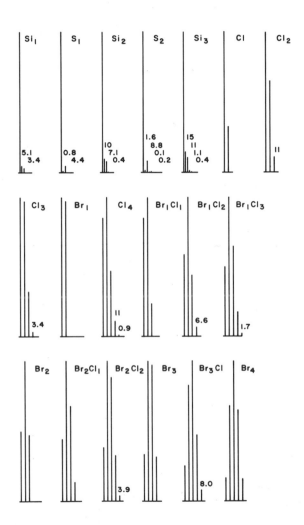

Figure 2B. Isotopic abundances for combinations of "A + 2" element compositions.

INTERPRETATION OF MASS SPECTRA

Third Edition

Organic Chemistry Series
Nicholas J. Turro, Editor

INTERPRETATION OF MASS SPECTRA

MASS SPECTRA

Third Edition

F. W. McLafferty
Cornell University

University Science Books
Mill Valley, California

University Science Books
20 Edgehill Road
Mill Valley, CA 94941

Library of Congress Catalog Card Number: 80-51179

ISBN 0-935702-04-0

Printed in the United States of America

10 9 8 7 6 5 4

to Tibby

CONTENTS

PREFACE TO THE THIRD EDITION

Since the last edition, the literature on organic mass spectrometry has more than doubled in size, and the need for interpretive expertise has grown by a far greater amount. Present worldwide instrument sales are estimated at nearly 1,000 mass spectrometers per year, and the number of unknown mass spectra which these can produce is inconceivable on the standards of the 1960s. The computerized gas chromatograph/mass spectrometer (GC/MS) is now an accepted primary tool in a wide variety of laboratories for the analysis of complex mixtures, and many of these are turning out 100 to 1,000 mass spectra per day. Liquid chromatography/MS and MS/MS promise a further large increase in the rate at which unknown mass spectra are produced. Although the time needed to identify an unknown can be reduced greatly by automated computer matching and interpretation procedures (Chapter 10), these are still only an aid to, not a replacement for, the skilled interpreter. My courses based on this book have now been given 50 times, to approximately 2,000 students and practicing chemists. Their interest and enthusiasm have been a tremendous encouragement in preparing this third edition, whose changes result from their suggestions as well as from the rapid progress in the field.

The major organizational change in this third edition is the separation of the basic and advanced material. The courses which I have presented during the last half-dozen years have been mainly divided into these two types, and several of my colleagues in other universities have suggested that a separation of the basic material would be helpful for the beginning student. The first six chapters, approximately one-third of the book, are recommended for this basic course; the remainder of the book should serve as a useful reference for the solution of actual unknown mass spectra, as well as advanced reading for the beginning student and a text for advanced courses. The largest addition to this part is the new Chapter 8 on "Detailed Mechanisms of Ion Fragmentation," which contains the advanced material, formerly in Chapter 4, expanded to reflect the greatly increased knowledge of such fragmentation reactions. Chapter 8 is a further attempt to relate such mechanisms to known physical-organic chemical principles and to classify them in a way that will make them easier to comprehend and to apply to unknown mass spectra. A short chapter on computer identification of unknown mass spectra recognizes the growing importance of such aids; in a few years such data analysis will be a real-time capability of modern MS computer systems, and this revolution in mass-spectral interpretation, unfortunately, could then demand yet another revision of this book.

Finally, I should emphasize that I remain convinced that the most important part of learning how to interpret unknown mass spectra is actually to practice interpreting mass spectra. Many unknowns which found little use have been removed, some new ones have been added, but many old favorites remain; such unknowns continue to be, in my opinion, the best teaching tool of this book. The student must attempt one or two unknowns for every 30 to 60 minutes of lecture or self-study. Try seriously to solve the unknown before checking the answers in Chapter 11; write down next to the spectral data your calculated elemental composition assignments, possible structures, and justifying mechanisms for later comparison with the book's reasoning. The solution of these unknowns should be fun; I continue to be amazed by the enthusiasm of the average student after more than fifteen years of teaching such courses.

Again, I am indebted to Dr. Willi Richter for a quote from Mao Tse-Tung that provides the *raison d'etre* for the third edition:

"We are obliged to organize the masses."

F. W. McLafferty

Ithaca, New York
September 1980

ACKNOWLEDGMENTS

The contributions of many people have made this book possible. The following helped to plan and teach the courses in which this material was developed: Drs. J. W. Amy, R. D. Board, E. Bonelli, A. L. Burlingame, M. M. Bursey, D. C. DeJongh, R. B. Fairweather, J. N. Gerber, M. L. Gross, I. Howe, K. L. Rinehart, J. W. Serum, S. Stallberg-Stenhagen, E. Stenhagen, G. E. VanLear, and Mr. W. E. Baitinger. The following postdoctoral fellows and students read the manuscript critically and worked the problems: M. A. Baldwin, M. P. Barbalas, P. F. Bente, III, R. D. Board, E. M. Chait, L. B. Dusold, W. F. Haddon, H. Hauer, K.-S. Kwok, J. G. Lawless, S. P. Levine, K. Levsen, E. R. Lory, D. J. McAdoo, D. C. McGilvery, I. K. Mun, G. M. Pesyna, R. M. Prinstein, M. Senn, J. W. Serum, T. W. Shannon, P. J. Todd, R. Venkataraghavan, T. Wachs, C. G. Warner, and P. C. Wszolek. Drs. G. M. Pesyna, S. L. McLafferty, and I. K. Mun drew the bar graphs; L. MacCaskill and J. Scriber drew the figures; L. M. Lawrence and C. B. Cook typed much of the manuscript; Dr. H. M. Rosenstock checked the ionization and appearance energy values; and Professor N. Turro checked the manuscript in detail. Most of this edition was written during a sabbatic leave at the University of Cambridge, England, and the Ecole Polytechnique, France; I am deeply indebted to Drs. D. H. Williams, R. D. Bowen, R. Raphael, J. M. Thomas, P. Knewstubb, M. Fetizon, P. J. Arpino, P. Longevialle, H. E. Audier, A. Milliet, G. Guiochon, and their colleagues for their warm hospitality, and to Churchill College, Cambridge, for an Overseas Fellowship.

GLOSSARY AND ABBREVIATIONS

$\overset{+}{\cdot}$	Radical cation, odd-electron ion (for example, $CH_4^{+\cdot}$).
⤸ (full arrow)	Transfer of an electron pair.
⤻ (fishhook)	Transfer of single electron.
[ ]	Relative abundance of the ion within the brackets.
α, alpha cleavage	$R\text{─}\!\!\!\!\!\backslash C_\alpha\text{─}\overset{\cdot+}{Y}$; cleavage of a bond on an atom adjacent to the atom bearing the odd electron (but *not* the bond to the latter atom).
A	Appearance energy; formerly appearance potential.
"A" element	Monoisotopic element (hydrogen is also considered to be an "A" element).
"A + 1" element	Element with an isotope whose mass is 1 amu above that of the most abundant isotope, but which is not an "A + 2" element.
"A + 2" element	Element with an isotope whose mass is 2 amu above that of the most abundant isotope.
A peak	Peak whose main elemental formula is composed of only the most abundant isotopes.
(A + 1) peak	The peak one mass unit above the A peak.
base peak	Peak representing the most abundant ion in the spectrum.
CA	Collisional activation.
CI	Chemical ionization.
cyclization, *rc*	Reaction in which a cyclized product (either the ion or neutral) is formed.

dalton	An atomic mass unit ($^{12}C = 12$ daltons).
daughter ion	The product of an ionic reaction.
displacement, rd	Reaction in which cyclization to form a new bond at a carbon (or other) atom results in the loss of another group attached to that carbon atom.
elimination, re	Reaction in which cyclization to form a new bond between two parts of an ion results in the loss of the actual group connecting these parts.
$E_o(M^{+\cdot} \to D^+)$	Critical energy for the reaction $M^{+\cdot} \to D^+$; also called activation energy, E_a.
$E_s(M^{+\cdot} \to D^+)$	Ion internal energy required so that half of $M^{+\cdot}$ ions will decompose to yield D^+ before leaving the ion source.
EE$^+$, even-electron ion	Ion in which the outer-shell electrons are fully paired; a "closed shell" ion.
EI	Electron ionization.
eV	Electron volt $= 96.487$ kJ/mol $= 23.06$ kcal/mol.
FI	Field ionization.
GC/MS	A gas chromatograph interfaced to a mass spectrometer.
I	Ionization energy; formerly ionization potential.
i	Inductive initiation of a reaction through electron withdrawal by the charge site.
isobaric	Of the same nominal mass but of different elemental compositions.
isotopic peak	Peak whose elemental composition contains an isotope not of highest natural abundance.
$k(E)$	The function describing the change in the rate constant, k, with change in the internal energy of the ion, E, for a particular ion-decomposition reaction.
LC/MS	A liquid chromatograph interfaced to a mass spectrometer.
m/z	The mass of the ion divided by its charge (usually unity); m/e has also been used.
m^*, metastable	The peak resulting from ion decompositions in a field-free drift region of the mass spectrometer.
mmu, millimass units	0.001 atomic mass unit.

$M^{\ddagger}$, molecular ion	The ionized molecule; "the molecular ion" is the peak representing the ionized molecule which contains only the isotopes of greatest natural abundance.
MI spectra	Metastable ion spectra.
MS/MS	Tandem mass analyzers used for the separation and identification of ions, such as those representing components in complex mixtures.
n-electrons	Nonbonding electrons.
nonisotopic peak	A peak whose elemental composition has only isotopes of highest natural abundance.
$OE^{\ddagger}$, odd-electron ion	Ion in which an outer-shell electron is unpaired; a radical ion.
Pa	Pascal (1 Pa = 0.0075 torr).
Parent, precursor	The decomposing ion in any reaction.
$P(E)$	The distribution function describing the probability for particular values of internal energy of an ion.
r, rearrangement	A reaction in which the molecular arrangement of the atoms in either the ionic or neutral product is not the same as that in the precursor ion.
r + db	Number of rings plus double bonds.
rc, rd, re, rH	Rearrangements involving cyclization, displacement, elimination, and hydrogen transfer, respectively.
relative abundance	The abundance (peak height) of an ion relative to the base peak in the spectrum (or, if so stated, relative to Σ_{ions}).
σ, sigma-electron ionization reaction	A simple cleavage reaction visualized as taking place through initial ionization at the sigma bond cleaved in the reaction.
Σ_{ions}	Total abundance of all ions in the spectrum.
Σ_{40}	Total abundance of all ions in the spectrum of mass 40 and above.
simple cleavage	An ion-decomposition reaction which involves cleavage of only a single bond.
z	The number of charges on an ion ("e" has also been commonly used for this definition).

INTERPRETATION OF MASS SPECTRA

Third Edition

1

INTRODUCTION

1.1 Appearance of the mass spectrum

Learning how to identify a simple molecule from its mass spectrum is much easier than identification from other types of spectra. *The mass spectrum shows the mass of the molecule and the masses of pieces from it.* Thus the chemist does not have to learn anything new—the approach is similar to an arithmetic brain-teaser. Try one and see.

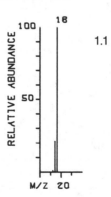

In the bar graph form of a spectrum (as in that for Unknown 1.1), the abscissa indicates the mass (actually m/z, the ratio of mass to the number of charges on the ions employed), and the ordinate indicates the relative abundance. If you need a hint, remember that the atomic weights of hydrogen and oxygen are 1 and 16, respectively. Check your answer (the solutions to the unknowns are given in Chapter 11).

Now try another simple spectrum, Unknown 1.2. Your structure will be correct if the molecule and its pieces have masses corresponding to those of the spectrum. (Make a serious attempt to solve each unknown before looking at the solution. This is a vital part of the book's instruction.)

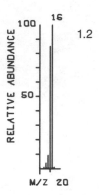

Unknown 1.3 contains carbon, hydrogen, and oxygen atoms. Obviously the possibilities for arranging these in a molecule of molecular weight 32 are limited. Compare any molecular structure possibilities with the major peaks of the spectrum.

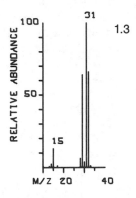

The tabular data for Unknowns 1.1, 1.2, and 1.3 include additional low-abundance peaks which can be particularly important for deducing isotopic compositions (Chapter 2). The mass spectrometer has a dynamic range much

Unknown 1.1

m/z	Rel. abund.
1	<0.1
16	1.0
17	21.
18	100.

Unknown 1.2

m/z	Rel. abund.
1	3.1
12	1.0
13	3.9
14	9.2
15	85.
16	100.
17	1.1

Unknown 1.3

m/z	Rel. abund.	m/z	Rel. abund.
12	0.3	28	6.3
13	0.7	29	64.
14	2.4	30	3.8
15	13.	31	100.
15.5	0.2	32	66.
16	0.2	33	1.0
17	1.0		

greater than the three orders of magnitude shown; that is, ion abundances of ≪ 0.1 per cent can be measured reproducibly. For the mass spectral data of the unknowns in this book, a reproducibility of ±10 per cent relative or ±0.2 absolute, whichever is the greater, will be used. With modern instruments, such reproducibility is achievable with relatively small sample sizes and fast measurement times.

The measured spectra of these compounds contained additional peaks (not shown) due to the "background" in the instrument. This arises from compounds that are desorbing from the walls of the instrument or are leaking in from various sources. To avoid confusion, such a "background spectrum" is usually run before the sample is actually introduced into the instrument. Such a spectrum is shown as Unknown 1.4. Can you identify any of the components? (For now, ignore anomalous peaks of only a few per cent relative abundance. Their significance will be discussed later.)

Unknown 1.4

m/z	Rel. abund.
14	4.0
16	0.8
17	1.0
18	5.0
20	0.3
28	100.
29	0.8
32	23.
34	0.1
40	2.0
44	0.10

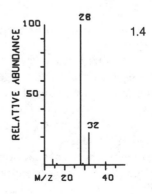

One of the components in its pure state gives the spectrum of Unknown 1.5. Can you identify it?

Unknown 1.5

m/z	Rel. abund.
12	8.7
16	9.6
22	1.9
28	9.8
29	0.1
44	100.
45	1.2
46	0.4

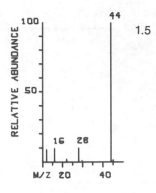

Figure 1A is an actual mass spectrum of the Martian atmosphere (after removal of CO and CO_2) from the US-NASA Viking mission.

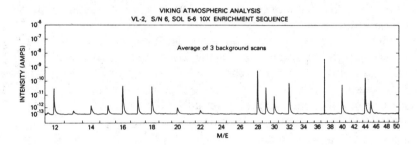

Figure 1A. Mass spectrum (log intensity scale) of gases in the atmosphere of Mars.

1.2 Formation of the mass spectrum

Of necessity this book gives little or no discussion of many important areas of mass spectrometry. Complete details of the method, techniques, and instrumentation are available in a number of excellent books (see the Bibliography). Compilations of references to the mass-spectrometry literature are also available from *Chemical Abstracts* and United Kingdom Chemical Information Service.

As in many chemical reactions used for analysis, the basic purpose of the mass spectrometer is to convert the sample into measurable products that are indicative of the original molecule. The products formed are also rather unusual: gaseous positive ions, whose masses and relative abundances are displayed in the mass spectrum. (Negative ions are generally less useful as molecular fragments because of lower specificity and sensitivity.) These are formed in the ion source, depicted schematically in Figure 1B. The background pressure (pressure without any sample) in the source is $\sim 10^{-5}$ Pa (7.5×10^{-8} torr), or 10^{-10} atm.

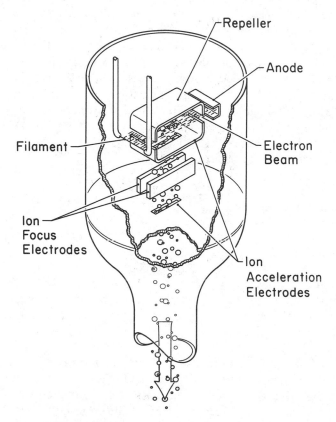

Figure 1B. Mass-spectrometer ion source.

Electron ionization (**EI**). For this method the "reagent" producing the ionic products is a beam of energetic (approximately 70 eV) electrons. These are "boiled off" an incandescent filament and travel through the ion chamber to an anode on the opposite side. The stream of vaporized sample molecules (at a pressure of approximately 10^{-2} Pa) entering the source interacts with the beam of electrons to form a variety of products, including positive ions. These are pushed out of the source by a relatively small "repeller" (or "draw-out") potential, and then are accelerated into the mass-analysis device. The bulk (~ 99.999 per cent) of the sample molecules and any other electron impact products are removed continuously by vacuum pumps on the ion-source housing.

Chemical Ionization (CI) (*Munson and Field 1966; Fales 1971*). Electron bombardment of methane or other gases at high pressures (100 Pa) in a special ion source produces an abundance of "reagent" ions such as CH_5^+ which will react with added sample molecules through a variety of ion-molecule reactions including charge exchange. Different types of molecules can be selectively ionized with specific positive and negative reagent ions. Heteroatom-containing molecules such as amines and ethers usually give abundant $(M + 1)^+$ ions; saturated hydrocarbons often yield $(M - 1)^+$ ions. Both types of ions are useful for indicating the molecular weight, since their fragmentation is greatly reduced in comparison to the molecular ions of EI spectra. Also, such diagnostic $(M + 1)^+$ ions can often be formed by increasing the sample pressure in the EI source (McLafferty 1957). Of similar utility is Field Ionization (FI) (Beckey 1977; Schulten 1979), another "soft ionization" technique (Section 6.1).

1.3 Mass analysis of ions

The ion beam from the source can be separated according to the respective masses (actually m/z, the ratio of the mass to the number of charges) of the ions by a variety of techniques. Magnetic deflection, quadrupole filter, time-of-flight, radio frequency, cyclotron resonance, and cycloidal focusing are separating techniques often employed; the first two are the most commonly used in commercial mass spectrometers. Excellent detailed discussions of these methods are available in the general references.

Magnetic sector. The single-focusing (focusing for direction) mass spectrometer in Figure 1C uses a large potential difference (1 to 10 kV) between the ion-acceleration electrodes (Figure 1B). Small potentials applied to the "repeller" and "ion-focus" electrodes are adjusted to maximize the ion current out of the source exit slit. This slit acts as the entrance focal point of the ion-optics system of the mass analyzer, allowing a diverging beam of entering ions of the same mass-to-charge ratio (m/z) to be focused on the analyzer exit slit.

The magnetic field acts as the mass analyzer. The m/z value of the ions which can pass through the exit slit depends on the radius (r, cm) of the ion path in the magnetic field, the field strength (B, gauss), and the ion-accelerating potential (V, volts) as defined by the fundamental equation (Beynon 1960)

$$m/z = 4.82 \times 10^{-5} B^2 r^2 / V. \qquad (1.1)$$

This can be derived on the basis that the ions of elementary charge e (as distinguished from z, the *number* of charges on the ion) have all been given

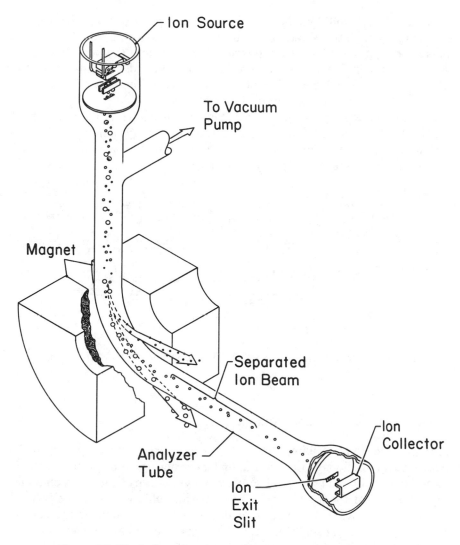

Figure 1C. Single-focusing, magnetic-sector mass spectrometer.

the same kinetic energy $eV = \frac{1}{2}mv^2$ (but different velocities, v) in acceleration, and that the force exerted by the magnet, Bev, must be equal to the centrifugal force, mv^2/r. Note that an ion whose $m/z = 100$ would require 1×10^{-6} sec to leave the ion source after formation (10-volt repeller, 5 mm distance), and 1×10^{-5} sec to travel 1 m after acceleration by 5 kV.

Quadrupole mass filter **(Figure 1D).** Ions from the source are injected into the end between the four rods. Mass separation is achieved by a combination of RF and DC fields with opposite polarities on the pairs of opposing rods. Path oscillations caused by the changing RF field tend to cause the lighter ions to strike the positive poles and the heavier to strike the negative pair, filtering out all ions but those of the desired mass. Dawson (1976) has compared this to keeping a ball from falling out of a saddle by inverting the saddle at the appropriate frequency.

An advantage of quadrupole over magnetic instruments is that the fields required to focus a particular mass can be changed very rapidly ($< 10^{-3}$ sec), which is especially valuable for computer-controlled measurements such as "multiple ion monitoring" (see Section 1.5).

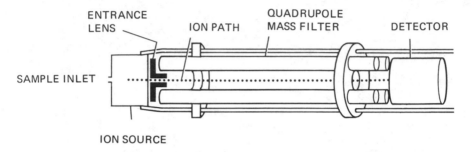

Figure 1D. Quadrupole mass spectrometer (courtesy of Hewlett-Packard).

1.4 Ion abundance measurement

The positive ions striking the collector produce a flow of neutralizing electrons proportional to the ion abundance, and this current can be measured accurately and with great sensitivity by modern electronic techniques. Amplification of the ion signal by an electron multiplier can make possible the detection of a *single ion* arriving at the collector. Although the efficiency of ionization and transmission in the mass spectrometer may yield only one ion at the collector for $\sim 10^5$ sample molecules introduced, useful spectra can be obtained from *nanogram* samples, and specific detection of *picogram* (10^{-12} g) samples is possible. Thus the method can be used in a variety of research and analytical problems for which most of the usual structural tools do not have sufficient sensitivity.

Changing the magnetic or electric fields that effect separation causes ions of different m/z values to reach the collector. Continuously changing the field while recording the ion signals on a strip chart produces a mass spectrum like that of Figure 1E. On-line computer systems produce sets of mass and abundance values directly by measuring the field and ion current at the top of each peak. Either system can record a complete mass spectrum in approximately one second, which is especially valuable for GC/MS (see below). However, the scan rate affects the accuracy of ion-abundance measurement, which is important for deducing elemental compositions from isotope ratios: check the performance of your instrument with known samples to be sure you have not sacrificed this accuracy unnecessarily.

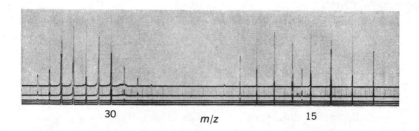

Figure 1E. Mass spectrum of Unknown 1.3 recorded with galvanometers of sensitivities ×1, ×3, ×10, ×30, ×100, from a magnetic-sector instrument. Peaks at m/z 18, 19, 20, 25, 26, and 27 are largely due to background.

1.5 Sample introduction systems

The sample must be vaporized so that the molecules will be separated from each other before ionization. Direct introduction into the ion source is preferred for the study of compounds of low volatility; samples are inserted with a probe through a vacuum lock into the ion source, where they are vaporized by heating (Figure 1F). With this system, one can obtain spectra

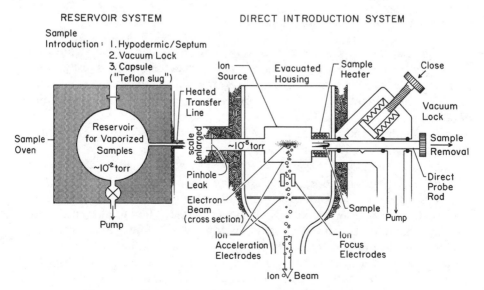

Figure 1F. Reservoir and direct-introduction sample systems.

of nonpolar molecules in the 1,000 to 2,000 molecular-weight range in nanogram amounts. *Warning:* it is very easy to use too much sample in the direct-introduction system; one crystal which is just large enough to see should be sufficient. The heating rate should also be controlled carefully to avoid volatilizing the sample too rapidly. More volatile samples can be introduced through a reservoir system. A large excess of sample (~0.1 mg) is vaporized into an evacuated, heated reservoir, from which the sample flows through a small orifice (molecular leak) into the ion source at a nearly constant rate. This makes spectral measurements more reproducible, which is helpful for quantitative analysis of multicomponent systems as well as for matching of unknown and reference spectra. The reservoir system should be all-glass and at constant temperature (no hot or cold spots) to avoid catalytic or thermal decomposition or condensation of the sample.

The mass spectrometer can also be directly coupled to a gas chromatograph (GC/MS; see Section 6.4 and McFadden 1973) so that the eluted components go directly to the ion source where their complete spectra are obtained "on the fly." By "multiple-ion monitoring," measuring two or more peaks sequentially, both high specificity and high sensitivity of detection can be achieved for compounds chosen in advance. Except for capillary columns it is usually necessary to reduce the sample pressure with a splitter or carrier-gas separator before the sample is introduced into the ion source. Methods for interfacing to the high-performance liquid chromatograph (LC/MS) have also been described (see Section 6.4, Dawkins and McLafferty 1978, and Arpino and Guiochon 1979).

1.6 Molecular-structure information

The main purpose of this book is to show how molecular-structure information can be derived from EI mass spectra. The spectrum of an average compound contains $~2^{50}$ bits of information, the mass and abundance values resulting from electron bombardment of the molecule. This information depends on the molecule's structure; the interpretation process attempts to convert the spectral information into molecular-structure predictions which are as complete and reliable as possible.

The 70-eV electron energies used are well above the $~10 \pm 3$ eV required for the ionization of molecules. The sample pressure in the ionization chamber is kept low enough (below 10^{-2} Pa) that secondary collisions of the ions with molecules or electrons will be negligible. The initial result of the electron interaction with the molecule is formation of the molecular ion by ejection of another electron. Part (sometimes all) of the molecular ions decompose further to yield the fragment ions of the spectrum. For example, the principal peaks in the spectrum of methanol (Unknown 1.3) are probably formed by the unimolecular processes in Equations 1.2:

$$CH_3OH + e^- \longrightarrow CH_3OH^{+\cdot} \ (m/z \ 32) + 2e^-$$
$$CH_3OH^{+\cdot} \longrightarrow CH_2OH^+ \ (m/z \ 31) + H\cdot$$
$$\longrightarrow CH_3^+ \ (m/z \ 15) + \cdot OH \tag{1.2}$$
$$CH_2OH^+ \longrightarrow CHO^+ \ (m/z \ 29) + H_2$$

The *dot* will be used to indicate a radical, so that the symbol $^{+\cdot}$ signifies a radical ion. Species containing such an unpaired electron are termed "odd-electron" (OE); similarly, an ion with only paired electrons is an "even-electron" (EE) ion. Note in Equations 1.2 that only one ion can be formed in the unimolecular decomposition of an ion, and that a radical product (neutral or ionic) must be formed from an $OE^{+\cdot}$ ion. It should be reemphasized that only *unimolecular* reactions are important; the sample pressure in the ion source is made low enough to avoid reactions between ions and molecules.

When one is interpreting the EI mass spectrum, identification of the molecular ion ($M^{+\cdot}$) determines the molecular weight and often the elemental composition of the molecule. If no $M^{+\cdot}$ is present, chemical or field ionization (CI or FI, Section 6.1) can often supply this information. The EI fragment ions indicate the pieces of which the molecule is composed, and the interpreter attempts to deduce how these pieces fit together in the original molecular structure. Spectra/structure correlations have now been published for a wide variety of complex molecules.

1.7 Standard interpretation procedure

In the game of "twenty questions" the most efficient approach usually is to classify the unknown first between major categories, e.g., "Is the person male?". Mass-spectral interpretation is similar; "blind alleys" can best be minimized if certain types of data in the spectrum are used before others. To learn to use the variety of information that is available in the mass spectrum, you should follow the outline of this book *step by step* in interpreting an unknown spectrum. This procedure is set forth inside the back cover in the form of a checklist to be used when you are interpreting an unknown.

This is a general, simplified approach applicable to the "average" mass spectrum. With experience the first several steps will be fast and largely automatic. While you are learning, however, each step should be done in this order, and your postulations, assignments, and conclusions from each step should be recorded, preferably on the spectrum. If more than one explanation appears possible for a particular spectral feature, be sure to note all possibilities.

Other sample information. It is important to incorporate all other available structural information (chemical, spectral, sample history) into the interpretation wherever appropriate. When a sample is submitted by another research worker to the mass-spectrometer laboratory for analysis, it is surprising how often other pertinent information is not transmitted. One of the strongest reasons to have the researcher interpret the mass spectra of his or her own samples is the importance of this information, for which that person will have the broadest and most thorough understanding. This is also one of the main incentives for the preparation of the book: mass spectrometry should be for the scientists, not just for mass spectrometrists.

Obtaining a Useful Spectrum. The interpretation procedure of this book assumes the unknown spectrum to be that of a pure compound. (If several compounds are present in the sample, the resulting spectrum will represent a linear superposition of the component spectra.) Even with good sample purity, care must be taken to avoid thermal or catalytic decomposition during introduction. Anomolous peaks can also arise from "background" due to leaks, previous samples ("memory"), or chromatographic column bleed. A background spectrum taken just before or after the sample run should be subtracted from the sample spectrum; however, note that the presence of the sample can affect the desorption rate of the background material. A log book of all samples run on the instrument is helpful in identifying contaminants. Unless scan times must be minimized, the spectrum should be scanned from m/z 12, or at least m/z 29, to well above the expected molecular weight of the sample, and even further if nonbackground peaks still are found. (Data for the spectra in this book will generally start at m/z 12, since there is little useful information below that value in mass spectra of organic molecules.) If CI or FI is available, use it to obtain an independent measurement of the molecular weight.

It is imperative that the measured m/z values be *correct* ($\pm$ <0.5 mass units). Common background peaks, such as those of air (Unknown 1.4), can serve as internal standards. If there is any doubt, add an internal mass standard, such as perfluorokerosene, C_nF_{2n+2}, and take a second scan. With computerized data systems it is usually quite convenient to check the accuracy of mass assignments. The reproducibility of peak-abundance measurements should also be checked. A multi-Cl/Br compound such as C_3Cl_6 provides a convenient standard, since its isotopic abundances are accurately predictable (Chapter 2).

Multiply Charged and Metastable Ions. Most peaks in a mass spectrum appear, conveniently, at integral mass numbers (for the exact mass differences, see Section 6.3). However, two types of peaks can appear at nonintegral masses, and therefore could lead to an erroneous mass-scale assignment. Multiply charged ions can appear at fractional masses; for example, the peak at m/z 15.5 in Figure 1E (Unknown 1.3) is actually a doubly charged ion of mass 31 (see Chapter 6).

In spectra from magnetic (not quadrupole) instruments, "metastable peaks" can also appear at fractional masses, but these are easily recognizable because they are characteristically wider and more diffuse than a regular peak. These arise from the decomposition of metastable ions after they are accelerated but before they are deflected in the magnetic field; for the decomposition $m_1 \rightarrow m_2$, the metastable decomposition product, m^*, will appear at $m/z = m_2^2/m_1$. In Figure 1.5 the broad peak at approximately m/z 26.8–27.3 arises from the metastable decomposition $m/z\ 31^+ \rightarrow 29^+$, and thus is centered at $m/z\ 29^2/31 = 27.1$ (see Section 6.3). Similar peaks can arise by "collisional activation" of the ions.

Before continuing to Chapter 2, try Unknowns 1.6 to 1.9, again ignoring the small peaks adjacent to large ones. Wait until you are really "stuck" before turning to Chapter 11 for help.

Unknown 1.6

m/z	Rel. abund.
12	0.9
13	3.6
24	6.1
25	23.
26	100.
27	2.2

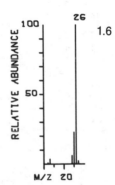

Unknown 1.7

m/z	Rel. abund.
12	4.2
13	1.7
13.5	0.9
14	1.6
26	17.
27	100.
28	1.6

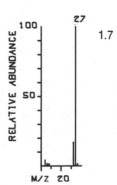

Unknown 1.8

m/z	Rel. abund.
14	17.
15.	100.
16	1.0
19	2.0
20	0.3
31	10.
32	9.3
33	89.
34	95.
35	1.1

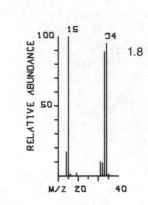

Unknown 1.9

m/z	Rel. abund.
12	3.3
13	4.3
14	4.4
16	1.7
28	31.
29	100.
30	89.
31	1.3

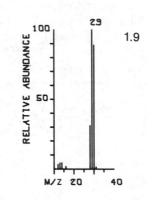

2

ELEMENTAL COMPOSITION

Mass-spectral peaks correspond to the molecule and pieces of it. In addition to the masses of these peaks, the elemental composition can be deduced. The most powerful technique for this uses exact mass measurement with a high-resolution mass spectrometer; measurement of the mass of a peak with sufficient accuracy defines its elemental composition unequivocally (Section 6.2). However, even with instrumentation of unit-mass resolution the presence of isotopes of known natural abundance makes possible a useful and simple method for deducing the elemental composition of many ions. The presence of a less abundant isotope gives the "isotopic peaks" that you were told to ignore as anomolous in solving the unknowns of Chapter 1. Even if you have high-resolution information on elemental compositions available to you, it is important to understand thoroughly the use of isotopic abundances, since this is basic to an understanding of mass spectra. For an introductory course, Sections 2.7 and 2.8 may be omitted.

2.1 Stable isotopes: classification according to natural abundances

A *chemically pure* organic compound will give a mixture of mass spectra because the elements that compose it are not isotopically pure. Recall the case of the element neon, which Sir J. J. Thompson (1913) showed gave not one peak at its chemical atomic weight of 20.2, but two peaks at masses 20.0 and 22.0, in relative abundances of 10:1. Of the common elements encountered in organic compounds (shown in Table 2.1, inside the front cover), many have more than one isotope of appreciable natural abundance. The isotopic abundances of other elements are shown in Table A.1 (in the Appendix); however, this book covers only the mass spectra of compounds containing the eleven elements of Table 2.1.

Unknown 2.1

m/z	Rel. abund.
35	12.
36	100.
37	4.1
38	33.

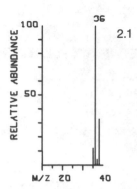

The effect of isotopes on a mass spectrum is illustrated in Unknown 2.1, which contains molecular ions at *m/z* 36 and 38. Their characteristic isotopic ratio of 3:1 should make the element easily recognizable from the data of Table 2.1. Note that the interpretation of *m/z* 36 as a fragment ion (loss of two H) from *m/z* 38 would be totally misleading. It is highly important that you become familiar with the data of Table 2.1. Note that the isotope of lowest mass is the most abundant for all these elements, quite fortunately. Note also that the isotopic abundances of the elements can be classified into three general categories: "A," those elements with only one natural isotope; "A + 1," those elements that have two isotopes, the second of which is one mass unit heavier than the most abundant isotope; and "A + 2," those elements that have an isotope that is two mass units heavier than the most abundant isotope. The "A + 2" elements are the easiest to recognize, and therefore you should look for these first. We will also use a similar classification for mass spectral peaks: an A peak is one whose main elemental formula is composed of only the most abundant isotopes; its (A + 1) peak is the peak one mass unit higher, and so forth.

2.2 "A + 2" elements: oxygen, silicon, sulfur, chlorine, and bromine

A second isotope makes an especially prominent appearance in the spectrum if it is more than one unit higher in mass than the most abundant isotopic species. Bromine and chlorine, and to a lesser extent silicon and sulfur, are striking common examples. The presence of these elements in an ion is often easily recognized from the "isotopic clusters" produced in the spectrum. Thus elements of Unknowns 2.2 and 2.3, like Unknown 2.1, can be recognized from the characteristic isotopic ratio of the ions separated by two mass units. (For now, ignore the small peaks next to the large ones.) An initial inspection of the bar graph is often an easy way to identify such isotopic clusters. For simplification, *only* the molecular ion data are included.

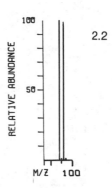

Unknown 2.2

m/z	Rel. abund.
94	100.
95	1.1
96	96.
97	1.1

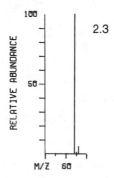

Unknown 2.3

m/z	Rel. abund.
64	100.
65	0.9
66	5.0

If there is more than one atom of these elements present in the molecule, the result is even more striking. For hydrogen bromide the isotopic molecular ions at m/z 80 and 82 ($H^{79}Br$ and $H^{81}Br$) are in the relative proportions of roughly 1:1. The mass spectrum of Br_2 shows prominent molecular ions at masses 158, 160, and 162, of relative abundances 1:2:1, due to the ions $^{79}Br_2$, $^{79}Br^{81}Br$ and $^{81}Br^{79}Br$, and $^{81}Br_2$, respectively. In a similar fashion, any ion structure containing three bromine atoms will exhibit four peaks at intervals of two mass units in the ratio 1:3:3:1, and the 3:1 isotopic ratio of $^{35}Cl/^{37}Cl$ yields three peaks in the ratio 9:6:1 for ion species containing two chlorine atoms (see Figure 2A).

The characteristic patterns resulting from combinations of the chlorine, bromine, sulfur, and silicon isotopes are illustrated by the bar graphs of Figure 2B (inside front cover), arranged in increasing order of the abundance ratio $[(A+2)]/[A]$. Numerical values for combinations of "A + 2" elements are given in Table A.2. These were calculated from the binomial expansion $(a+b)^n = a^n + na^{n-1}b + n(n-1)a^{n-2}b^2/2! + n(n-1)(n-2)a^{n-3}b^3/(3!) + \cdots$. Thus for a peak containing four chlorine atoms, the abundance of the $(A+4)$ relative to A should be $4 \cdot 3 \cdot 1^2 \cdot 0.325^2/2 = 0.635$.

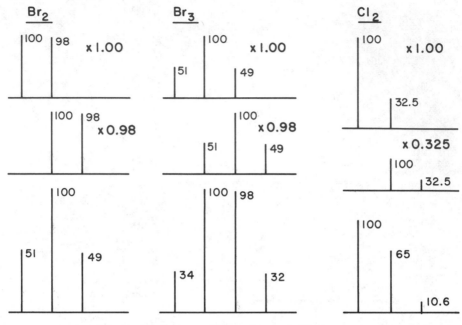

Figure 2A. Patterns of bromine and chlorine peaks.

Oxygen and Abundance Accuracies. The $(A + 2)$ relative abundance of oxygen is very low (0.2 per cent), and thus high abundance accuracy is necessary to deduce the number of oxygen atoms (for example, Unknown 2.3). For the unknowns in this book, a possible relative error of ± 10 per cent or absolute error of ± 0.2 (on the scale that the abundance of the highest peak $= 100$), whichever is greater, in the value of $[(A + n)]/[A]$ will be assumed, resulting in an uncertainty of $\geq O_1$ in calculating the number of oxygens. Further, there often is a substantial probability that several atoms of the "$A + 1$" element carbon are present, producing a small contribution to the $A + 2$ peak; so it is usually necessary to calculate the number of oxygen atoms after the "$A + 1$" elements as well as the other "$A + 2$" elements have been identified.

Absence of "A + 2" Elements. The value of "negative information" should not be overlooked. Another reason for checking first for the presence of "$A + 2$" elements is that one often finds them to be absent. Consider any peak A (fragment ion as well as M^+) in a mass spectrum and the peak two mass units higher (the "$A + 2$" peak); if $[(A + 2)]/[A] < 3$ per cent, the peak A *cannot* contain the most abundant isotope of the elements Si, S, Cl, or Br (if A contains more than one ion formula, this ratio determines the proportion of ions free of these elements). Table 2.1 shows that if peak A contains one ^{28}Si

atom, $[(A + 2)]/[A] \geq 3.4$ per cent, since in nature 3.4 ^{30}Si atoms must be found with every 100 atoms of ^{28}Si. The ratio $[(A + 2)]/[A]$ can be greater than 3.4 per cent if ions of some other composition also contribute to the $(A + 2)$ peak.

2.3 "A + 1" elements: carbon and nitrogen

The three "A + 1" elements in Table 2.1 are hydrogen, carbon, and nitrogen, but the ^{2}H/^{1}H ratio is so low that we shall consider hydrogen to be an "A" element. Each element exhibits its isotopic abundances independently; thus the carbon atom in the CH_3Br molecule of Unknown 2.2 contributes 1.1 per cent of the abundance of the m/z 94 peak ($^{12}CH_3{}^{79}Br^+$) to the m/z 95 peak ($^{13}CH_3{}^{79}Br^+$), and of the m/z 96 ($^{12}CH_3{}^{81}Br^+$) to the 97 ($^{13}CH_3{}^{81}Br^+$). Unknown 2.3, which contains no "A +1" elements, shows $[m/z\ 65]/[m/z\ 64]$ close to that expected for one carbon (but actually due to $^{33}S^{16}O_2{}^+$ and $^{32}S^{16}O^{17}O^+$); that gives us another reason why "A + 2" elements should be identified first.

Increasing the number of carbon atoms in an ion increases the probability that one of these atoms will be a ^{13}C isotope; $[(A + 1)]/[A]$ for a C_{10} ion will thus exhibit ten times the probability of C_1, or 10×1.1 per cent $= 11$ per cent. This fact provides a way to deduce the number of carbon atoms, which is obviously of key importance in interpreting the spectra of organic compounds. (At this stage, do not worry about detecting nitrogen as an "A + 1" element; the "nitrogen rule" in Section 3.3 will be helpful for this.) Table 2.2 (inside the front cover) tabulates the probability that an ion of a specified number of ^{12}C atoms will contain one ^{13}C atom. The factor of 1.1 per cent per carbon atom varies slightly (~2 per cent relative) with the organic source, such as petroleum (low value) versus a contemporary plant source; we will ignore the small contribution of deuterium from the usual hydrogen content of the ion. [The 1.1 per cent is equivalent to $^{13}C/^{12}C = 1.08$ per cent with 1.5 H atoms per C atom; ± 0.5 H/C changes the intensity of $(A + 1)$ by only ± 0.7 per cent of its value, $(A + 2)$ by ± 1.5 per cent of its value.]

In Unknown 2.4, calculate (after checking for "A + 2" elements) the maximum number of carbon atoms in the ions of m/z 43 and 58. The results indicate that m/z 43 peak is formed from the m/z 58 by the loss of what group?

Note that many small peaks in Unknown 2.4 were relatively unimportant for deriving its structure, such as peaks which can be formed from abundant peaks by the loss of hydrogen or doubly charged ions. In other unknowns of this book, unimportant peaks will often be omitted. For Unknown 2.5, find the elemental formula of the molecular ion; this is the largest ("base") peak in the spectrum, indicating high stability.

Unknown 2.4

m/z	Rel. abund.	m/z	Rel. abund.
12	0.1	40	1.6
13	0.3	41	27.
14	1.0	42	12.
15	5.3	43	100.
25	0.5	44	3.3
25.5	0.4	48	0.1
26	6.1	49	0.4
26.5	0.1	50	1.2
27	37.	51	1.0
27.5	0.1	52	0.3
28	32.	53	0.7
29	44.	54	0.2
30	1.0	55	0.9
36	0.1	56	0.7
37	1.0	57	2.4
38	1.8	58	12.
39	12.	59	0.5

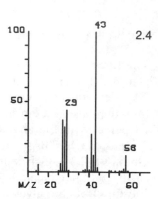

Double C-13. The Unknown 2.5 spectrum also contains a minor, but important, peak at *m/z* 80 of 0.18 per cent abundance. Assigning this as the ^{18}O contribution of one ^{16}O atom in *m/z* 78 would mean that the latter contains no more than *five* carbon atoms, whereas the *m/z* 79 abundance indicates six. The peak at *m/z* 80 actually arises from $C_6H_6^+$ ions containing two ^{13}C atoms and four ^{12}C atoms. The abundance of such $(A+2)^+$ ions relative to the A^+ ion depends on the number of carbon atoms, just as $[(A+1)^+]/[A^+]$ does: these abundances are also tabulated in Table 2.2.

Unknown 2.5

m/z	Rel. abund.	m/z	Rel. abund.
12	0.2	53	0.8
13	0.4	60	0.2
14	0.4	61	0.4
15	1.0	62	0.8
24	0.4	63	2.9
25	0.8	64	0.2
26	3.2	72	0.4
27	2.6	73	1.0
36	0.9	74	3.9
37	3.8	75	2.2
39	13.	76	7.0
40	0.4	77	15.
50	16.	78	100.
51	19.	79	6.8
52	20.	80	0.2

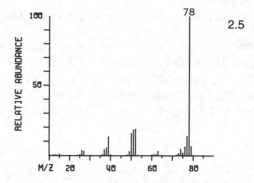

Of course $(A+2)^+$ isotopic contributions can also arise from the "$A+2$" elements. After one has assigned the number of carbon and nitrogen atoms from the $(A + 1)^+$ abundance value, the $(A + 2)^+$ value must be rechecked for the detection of oxygen; although the $^{18}O/^{16}O$ value of 0.2 per cent is small, the abundance accuracies assumed in this book (the greater of ±0.2 absolute or ±10 per cent relative) usually allow one to calculate the oxygen to within ±1 atoms. For Unknown 2.6, deduce the elemental compositions of the m/z 72 and 55 peaks; do these suggest its molecular structure?

Unknown 2.6

m/z	Rel. abund.
25	5.2
26	38.
27	74.
27.5	0.3
28	12.
29	4.3
31	0.5
41	1.2
42	1.3
43	5.8
44	14.
45	32.
46	2.5
52	1.4
53	6.0
54	2.3
55	74.
56	2.6
57	0.2
71	4.3
72	100.
73	3.5
74	0.5

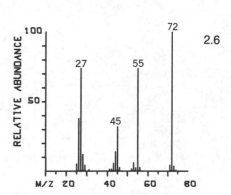

2.4 "A" elements: hydrogen, fluorine, phosphorus, and iodine

After the number of each "A + 2" and "A + 1" element has been fixed (or estimated, depending on the experimental accuracy), the balance of the mass of the peak must be due to the monoisotopic "A" elements. The assignment of the total elemental composition (or of the several possible compositions) is then completed by using numbers of these elements consistent with rules of bonding.

Try this partial spectrum:

m/z	Rel. abund.
127	100.
128	0.0
129	0.0

No "A + 2" or "A + 1" elements are possible. All the "A" elements but phosphorus are monovalent; so this grouping must be due to a single one of these (which?), or a pair, or a combination of these with phosphorus (thus H_3P_4 is a conceivable assignment).

The fragment ions of two further examples involve the same masses:

m/z	Rel. abund.	Rel. abund.
69	100.	100.
70	1.1	0.0
71	0.0	0.0

What is the maximum number of "A + 2" elements? Of "A + 1" elements? For the data in the first column, the presence of one carbon should be an obvious possibility; did you think of N_3? Compositions that satisfy the data are CF_3 and PF_2.

Try these procedures with Unknowns 2.7 and 2.8.

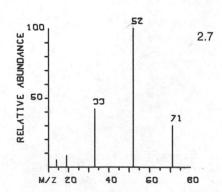

Unknown 2.7

m/z	Rel. abund.
14	5.2
19	8.4
33	36.
52	100.
53	0.4
71	30.

Unknown 2.8

m/z	Rel. abund.
12	0.36
19	0.2
31	1.8
35	3.0
37	1.0
42.5	1.7
43.5	0.6
47	0.5
49	0.2
50	6.3
66	0.3
69	100.
70	1.2
85	18.
86	0.2
87	5.9
104	0.7
106	0.2

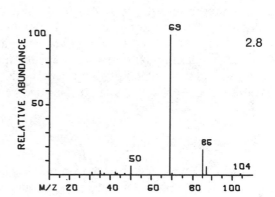

2.5 Rings plus double bonds

Because of the valences of the elements involved, the total number of rings and double bonds in a molecule of the formula $C_xH_yN_zO_n$ will be equal to $x - \frac{1}{2}y + \frac{1}{2}z + 1$. For ions, the calculated value may end in "$\frac{1}{2}$," and this fraction should be subtracted to obtain the true value. How to use this should be more obvious from inspection of the examples in Box 2.1. The value 4 found for pyridine represents the ring and three double bonds of this molecule. The 5.5 calculated for the benzoyl ion represents the ring, the three

Box 2.1 Rings plus Double Bonds ($r + db$)

For the general formula $C_xH_yN_zO_n$
   (more general case $I_yII_nIII_zIV_x$, where I = H, F, Cl, Br, I; II = O, S; III = N, P; and IV = C, Si, etc.)

Total rings plus double bonds = $x - \frac{1}{2}y + \frac{1}{2}z + 1$

For an even-electron ion (see Section 3.2), the true value will be followed by "$\frac{1}{2}$"

Examples:
C_5H_5N: rings plus double bonds = $5 - 2.5 + 0.5 + 1 = 4$
   For example, pyridine$\overset{+}{\cdot}$ (odd-electron)
C_7H_5O: rings plus double bonds = $7 - 2.5 + 1 = 5.5$
   For example, $C_6H_5CO^+$, benzoyl (even-electron)

double bonds of benzene, and the double bond of the carbonyl group. Calculate the number of rings and double bonds for the empirical formulas that you found for m/z 43 and 58 in Unknown 2.4.

If other elements are present, these are counted as additional atoms of the element C, H, N, or O to which they correspond in valence. Thus, the number of silicon atoms should be added to the number of carbon atoms, the number of halogen atoms to the number of hydrogen atoms, and the number of phosphorus atoms to the number of nitrogen atoms. Note also that this is based on the lowest valence state of the elements and does not count double bonds formed to elements in higher valence states. Thus, the formula indicates one double bond in CH_3NO_2 (nitromethane), no double bonds in $CH_3SO_2CH_3$ (dimethylsulfone), and a *negative* value in H_3O^+.

2.6 Exercises

Extensive experience has shown that *practice* is necessary to develop one's ability to calculate elemental compositions from isotopic abundances. (Perhaps this is because the chemist must become accustomed to the fact that the mass spectrum shows his carefully purified compound to be a mixture of isotopically different molecules.) It is important that you understand this procedure, since it is the key primary step in interpreting an unknown spectrum. To avoid confusion, follow the stepwise procedure below while you are learning. Table 2.3 illustrates this procedure with data from the $M^{\ddag}$ region of the spectrum of *t*-butylthiophene, $C_8H_{12}S$.

Table 2.3. *The $M^{\ddag}$ region of the spectrum of t-butylthiophene.*

m/z	Rel. abund.	Normalize	S_1	O_1	C_8	C_9	$^{13}CS_1$
139	0.5	2.					
140	25.	100.	100.	100.	100.	100.	
141	2.5 ± 0.25	$10. \pm 1.0$	0.8	0.0	8.8	9.9	8.8
142	1.2 ± 0.2	4.8 ± 0.8	4.4	0.2	0.2	0.3	0.1
143	0.1 ± 0.2	0.4 ± 0.8					0.4

(i) Insert the expected experimental accuracies on the original data, in the "Relative abundance" column (*not* on the normalized data of Step ii).

(ii) Normalize all data (including experimental accuracies) of the peak group; i.e., multiply all by the factor needed to set [A] equal to 100 per cent.

(iii) Find all possible "A + 2" elements (the value for oxygen will not be accurate), and show their expected abundance contributions in separate columns (S_1 and O_1 in the example; actually at this stage up to three oxygens can be considered).

(iv) Assign the possible number of carbons, showing these in similar columns. Subtraction of the ^{33}S contribution to m/z 141 places the carbon-isotope contribution in the range 8.2 to 10.2 per cent, leading to C_8 and C_9 as possibilities. However, S_1 and C_9 total 140 mass units, equal to that of peak A; C_9S is a highly unlikely ion formula, making C_8 the favored assignment. The column headed $^{13}CS_1$ shows that every 8.8 ions of $^{12}C_7{}^{13}C^{32}S$ will be accompanied by 4.4 per cent $\times$ 8.8 = 0.4 ions of $^{12}C_7{}^{13}C^{34}S$, predicting a corresponding contribution to m/z 143.

(v) Assign the "A" elements by difference. The postulated C_8S formula accounts for $96 + 32 = 128$ of the 140 mass units of peak A; H_{12} is the obvious choice for the remainder.

Using this procedure, can you assign compositions and values for rings plus double bonds to the fragmentary spectra shown in Unknowns 2.9 through 2.14? The peaks of Unknowns 2.9, 2.10, and 2.12–2.14 contain the molecular ion $(M^{\ddot{+}})$; peaks in Unknown 2.11 arise by loss from $M^{\ddot{+}}$ as indicated.

Unknown 2.9

m/z	Rel. abund.
130	1.0
131	2.0
132	100.
133	9.9
134	0.7

Unknown 2.10

m/z	Rel. abund.
73	26.
74	19.
75	41.
76	80.
77	2.7
78	<0.1

Unknown 2.11

m/z	Rel. abund.	
84	0.0	
85	40.	$(M - 19)^+$
86	2.0	
87	1.3	
88	0.0	

Unknown 2.12

m/z	Rel. abund.
179	2.2
180	1.1
181	7.4
182	55.
183	8.3
184	0.6

Unknown 2.13

m/z	Rel. abund.
171	0.1
172	98.
173	6.7
174	100.
175	6.5
176	0.5

Unknown 2.14

m/z	Rel. abund.
128	0.0
129	30.
130	100.
131	31.
132	98.
133	12.
134	32.
135	1.7
136	3.4
137	0.0

2.7 Interfering abundance contributions

Maximum Number of Atoms of the Element. In calculating the elemental composition of a peak A above, we assumed that the only contributions to the $(A + 1)$ and $(A + 2)$ peaks were from the less abundant isotopes of the elements in A. However, this is not necessarily true; there could be important additional contributions to the $(A + 1)$ and $(A + 2)$ peaks (and, occasionally, to peak A) for which corrections must be made. For the $(A + 1)$ and $(A + 2)$ peaks, these could arise from other fragment ions (or impurities, background, or ion-molecule reactions); for the $(A + 1)$ peak these could also arise from heavy isotope contributions of "$A + 2$" elements, such as from the $(A - 1)$ peak. Without such corrections, the calculations thus give the *maximum* number of atoms of each type in the elemental composition.

Interference from other fragment ions. In Unknown 2.4 calculating the number of carbon atoms of m/z 41 from the relative abundance of ions in m/z 42 gives a ridiculously high number of carbon atoms; $[42^+]/[41^+] = 12/27 = 44$ per cent, corresponding to 40 carbon atoms in Table 2.2. Although this is correct for the *maximum* number of C atoms, it is hardly a helpful calculation; the reason for this value is that most of the m/z 42 peak arises from $^{12}C_3{}^1H_6{}^+$, another fragment ion that contains only the most abundant isotopes. This peak is part of a group of peaks, each separated from its neighbor by one mass unit, of the formulas C_3H_{0-7}. Unknown 2.4, which is the spectrum of n-C_4H_{10}, has several other peak groups which correspond to the fragment ions C_1H_{0-3}, C_2H_{0-5}, and C_4H_{0-10}. The only A peaks for which the corresponding $(A + 1)$ and $(A + 2)$ peaks do not contain interfering fragment ions are m/z 15, 29, 43, and 58, corresponding to $CH_3{}^+$, $C_2H_5{}^+$, $C_3H_7{}^+$, and $C_4H_{10}{}^+$.

Thus the most useful isotopic-abundance calculations are generally those that use the significant peaks which are at the high-mass end of peak groups. Can you deduce the elemental compositions of any other ions in Unknowns 2.5 and 2.6?

Interference from "$A + 2$" elements of lower mass peaks. The full mass spectrum of methyl bromide (Unknown 2.2) also contains fragment ions from the loss of H atoms: CBr^+, 3 per cent; $CHBr^+$, 2 per cent; and CH_2Br^+, 4 per cent (with CH_3Br^+, 100 per cent). Each species must give a 1:1 pair of ^{79}Br, ^{81}Br peaks, and there must be a 1.1 per cent ^{13}C peak for each ^{12}C peak (see Figure 2C). The resulting mass spectrum (bottom in Figure 2C) is a linear superposition of these ion contributions. For the base peak, m/z 94, the $[(A + 1)]/[A]$ ratio of 5.0 per cent indicates a maximum of four or five carbon atoms; to correct for the $CH_2{}^{81}Br^+$ contribution to this peak, the

$CH_2{}^{79}Br^+$ abundance must be calculated by correcting $[m/z\ 93]$ for its $C^{81}Br^+$ component, based on $[C^{79}Br^+]$. However, it is much simpler to calculate the "A + 1" elements from the $[m/z\ 97]/[m/z\ 96]$ ratio, since m/z 94 and 96 must contain the same number of carbon atoms. Because the

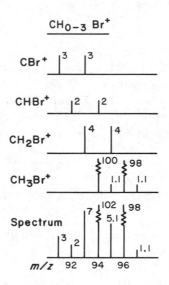

Figure 2C. Isotopic peaks in the spectrum of CH_3Br.

lower-mass fragment ions cannot contribute to this (A + 1) peak, the $[(A + 1)]/[A]$ ratio of 1.1 per cent correctly indicates the presence of one carbon atom. Again, the most useful isotopic calculations generally are those which use the significant peaks at the high-mass end of peak groups. Here a peak's significance is determined by the abundance accuracy of the peaks necessary for the calculation. In Unknown 2.14 the needed m/z 137 peak has an abundance of 0.0 ±0.2 per cent, so that $[m/z\ 137]/[m/z\ 136] = <6$ per cent, or 0–5 carbon atoms. If you cannot sort out the overlapping isotopic contributions, see the breakdown in Figure 2D.

In summary, if n "A + 2" elements are present, the (A + 1) peak [and possibly other $(A + 2n + 1)$ peaks] will contain isotopic contributions from the (A − 1) peak [and possibly other $(A − 2n − 1)$ peaks]. In calculating the number of carbon atoms using $[(A + 1)]/[A]$, correction should be made for such isotopic interferences. Alternatively, use $[(A + 2n + 1)]/[A + 2n)]$ corresponding to the peaks of highest mass, since the (A − 1) peak cannot make an isotopic contribution to these.

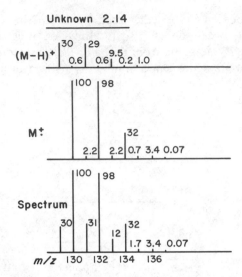

Figure 2D. Isotopic peaks in the spectrum of C_2HCl_3.

Erroneously low values for an element. Occasionally a significant proportion of the peak A abundance will be due to the isotopic contribution of an "A + 1" element in the (A − 1) peak, or an "A + 2" element in the (A − 2) peak. For example, the spectrum of *n*-butylbenzene shows the group of peaks:

Figure 2E.

m/z	Rel. abund.
91	100.
92	55.
93	3.9
94	0.1

The ratio $[m/z\ 93]/[m/z\ 92] = 7.1$ per cent, equivalent to 6.4 carbon atoms, although m/z 92 is mainly $C_7H_8^+$. Actually, 7.7 per cent of m/z 92 and 0.25 per cent of 93 are due to $^{12}C_6{}^{13}CH_7^+$ and $^{12}C_5{}^{13}C_2H_7^+$, respectively; thus $^{12}C_6{}^{13}CH_8^{+\cdot} = 3.65$ per cent and $^{12}C_7H_8^{+\cdot} = 47.3$ per cent, and their ratio is 7.7 per cent, as expected for seven carbons.

2.8 Deducing elemental compositions

Above we derived peak abundances in the methyl bromide spectrum from knowledge of abundances of particular ion compositions; however, for an unknown spectrum we face the opposite problem. The suggestions that follow should increase your efficiency in arriving at all reasonable solutions to such a problem.

1. Highest mass peaks. Use first those peaks which should yield composition assignments of high accuracy as well as high utility. The least isotopic contamination to the A, $(A + 1)$, *etc.*, peaks should be found in the peak group (compositions differing only in numbers of hydrogens) of highest mass (especially if it contains $M^{\ddagger}$). Further, for any particular peak group, the high-mass peaks should contain the least isotopic contamination.

2. Highest abundances. Within these limits, the accuracy of elemental composition calculations obviously should improve with increasing peak abundances.

3. Selection of the A peak. The desired A peak of the peak group is the one of highest mass which contains only the most abundant isotopes (such as ^{12}C, ^{16}O, ^{35}Cl); it is also called the "nonisotopic" peak. As the first candidate for A, try the largest peak in the group; if the second largest is at a mass higher than $(A + 2)$, try it instead as peak A. Next, if $[(A - 2)]/[A] > 30$ per cent, check the possible Cl/Br isotopic patterns (Figure 2B and Table A.2) to see if these are isotopic peaks of the same elemental formula. For example, the largest peak of the Br_4 isotopic cluster is actually the $(A + 4)$ peak. For the molecular-ion region, note that $(M - 2)^{\ddagger}$ is usually much less abundant than either $M^{\ddagger}$ or $(M - 1)^+$; occasionally, however, $(M - 1)^+$ is more abundant than $M^{\ddagger}$. Next, calculate the possible elemental compositions for this A peak; if an assignment cannot be found that accounts for the abundances of the higher mass peaks within experimental error, at least one of these must contain nonisotopic ions, and thus should be used as peak A. As an example, try the m/z 91–94 data from the *n*-butylbenzene spectrum in Figure 2E. Because m/z 91 cannot contain more than seven carbon atoms, no elemental composition for this A peak can account for the abundance of m/z 92; it thus must be used as peak A.

4. Other A peaks. See if useful information on elemental compositions can be obtained from lower mass peaks in each group of peaks differing by one mass unit. It is possible that their compositions differ by more than just the number of hydrogen atoms. Even "negative information" (Section 2.2) can be helpful.

5. Consistency of composition assignments. Elemental formulas which predict the observed isotopic abundances within experimental error are not necessarily correct. The elemental composition of $M^{\ddagger}$ (if present) gives the maximum number of each element that can be found in any lower mass peak.

Similarly, the fragment-ion compositions will usually show some mutual consistencies; use these to infer the more probable composition(s) if more than one agrees within experimental error. Common differences found between the compositions of molecular and fragment ions are given in Table A.5 and are discussed in Section 3.5. Common structural assignments for fragment ions are given in Tables A.6 and A.7 (see Section 5.2). Table A.8 gives compositional assignments for molecular ions containing the eleven common elements; use this to check if you have considered all possibilities.

3

THE MOLECULAR ION

The molecular ion, M^{+}, provides the most valuable information in the mass spectrum; its mass and elemental composition show the molecular boundaries into which the structural fragments indicated in the mass spectrum must be fitted. Unfortunately, for some types of compounds the molecular ion is not sufficiently stable to be found in appreciable abundance in the EI spectrum. An increasingly large proportion of mass-spectrometry facilities also have a "soft ionization" technique such as chemical or field ionization (CI or FI, Section 6.1) available. Such data should be used for molecular weight assignment wherever possible. However, even with CI or FI evidence the unknown spectrum should still be examined as described in this chapter, since this should lead to useful structure information as well as verification of the M^{+} assignment.

By convention, mass spectrometrists calculate the molecular weight (m/z of *"the"* molecular-ion peak) on the basis of the mass of the most abundant isotope of each of the elements present. For benzene (C_6H_6), which has substantial m/z 79 and m/z 80 peaks, *the* molecular ion is considered to be at mass 78 (C = 12, H = 1). M^{+} for the molecule Br_2 is considered to be 158, twice the mass of the most abundant isotope, ^{79}Br, although in the mass spectrum of Br_2 the most abundant ion is at m/z 160 (Section 2.2). Within these constraints, in the EI mass spectrum of a pure compound the molecular ion, if present, must be found at the highest value of m/z in the spectrum. There are further tests which must be used to ascertain if this peak does not represent the molecular ion, although these tests cannot demonstrate the converse.

3.1 Requirements for the molecular ion

The following are necessary, *but not sufficient,* requirements for the molecular ion in the mass spectrum of a pure sample, free of extraneous peaks such as those from background and ion-molecule reactions.

1. It must be the ion of highest mass in the spectrum.
2. It must be an odd-electron ion (Section 3.2).
3. It must be capable of yielding the important ions in the high-mass region of the spectrum by loss of logical neutral species (Section 3.5).

If the ion in question fails any of these tests, it cannot be the molecular ion; if it passes all these tests, it may or may not be the molecular ion.

3.2 Odd-electron ions

Ionization of the sample molecule occurs through the loss of an electron, and therefore the molecular ion is a *radical* species. Such an ion, either molecular or fragment, with an unpaired electron is called an "odd-electron" (OE) ion, and is designated by the symbol $\overset{+}{\cdot}$. It is often useful and convenient in explaining and classifying ion-decomposition reactions to distinguish between such radical ions and "even-electron" (EE) ions, those in which the outer-shell electrons are fully paired; the symbol $^+$ will be used only to refer to even-electron ions.

This concept can be visualized in its simplest form with structures (Equations 3.1 to 3.3) that include the outer-shell electrons. Usually a number of canonical resonance forms can be drawn to approximate the electron distribution in the ion. Note that the symbolism $\overset{+}{\cdot}$ is meant only to indicate an ion with an *unpaired* electron, *not* an electron *in addition* to those the formula represents; adding an electron to CH_4 would give the negative ion, $CH_4{}^-$.

$$\begin{array}{c} H \\ H{:}\overset{\cdot\cdot}{C}{:}H \\ H \end{array} \longrightarrow \begin{array}{c} H \\ H{:}\overset{\cdot}{C}{\overset{+}{:}}H \\ H \end{array} \quad \text{or} \quad CH_4{}^{\overset{+}{\cdot}} \qquad (3.1)$$

$$R{:}\overset{\cdot\cdot}{\underset{\cdot\cdot}{O}}{:}R \longrightarrow R{:}\overset{\overset{+}{\cdot}}{\underset{\cdot\cdot}{O}}{:}R \quad \text{or} \quad R{\overset{+}{:}}\overset{\cdot\cdot}{\underset{\cdot}{O}}{:}R \quad \text{or} \quad ROR^{\overset{+}{\cdot}} \qquad (3.2)$$

$$H_2C{::}CH_2 \longrightarrow H_2C{\overset{+}{:}}{:}CH_2 \quad \text{or} \quad H_2C{=}CH_2{}^{\overset{+}{\cdot}} \qquad (3.3)$$

If you can establish the elemental composition for the proposed molecular ion, the rings-plus-double-bonds formula will show immediately if the ion is an odd-electron species. For the general formula $C_xH_yN_zO_n$, the value of $x - \frac{1}{2}y + \frac{1}{2}z + 1$ will be a whole number for any odd-electron ion; examples are given in Box 2.1.

3.3 The nitrogen rule

For most elements encountered in organic compounds, there is a fortunate correspondence between the mass of an element's most abundant isotope and its valence; either both are even-numbered, or both are odd-numbered, with nitrogen as the major exception. This leads to the so-called "nitrogen rule," which can be stated as follows:

If a compound contains an even number of nitrogen atoms, its molecular ion will be at an even mass number. As examples, the following molecules yield even-mass molecular ions: H_2O, m/z 18; CH_4, m/z 16; C_2H_2, m/z 26; CH_3OH, m/z 32; $CClF_3$, m/z 104; C_6H_5OH, m/z 94; $C_{17}H_{35}COOH$, m/z 284; cholesterol, $C_{27}H_{46}O$, m/z 386; H_2NNH_2, m/z 32; and aminopyridine, $C_5H_6N_2$, m/z 94. An odd number of nitrogen atoms causes $M^{\ddot{+}}$ to be at an odd mass number: NH_3, m/z 17; $C_2H_5NH_2$, m/z 45; and quinoline, C_9H_7N, m/z 129. Thus if the ion of highest m/z is at an odd mass number, to be the molecular ion it must contain an odd number of nitrogen atoms.

This relationship applies to all ions, not just $M^{\ddot{+}}$; thus the nitrogen rule can also be stated as follows:

An odd-electron ion will be at an even mass number if it contains an even number of nitrogen atoms. Similarly, an even-electron ion containing an even number of nitrogen atoms will appear at an odd mass number. You may find this confusing at first; until you get used to working with this, it will probably be easier to derive this from the first statement, remembering that $M^{\ddot{+}}$ is an odd-electron ion.

Unknown 3.1. Indicate whether ions of the following formulas are odd-electron or even-electron: C_2H_4, C_3H_7O, C_4H_9N, C_4H_8NO, C_7H_5ClBr, C_6H_4OS, $C_{29}F_{59}$, H_3O, and C_3H_9SiO. Which of these ions will appear at even mass numbers?

3.4 Relative importance of peaks

$OE^{\ddot{+}}$ ions have a special mechanistic significance, as discussed in Section 4.4. Because of this you should *indicate all important $OE^{\ddot{+}}$ ions,* marking these directly on the spectrum. This is the next step in the "Standard Interpretation Procedure" (inside back cover). The "importance" of a peak (Section 2.6), after one has corrected its abundance for contributions of ions containing less common isotopes, generally increases with (i) increasing abundance, (ii) increasing mass in the spectrum, and (iii) increasing mass in the peak group (particularly the most or the second-most number of hydrogen atoms for an $OE^{\ddot{+}}$ peak).

Important $OE^{\ddot{+}}$ ions are even less probable in the lower-mass end of the spectrum; thus abundant even-*mass* peaks in this region are usually due to ions containing an odd number of nitrogen atoms, such as CH_4N^+, m/z 30. Similar reasoning leads to a corollary of the nitrogen rule:

A scarcity of important even-mass ions, especially at lower m/z values, indicates an even-mass molecular weight. However, the reverse is not always true; the presence of abundant even-mass ions does not necessarily indicate an odd-mass $M^{\ddagger}$. This rule is helpful for the mass spectrum of neopentane, molecular weight 72, which shows no molecular ion (Figure 3A).

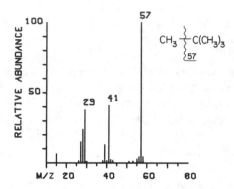

Figure 3A. Mass spectrum of neopentane.

3.5 Logical neutral losses

There are only a limited number of neutral fragments of low mass which are commonly lost in decompositions of molecular ions. The presence of an "important" ion separated from the highest mass ion by an anomalous mass or elemental formula will indicate that the latter ion is not the molecular ion. Presence of an abundant (as compared to its neighboring ions) ion 5 mass units below the ion of highest m/z would have to represent the loss of five hydrogen atoms—a highly unlikely decomposition. Small neutral fragments lost from the molecular ion are commonly those attached by a single bond. For example, if Figure 3A had been obtained from an unidentified pure compound, m/z 57 might be considered as $M^{\ddagger}$ of $CH_3CH=CHNH_2$, with m/z 41 and 42 as losses of NH_2 and CH_3, respectively. However, a significant $(M - CH_2)^{\ddagger}$ peak is very rare. Thus, the presence of a nonisotopic ion of m/z 43 indicates that the m/z 57 ion cannot be the molecular ion, even though it is the highest-mass nonisotopic peak in the spectrum. Such a mass difference of CH_2 is commonly encountered when two such homologous ions are produced by decomposition or a larger ion, here, $C_5H_{12}^{\ddagger}$ producing $C_4H_9^+$ and $C_3H_7^+$; this can also be due, however, to the molecular ion of an unexpected impurity which is a homolog of the unknown compound.

Mass losses of 4 to 14 and 21 to 25 that give important peaks are highly unlikely. Remember these ranges; Table A.5 presents a more complete list plus neutral fragments which are commonly lost. Maximum expected relative abundances are given by Speck (1978); for example, $(M-2)^{+\cdot}$ is usually much less abundant than either M^+ or $(M-1)^+$ (Section 2.6).

If the elemental composition of the fragment lost can be deduced, this gives an even more powerful test. For example, the presence of a major $(M-15)^+$ ion is common but a major $(M-NH)^{+\cdot}$ ion is probably an anomalous ion; the loss of 35 is logical only if chlorine is present.

Can the ion of highest mass be the molecular ion if the following are the major ions of high mass in the spectrum?

Unknown 3.2. $C_{10}H_{15}O$, $C_{10}H_{14}O$, $C_9H_{12}O$, $C_{10}H_{13}$, $C_8H_{10}O$

Unknown 3.3. C_9H_{12}, C_9H_{11}, C_9H_9, C_8H_9, C_8H_7, C_7H_7

Identifying and testing the molecular ion are important keys to Unknowns 3.4 and 3.5. *Hint:* In Unknown 3.4, use $[64^+]/[63^+]$ and $[99^+]/[98^+]$ to calculate the number of carbon atoms, since m/z 62 and 97 also contain an isotopic contribution from an "A + 2" element in m/z 60 and 95, respectively.

Unknown 3.4

m/z	Rel. abund.
12	2.7
13	3.0
14	0.6
24	4.0
25	15.
26	34.
27	0.7
31	0.3
35	7.0
36	1.9
37	2.3
38	0.7
47	6.5
47.5	0.2
48	5.9
49	4.2
50	1.8
51	0.7
59	2.6
60	24.
61	100.
62	9.9
63	32.
64	0.7
95	1.5
96	67.
97	2.4
98	43.
99	1.0
100	7.0
101	0.1

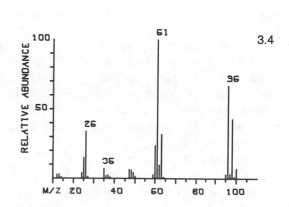

Unknown 3.5

m/z	Rel. abund.
31	42.
32	0.5
35	2.5
37	0.7
43	0.5
47	1.7
49	0.6
50	11.
69	57.
70	0.7
85	100.
86	1.1
87	33.
88	0.4
100	2.8
119	52.
120	1.2
135	24.
136	0.5
137	7.7

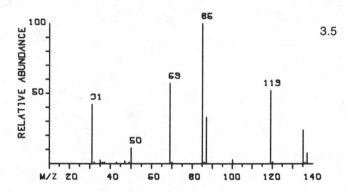

3.5

3.6 Molecular-ion abundance versus structure

The abundance of the molecular ion, $[M^+]$, depends mainly on its stability and the amount of energy needed to ionize the molecule (Table A.3). Particular structural features tend to show characteristic values of these properties, so that the magnitude of $[M^+]$ provides an indication of the structure of the molecule. Table A.4 gives typical $[M^+]$ values for a number of types of compounds, listed in order of decreasing abundance. Figures 3B to 3Y present spectra that also show such trends. (These figures will be referred to often later in the book.)

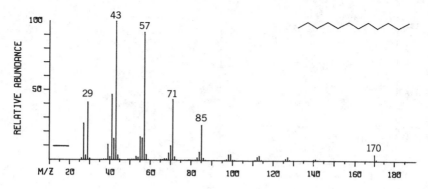

Figure 3B. Mass spectrum of *n*-dodecane.

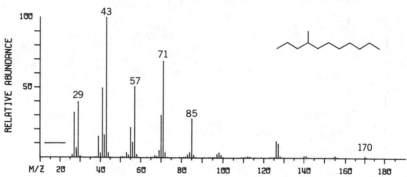

Figure 3C. Mass spectrum of 4-methylundecane.

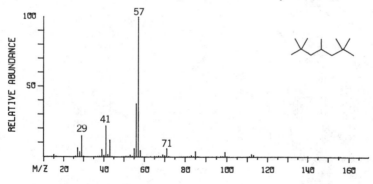

Figure 3D. Mass spectrum of 2,2,4,6,6-pentamethylheptane.

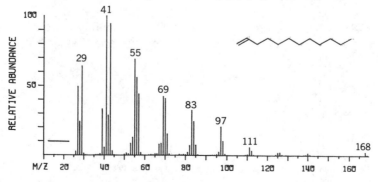

Figure 3E. Mass spectrum of 1-dodecene.

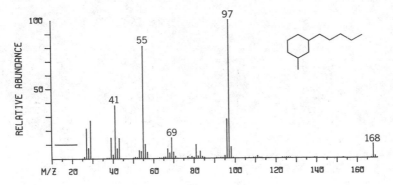

Figure 3F. Mass spectrum of 1-methyl-3-*n*-pentylcyclohexane.

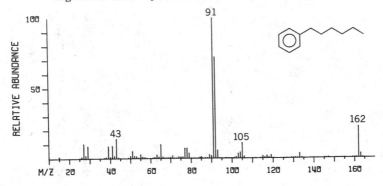

Figure 3G. Mass spectrum of 1-phenyl-*n*-hexane.

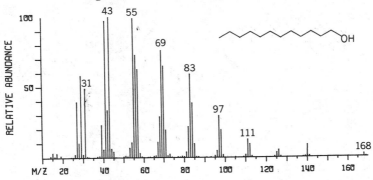

Figure 3H. Mass spectrum of 1-dodecanol.

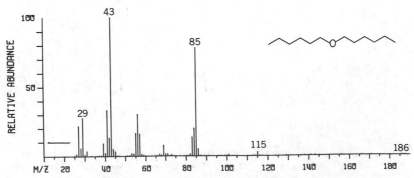

Figure 3I. Mass spectrum of *n*-hexyl ether.

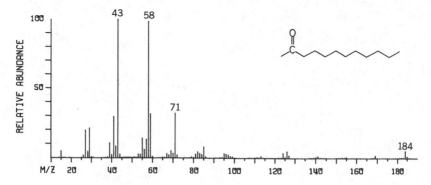

Figure 3J. Mass spectrum of 2-dodecanone.

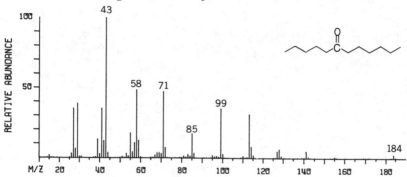

Figure 3K. Mass spectrum of 6-dodecanone.

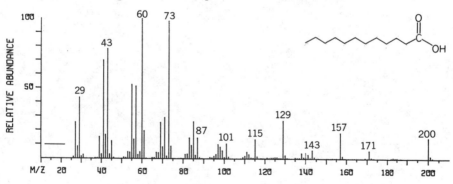

Figure 3L. Mass spectrum of *n*-dodecanoic acid.

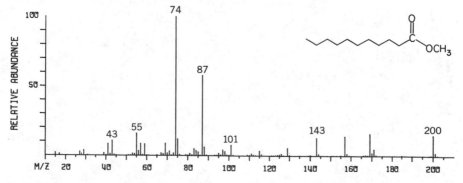

Figure 3M. Mass spectrum of methyl *n*-undecanoate.

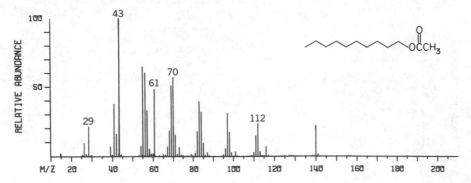

Figure 3N. Mass spectrum of *n*-decyl acetate.

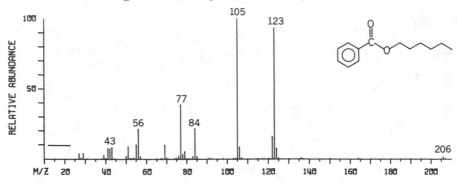

Figure 3O. Mass spectrum of *n*-hexyl benzoate.

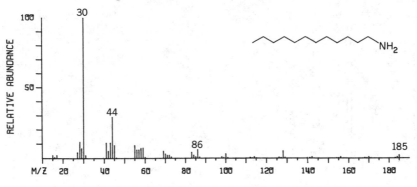

Figure 3P. Mass spectrum of *n*-dodecylamine.

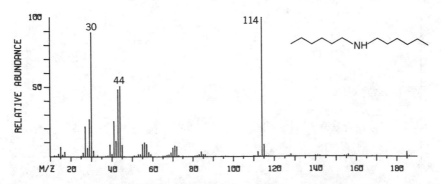

Figure 3Q. Mass spectrum of *bis*-(*n*-hexyl)amine.

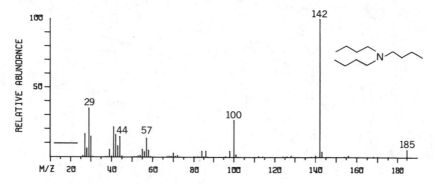

Figure 3R. Mass spectrum of *tris*-(*n*-butyl)amine.

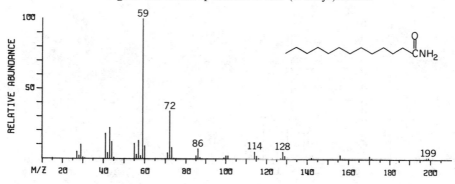

Figure 3S. Mass spectrum of dodecanamide.

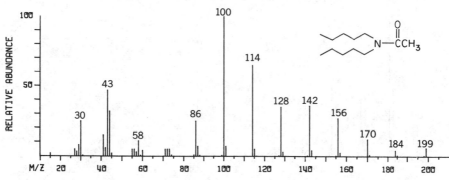

Figure 3T. Mass spectrum of N,N-dipentylacetamide.

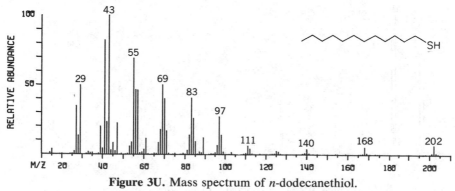

Figure 3U. Mass spectrum of *n*-dodecanethiol.

41

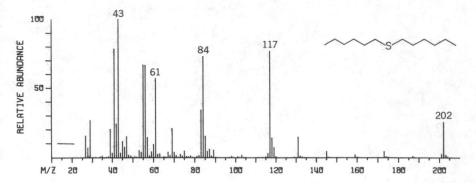

Figure 3V. Mass spectrum of *bis*-(*n*-hexyl)sulfide.

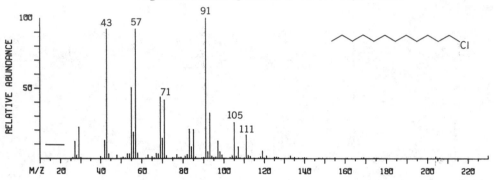

Figure 3W. Mass spectrum of 1-chlorododecane.

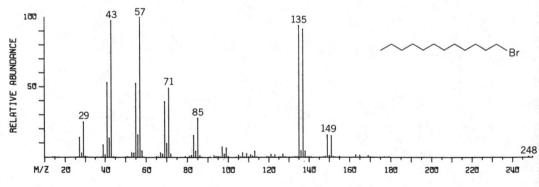

Figure 3X. Mass spectrum of 1-bromododecane.

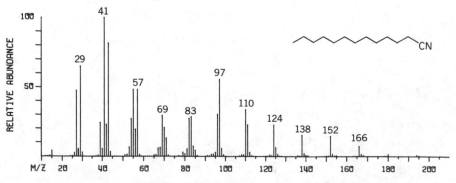

Figure 3Y. Mass spectrum of *n*-dodecyl cyanide.

In general, the chemical stability of the molecule parallels the stability of M^{+}, and so is reflected in the abundance of M^{+}; $[M^{+}]$ usually increases with increased unsaturation and number of rings, as illustrated by the striking abundance of M^{+} in strychnine (Figure 3Z). The effect of molecular weight is less clear-cut; increasing the chain length up to C_6 or C_8 generally decreases $[M^{+}]$ substantially (see Table A.4), but often $[M^{+}]$ increases again for longer straight chains. Chain-branching substantially decreases M^{+} stability and thus its abundance.

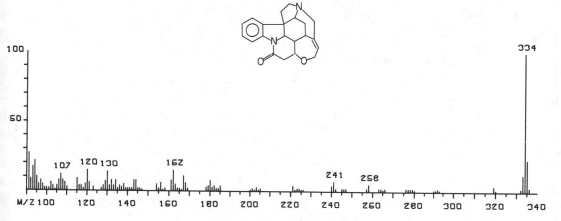

Figure 3Z. Mass spectrum of strychnine.

If less energy is required to ionize the molecule (that is, if it has a lower ionization energy), more molecular ions of lower internal energy ("cool ions") can be formed, and $[M^{+}]$ will tend to be higher. As shown in Table A.3, the ease of ionization of the outer-shell nonbonding electrons on heteroatoms increases in going down a column or to the left in a row of the periodic table. This accounts for the dramatic increase in $[M^{+}]$ for mercaptans in comparison to the corresponding alcohols (Figure 3H and 3U); primary amines (Figure 3P) show a smaller, though significant, increase in $[M^{+}]$ versus the corresponding alcohols.

Try one more unknown, 3.6, before tackling the next important area, Mechanisms.

Unknown 3.6

m/z	Rel. abund.
35	41.
37	13.
41	1.2
42	0.8
47	40.
48	0.5
49	13.
58.5	4.7
59.5	4.5
60.5	1.4
70	1.4
72	0.9
82	29.
83	0.3
84	19.
85	0.2
86	2.9
117	100.
118	1.0
119	96.
120	1.0
121	30.
122	0.3
123	3.1

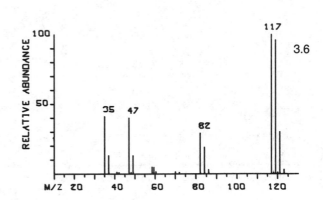

4

BASIC MECHANISMS OF ION FRAGMENTATION

Our initial statement (Chapter 1) that "the mass spectrum shows the mass of the molecule and the masses of pieces from it" is oversimplified, in that it neglects the important second dimension of a mass spectrum: ion *abundance*. The abundance of a specific fragment ion relative to the abundances of the molecular ion and other fragment ions can be a very useful indication of the *structure of that fragment and its environment in the molecule*. However, to make use of this information, we must understand the factors that control fragment-ion abundances. A preliminary, somewhat empirical, discussion of these factors will be given here. A more detailed treatment of mechanisms appears in Chapter 8, and the basic processes involved in forming mass spectra will be examined in Chapter 7. It is important to emphasize, however, that there are serious limitations to our present knowledge of mass-spectral reactions. To recognize for more complex molecules what alternative structural possibilities could give rise to the observed mass-spectral features, we must study the spectra of closely related molecules.

Such ion-decomposition reactions can be viewed as another field of chemistry, but fortunately for most chemists studying this book, there are many close similarities to pyrolytic, photolytic, radiolytic, and other energetic reactions, and there are even many general similarities to condensed-phase (solution) organic reactions. The largest points of difference are that ionic and often radical species are involved in each reaction in the mass spectrometer, and their combined effects sometimes appear unusual to the organic chemist. Chemists may also question the reliability of structural information based on rearrangement reactions. However, the wide variety of detailed studies which have now been published offer persuasive evidence that these ion-decomposition reactions take place by means of chemically reasonable processes.

4.1 Unimolecular ion decompositions

Mass-spectral reactions are *unimolecular;* the sample pressure in the ion source is kept sufficiently low that bimolecular ("ion-molecule") or other collision reactions are usually negligible. Molecular ions are formed with a wide range of internal energies. Those that are sufficiently "cool" will not decompose before collection, and will appear as $M^{\ddagger}$ in the spectrum. If sufficiently excited, the $M^{\ddagger}$ ions can decompose by a variety of energy-dependent reactions, each of which results in the formation of an ion and a neutral species; this primary product ion may have sufficient energy to decompose further. In the mass spectrum of ABCD (Equation 4.1), the abundance of BCD^{+} will depend on the average rates of its formation and decomposition, whereas $[BC^{+}]$ will depend on the relative rates of several competitive and consecutive reactions. Isomerization is a possible uni-molecular reaction; formation of the $OE^{\ddagger}$ ion $AD^{\ddagger}$ involves *rearrangement* (r) of the atom-bonding relationships of $ABCD^{\ddagger}$.

$$
\begin{aligned}
ABCD \xrightarrow{-e} ABCD^{\ddagger} &\longrightarrow A^{+} + BCD\cdot \\
&\longrightarrow A\cdot + BCD^{+} \\
&\qquad\qquad \hookrightarrow BC^{+} + D \qquad (4.1)\\
&\longrightarrow D\cdot + ABC^{+} \\
&\qquad\qquad \hookrightarrow A + BC^{+} \\
&\xrightarrow{r} AD^{\ddagger} + BC
\end{aligned}
$$

4.2 Basic factors that influence ion abundance

In general, the most abundant fragment-ion peaks in the mass spectrum correspond to reactions that form the most stable products (ion and neutral). *Steric effects* can also be of key importance, but a general discussion of these is postponed until Chapter 8.

Stability of the product ion. The most important general factor affecting the abundance of a product ion is its stability. Important types of ion stabilization include *electron sharing* involving a nonbonding orbital of a heteroatom, such as in the acetyl ion $CH_3-\overset{+}{C}=O \longleftrightarrow CH_3-C\equiv\overset{+}{O}$ (the latter is iso-electronic with $CH_3-C\equiv N$), and *resonance stabilization,* such as in the allyl cation $CH_2=CH-\overset{+}{C}H_2 \longleftrightarrow \overset{+}{C}H_2-CH=CH_2$.

Stevenson's rule (1951). Cleavage of a single bond in an odd-electron ion can lead to two sets of ion and radical products; $ABCD^{\ddagger}$ can give $A^{+} + \cdot BCD$

or $A \cdot + BCD^+$. The fragment with the higher tendency to retain the unpaired electron should have the higher ionization energy (I, Table A.3). Thus there should be a higher probability for forming the fragment *ion* of lower I value; because this ion usually is also the more stable, it should be the more abundant of the complementary pair of ions resulting from this bond cleavage (Audier 1969; McLafferty *et al.* 1970; Harrison *et al.* 1971).

Loss of the largest alkyl. A notable exception, in which abundance decreases with increasing ion stability, is the widespread preference for the loss of the *largest* alkyl radical at a reactive site:

$$
\begin{array}{c}
CH_3 \\
| \\
C_2H_5{-}CH{-}C_4H_9^{\cdot +}
\end{array}
\longrightarrow
\begin{array}{c}
CH_3 \\
| \\
[C_2H_5\overset{+}{C}H]
\end{array}
>
\begin{array}{c}
CH_3 \\
| \\
[{+}CHC_4H_9]
\end{array}
> \qquad (4.2)
$$

$$
[C_2H_5\overset{+}{C}HC_4H_9] >
\begin{array}{c}
CH_3 \\
| \\
[C_2H_5\underset{+}{C}C_4H_9]
\end{array}
$$

Stability of the neutral product. The importance of a reaction pathway can also be increased by a favorable product site for the unpaired electron. Such radical stabilization is improved on carbon by delocalization (e.g., allyl radical) or increased branching (e.g., *tert*-butyl radical); electronegative sites such as oxygen (e.g., alkoxyl radical) are also favorable. The neutral product can also be a molecule; small stable ones such as H_2, CH_4, H_2O, C_2H_4, CO, NO, CH_3OH, H_2S, HCl, $CH_2{=}C{=}O$, and CO_2 are often favored (Table A.5).

Unknown 4.1. Predict the most abundant product ion in the mass spectrum of

$$H_2N{-}\langle\rangle{-}CH_2{-}CH_2{-}\langle\rangle$$

4.3 Reaction initiation at radical or charge sites

To predict preferred decomposition pathways, we will use the simplistic assumption that the reactions are initiated at the favored sites for the unpaired electron and for the positive charge in the decomposing ion. Such a site is viewed as providing the driving force for specific types of reactions which are characteristic of the chemical nature of the site. Although this only gives, at best, an approximation of the actual electronic displacements, it provides a convenient way to correlate (and for you to remember) a large number of the reactions of diverse structural moieties.

The most favored radical and charge sites in the molecular ion are assumed to arise from loss of the molecule's electron of lowest ionization energy (I, Table A.3). Relative energy requirements are similar to those for the electronic transitions affecting ultraviolet spectra; favorability for ionization generally is on the order of $\sigma- < \pi- < n$-electrons.

$$\text{Sigma } (\sigma): RH_2C:CHR' \xrightarrow{-e^-} RH_2C^+CH_2R'$$

$$\text{Pi } (\pi): RHC::CHR' \xrightarrow{-e^-} RHC:^+CHR'$$

$$\text{Non-bonding } (n): R-\ddot{O}-R' \xrightarrow{-e^-} R-\overset{+\cdot}{O}-R'$$

The symbol $\overset{+}{\cdot}$ at the end of the molecule signifies an odd-electron ion without designating the radical site. Use of either $\cdot$ or $^+$ within the molecule, as in $CH_3\overset{+}{\cdot}CH_3$, implies localization of the radical or the charge.

4.4 Reaction classifications

Decompositions involving only the cleavage of a single bond in the odd-electron ($OE^{+\cdot}$) molecular ion must produce an even-electron (EE^+) fragment ion and a neutral radical species:

$$CH_3CH_2^+CH_3 \longrightarrow CH_3CH_2^+ + \cdot CH_3$$
$$\longrightarrow CH_3CH_2\cdot + CH_3^+ \tag{4.3}$$

The two fragments compete for the charge and the unpaired electron; the abundances of the two ionic products from this type of reaction would be equal only by coincidence (Stevenson's Rule, Section 4.2).

In contrast, an $OE^{+\cdot}$ ion is formed from $M^{+\cdot}$ by cleavage of *two* bonds between the products. Rearrangements and reactions involving fragmentation of a ring are two ways in which abundant $OE^{+\cdot}$ ions can be produced:

$$H_2C-CHOH^{+\cdot} \longrightarrow H_2C=CHOH^{+\cdot} + H_2C=CH_2 \text{ (charge retention)}$$
$$H_2C-CH_2 \longrightarrow H_2C=CHOH + H_2C=CH_2^{+\cdot} \text{ (charge migration)}$$

$$H \quad OH^{+\cdot} \longrightarrow HOH^{+\cdot} + H_2C=CH_2 \text{ (charge retention)}$$
$$H_2C-CH_2 \longrightarrow HOH + H_2C=CH_2^{+\cdot} \text{ (charge migration)} \tag{4.4}$$

Again, the two fragments compete for the charge and radical. In a similar fashion, cleavage of three bonds of $M^{+\cdot}$ produces an EE^+ ion:

$$CH_3CH_2\!\!-\!\!\overset{H}{\underset{H}{\overset{\diagup}{\underset{\diagdown}{}}}}\!\!CH_2C\overset{\diagup}{\underset{\diagdown}{H}}\Bigg]^{\overset{+}{\cdot}} \longrightarrow C_2H_3^+ + CH_3CH_2\cdot + H_2 \qquad (4.5)$$

The decompositions of even-electron ions are strongly influenced by the preference for formation of an EE^+ ion and EE^0 neutral (the "even-electron rule," Karni and Mandelbaum 1980). Production of an $OE^{\overset{+}{\cdot}}$ ion must be accompanied by formation of a radical neutral species, involving the energetically unfavorable separation of an electron pair:

$$
\begin{aligned}
CH_3CH_2CH_2^+ &\longrightarrow CH_3^+ + CH_2\!\!=\!\!CH_2 \ \text{(charge migration)} \\
&\longrightarrow CH_3\cdot + CH_2CH_2^{\overset{+}{\cdot}} \ \text{(unfavored)} \\
H_2C\!\!-\!\!\overset{+}{C}H &\longrightarrow CH_2\!\!=\!\!\overset{+}{C}H + CH_2\!\!=\!\!CH_2 \ \text{(charge retention)} \\
H_2C\!\!-\!\!CH_2 &\longrightarrow CH_2\!\!=\!\!\dot{C}H + CH_2\!\!=\!\!CH_2^{\overset{+}{\cdot}} \ \text{(unfavored)}
\end{aligned}
\qquad (4.6)
$$

The tendency for such further EE^+ ion decomposition by either charge migration or charge retention depends on the charge stabilization in the product ion relative to that in the precursor ion, as well as on the stability of the neutral product.

The seven main types of cation fragmentations discussed in this chapter are those not in brackets in Table 4.1. Odd-electron ion decompositions involving the cleavage of one bond are discussed in Sections 4.5 to 4.7, of two bonds in 4.8 and 4.9, and of three bonds in 4.11; EE^+ decompositions cleaving one and two bonds are covered in Sections 4.7 and 4.10, respectively. Note that $OE^{\overset{+}{\cdot}}$ formation is favored *only* for the cleavage of two bonds in

Table 4.1 *Types of Ion Decompositions*

Precursor ion	Number of bonds cleaved	Product ion[a]	
		Charge retention	Charge migration
$OE^{\overset{+}{\cdot}}$ ($M^{\overset{+}{\cdot}}$)	1	EE^+ (α)	EE^+ (i)
$OE^{\overset{+}{\cdot}}$ ($M^{\overset{+}{\cdot}}$)	2	$OE^{\overset{+}{\cdot}}$ ($\alpha\alpha$)	$OE^{\overset{+}{\cdot}}$ (αi)
$OE^{\overset{+}{\cdot}}$ ($M^{\overset{+}{\cdot}}$)	3	EE^+ ($\alpha\alpha\alpha$)	$[EE^+$ ($\alpha\alpha i$)$]^b$
EE^+	1	$[OE^{\overset{+}{\cdot}}]^b$	EE^+
EE^+	2	EE^+	$[OE^{\overset{+}{\cdot}}]^b$

[a]Designations "α" and "i" are alpha and inductive cleavages, respectively, as explained in the text. Two i reactions lead to the same charge behavior (retention or migration) as two α reactions. Brackets indicate products of reactions discussed in Chapter 8.
[b]Not favored.

an $OE^{\ddagger}$ precursor; this is why you should mark important $OE^{\ddagger}$ ions (Section 3.4).

Summary. The ion-decomposition mechanisms discussed in the rest of this chapter are summarized inside the back cover.

4.5 Sigma-bond dissociation (σ)

$$\text{Alkanes:} \quad R^{\ddagger}CR_3 \xrightarrow{\sigma} R\cdot + \overset{+}{C}R_3 \qquad (4.7)$$

If the electron lost in ionization comes from a single bond, cleavage at this location will of course be favored. The more abundant ionized fragment will be the one better able to stabilize the positive charge. For saturated hydrocarbons such single-bond ionization is the lowest energy process; this can then account for favored alkane fragmentation at carbon atoms which are more highly substituted, and which therefore should be more easily ionized. (The percentage values indicate the ion abundance relative to the base peak of the spectrum.)

$$(CH_3)_3C-CH_2CH_3 \xrightarrow{-e^-} (CH_3)_3C^{\ddagger}CH_2CH_3 \xrightarrow{\sigma}$$
$$(CH_3)_3C^+ + \cdot CH_2CH_3 \qquad (4.8)$$
$$100\%$$

Also compare the abundances of the homologous $C_nH_{2n+1}{}^+$ ions (an "EE^+ ion series," Section 5.2) in Figures 3.2 to 3.4, and in the spectrum of pristane, Figure 4A. However, for σ-ionization at the same carbon atom, such as that leading to the formation of the key m/z 113 and 183 peaks in Figure 4A, the rule for loss of the largest alkyl radical applies; $[(M - C_6H_{13})^+] < [(M - C_{11}H_{23})^+]$. The abundant $C_nH_{2n+1}{}^+$ ions such as m/z 43 and 57 formed by secondary fragmentations will be discussed in Section 5.1.

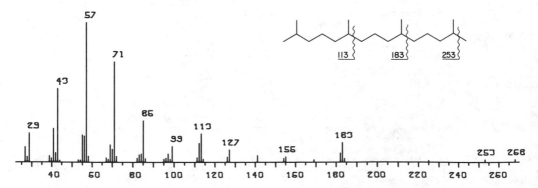

Figure 4A. Mass spectrum of pristane (2,6,10,14-tetramethylpentadecane).

4.6 Radical site initiation (α-cleavage)

Equations 4.9 to 4.12 summarize the reactions discussed in this section.

$$\text{Saturated site:} \quad R\overset{\frown}{-}CR_2\overset{\curvearrowright}{\overset{\cdot+}{Y}}R \xrightarrow{\quad\alpha\quad} R\cdot + CR_2{=}\overset{+}{Y}R \tag{4.9}$$

$$\overset{+}{Y}R\overset{\frown}{-}CH_2\overset{\curvearrowright}{\overset{\cdot}{C}}H_2 \xrightarrow{\quad\alpha\quad} Y\overset{\cdot+}{R} + CH_2{=}CH_2 \tag{4.10}$$

$$\begin{matrix}\text{Unsaturated} \\ \text{heteroatom:}\end{matrix} \quad R\overset{\frown}{-}CR{\overset{\curvearrowright}{\equiv}}\overset{\cdot+}{Y} \xrightarrow{\quad\alpha\quad} R\cdot + CR{\equiv}Y^+ \tag{4.11}$$

$$\begin{matrix}\text{Alkene} \\ \text{(allylic cleavage):}\end{matrix} \quad R\overset{\frown}{-}CH_2\overset{\curvearrowright}{\overset{\cdot+}{C}}H{\cdot+}CH_2 \xrightarrow{\quad\alpha\quad}$$
$$R\cdot + CH_2{=}CH{-}\overset{+}{C}H_2 \tag{4.12}$$

Reaction initiation at the radical site arises from its strong tendency for *electron pairing;* the odd electron is *donated to form a new bond* to an adjacent atom. This is accompanied by cleavage of another bond to that α-atom; thus this is commonly called the "α-cleavage reaction":

$$CH_3\overset{\frown}{-}CH_2\overset{\curvearrowright}{\overset{\cdot+}{O}}C_2H_5 \xrightarrow{\quad\alpha\quad} CH_3\cdot + CH_2{=}\overset{+}{O}C_2H_5 \ (\leftrightarrow \overset{+}{C}H_2{-}OC_2H_5)$$
$$50\% \ (COH_3{}^+ = 100\%) \tag{4.13}$$

(see Figure 3I). A "fishhook" half-arrow indicates transfer of a *single* electron: the doubly barbed arrow indicates transfer of an electron pair.

The driving force is like that underlying the high reactivity of neutral radicals such as dimerization and hydrogen abstraction. The tendency for the radical site to initiate a reaction in competition with the charge site generally parallels the radical site's tendency to donate electrons: N > S, O, π, R$\cdot$ > Cl, Br > H, where π signifies an unsaturated site and R$\cdot$ an alkyl radical. (This ordering does not mean, however, that a chlorine atom cannot initiate a radical-site reaction in the absence of stronger driving forces.) The donating ability of a particular site is affected by its molecular environment; for predicting this, conventional resonance and inductive effects are generally applicable.

Ionization of an aliphatic ether (4.13) should occur preferentially by loss of an n-electron of the oxygen. Donation of the unpaired electron to the adjacent C—O bond is followed by transfer of an electron from another bond of this α-carbon atom. The resulting one-elctron bond then cleaves to give the alkyl radical and the resonance-stabilized oxonium ion; the greater the double-bond character of this ion, the lower will be the critical (activation) energy of the reaction. Note that only the radical site moves; the charge site remains on the oxygen.

Unknown 4.2. Predict which will be the most abundant product ion in the mass spectrum of 1-hydroxy-2-aminoethane (Harrison *et al.* 1971).

α-Cleavage at the carbonyl group can be visualized similarly with formation of the stable acylium ion as in reaction 4.14 (see Figures 3J and 3K).

$$
\begin{array}{c}
C_2H_5 \\
\diagdown \\
\overset{+\cdot}{C}{=}O \quad \xrightarrow{\ \alpha\ } \quad C_2H_5\cdot \ + \ C_2H_5{-}C{\equiv}\overset{+}{O} \\
\diagup \\
C_2H_5
\end{array}
\qquad (4.14)
$$

$$100\%$$

A reaction initiated by an olefinic double bond (or phenyl π-system) yields an allyl (or benzyl) ion (reaction 4.15); here the radical site can be on either atom of the double bond. An allylic-type cleavage should be less favored at

$$
CH_3{-}CH_2{-}CH\overset{\cdot}{\underset{\cdot}{+}}CH_2 \quad \xrightarrow{\ \alpha\ }
$$
$$
CH_3\cdot \ + \ CH_2{=}CH{-}\overset{+}{C}H_2 \ (\leftrightarrow \overset{+}{C}H_2{-}CH{=}CH_2)
$$
$$100\%$$
$$(4.15)$$

$$
CH_3{-}CH_2{-}\overset{+\cdot}{C_6H_5} \quad \xrightarrow{\ \alpha\ } \quad CH_3\cdot \ + \ CH_2{=}\overset{+}{C}_6H_5 \ (\leftrightarrow \overset{+}{C}H_2{-}C_6H_5)
$$
$$100\%$$

a carbonyl double bond because of poor resonance stabilization of the resulting ion:

$$
CH_3{-}CH_2{-}C(C_2H_5)\overset{\cdot}{\underset{\cdot}{+}}O \xrightarrow{\ \times\ }
$$
$$
CH_3\cdot \ + \ CH_2{=}C(C_2H_5){-}\overset{+}{O} \longleftrightarrow \overset{+}{C}H_2{-}C(C_2H_5){=}O \qquad (4.16)
$$
$$0.5\% \ (C_3H_5O^+ = 100\%)$$

Unknown 4.3. Abundant loss of H is an uncommon fragmentation pathway. Postulate a structure for Unknown 4.3 for which such a loss is reasonable.

Loss of the largest alkyl radical. This rule (Section 4.2) is generally applicable to α-cleavage reactions. Note that there are two α-bonds for the carbonyl

$$
\text{group } {-}\overset{\overset{\displaystyle O}{\|}}{C}{-}, \text{ three for an alcohol } HO\overset{|}{\underset{|}{C}}{-} \text{ or primary amine } H_2N\overset{|}{\underset{|}{C}}{-}, \text{ six}
$$

$$
\text{for an ether } {-}\overset{|}{\underset{|}{C}}{-}O{-}\overset{|}{\underset{|}{C}}{-}, \text{ and nine for a tertiary amine } N{\equiv}\overset{|}{\underset{|}{C}}{-})_3.
$$

This rule predicts that the spectrum of 3-methyl-3-hexanol (Figure 4B) should show characteristic peaks at m/z 73, 87, and 101, with abundances decreasing in the order shown in reaction 4.17.

Unknown 4.3

m/z	Rel. abund.	m/z	Rel. abund.	m/z	Rel. abund.
15	1.1	52	1.4	70	1.8
26	3.3	53	8.7	71	3.8
27	8.0	54	0.3	72	0.5
37	4.1	57	3.9	81	0.9
38	4.9	58	6.6	82	1.0
39	13.	59	5.1	83	0.5
40	0.4	60	0.7	95	1.0
45	21.	61	1.6	96	0.6
46	0.9	62	1.9	97	100.
47	1.9	63	3.0	98	56.
48	0.8	64	0.5	99	7.6
49	3.1	65	2.4	100	2.4
50	3.6	68	0.7		
51	4.0	69	6.2		

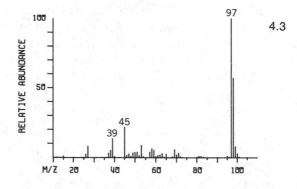

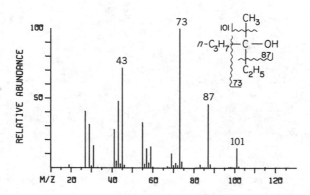

Figure 4B. Mass spectrum of 3-methyl-3-hexanol.

$$C_3H_7 - \overset{\overset{\displaystyle CH_3}{|}}{\underset{\underset{\displaystyle C_2H_5}{|}}{C}} \overset{+}{\not{O}H} \xrightarrow{\ \alpha\ }$$

(4.17)

$$[C_2H_5C(CH_3)\!=\!\overset{+}{O}H] > [C_3H_7C(CH_3)\!=\!\overset{+}{O}H] >$$
$$m/z\ 73,\ 100\% \qquad\qquad m/z\ 87,\ 50\%$$

$$[C_3H_7C(C_2H_5)\!=\!\overset{+}{O}H]$$
$$m/z\ 101,\ 10\%$$

The overwhelming driving force provided by the nitrogen atom's electron-donating ability makes the α-cleavage reaction dominant in the spectra of aliphatic amines. The only α-groups in *tert*-butylamine are methyls, and this loss yields the major peak of the spectrum (Figure 4C); the righthand ordinate shows that this represents 58 per cent of all ions collected. There are two α-methyls and four α-hydrogens in diethylamine (Figure 4D); methyl loss is still dominant, yielding 30 per cent of the total ions, versus 6 per cent for loss of H (the large m/z 30 peak is discussed in Section 4.10). For *n*-alkylamines ($R-NH_2$, $R = CH_3$ to $n\text{-}C_{14}H_{29}$), the m/z 30 is the most abundant ion. The replacement of a terminal hydrogen atom in *n*-dodecane by an amine group causes the profound spectral change shown between Figures 3B and 3P.

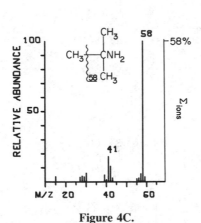

Figure 4C.

Mass spectrum of *tert*-butylamine.

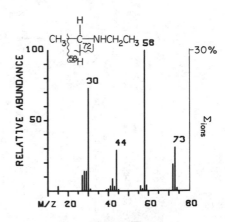

Figure 4D.

Mass spectrum of diethylamine.

Test your understanding of these concepts with Unknown 4.4 to 4.8, which are spectra of isomeric $C_4H_{11}N$ alkylamines.

Unknowns 4.4 to 4.8

m/z	Relative abundance 4.4	4.5	4.6	4.7	4.8
15	0.7	0.5	3.1	1.3	1.9
27	2.9	2.7	4.6	10.	0.8
28	4.6	5.1	9.1	11.	4.2
29	2.2	2.2	3.6	8.1	9.1
30	100.	100.	29.	13.	2.9
31	2.1	2.2	1.3	0.3	4.1
32	0.3	0.3	1.8	0.0	0.4
33	0.0	0.0	0.0	0.0	1.1
39	1.9	0.2	4.2	1.2	2.0
40	0.4	0.5	1.1	2.1	0.8
41	2.9	2.8	7.4	4.5	9.4
42	1.7	0.4	7.4	28.	6.0
43	1.2	5.8	8.8	7.2	3.1
44	2.0	1.3	1.6	25.	100.
45	0.4	1.6	0.2	1.8	2.8
56	1.1	2.1	8.1	7.3	2.3
57	0.2	1.1	4.2	5.0	1.6
58	0.3	1.9	100.	100.	10.
59	0.0	0.1	3.9	3.9	0.4
71	0.0	0.0	0.6	1.0	0.4
72	1.0	1.3	9.6	17.	2.3
73	10.	10.	11.	23.	1.2
74	0.8	0.5	1.3	1.1	0.1

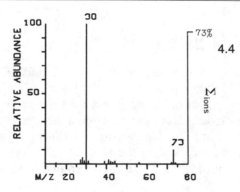

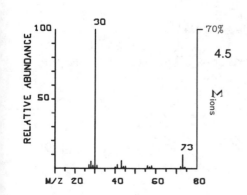

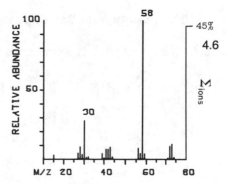

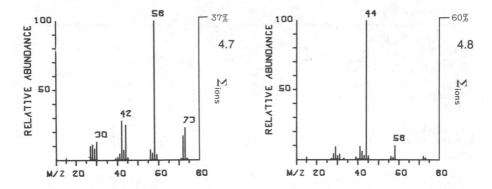

First write down the eight possible amines of molecular weight 73. Start with the four spectra that have a base peak at m/z 58; which of your structures will show a ready loss of 15? (Thus, which have an α-methyl group?) You should have

$$\underset{\substack{| \\ (CH_3)_2CNH_2,}}{CH_3} \quad \underset{\substack{| \\ C_2H_5CHNH_2,}}{CH_3} \quad \underset{\substack{| \\ CH_3CHNHCH_3,}}{CH_3}$$

$$\underset{\substack{| \\ CH_2NHC_2H_5,}}{CH_3} \quad \text{and} \quad \underset{\substack{| \\ CH_2N(CH_3)_2}}{CH_3}$$

The spectra of two of these are Figures 4C and 4D. The other isomeric C_4 amines of Unknowns 4.4 through 4.8 contain the remaining three α-methylamines; yet in only two is mass 58 the base (largest) peak. Why? This should be obvious from the possible molecular structures.

To assign structures to Unknowns 4.4 through 4.8, try *predicting* the major ions in the spectra of all the isomeric C_4 alkylamines. The spectrum of one isomer is not included in these unknowns. Which one?

4.7 Charge-site initiation (inductive cleavage, *i*)

Equations 4.18 to 4.21 summarize the reactions discussed in this section.

$$OE^{\ddagger}: R\overset{\frown}{\underset{}{Y}}-R \xrightarrow{\ i\ } R^+ + \cdot YR \tag{4.18}$$

$$\underset{\substack{R \\ | \\ R}}{\overset{}{C}}=\overset{+\cdot}{Y} \left(\longleftrightarrow \underset{\substack{R \\ | \\ R}}{\overset{}{\overset{+}{C}}}-\overset{..}{Y}: \right) \xrightarrow{\ i\ } R^+ + R-\overset{.}{C}=Y \tag{4.19}$$

$$\text{EE}^+: \quad \text{R}\overset{\curvearrowleft}{\underset{}{-}}\overset{+}{\text{Y}}\text{H}_2 \xrightarrow{\quad i \quad} \text{R}^+ + \text{YH}_2 \tag{4.20}$$

$$\text{R}\overset{\curvearrowleft}{\underset{}{-}}\overset{+}{\text{Y}}{=}\text{CH}_2 \xrightarrow{\quad i \quad} \text{R}^+ + \text{Y}{=}\text{CH}_2 \tag{4.21}$$

Initiation of a cleavage reaction by the positive charge involves *attraction of an electron pair*. The tendency for the formation of R^+ from RY is: halogens $>$ O, S $\gg$ N, C; for elements of the same row of the periodic table this tendency parallels the inductive effect (i) of Y. However, stabilization of the charge is generally more important than that of the radical for determining reaction products (Section 4.2) in both the formation and decomposition of the precursor; because these i cleavages require *migration of the charge*, they are generally *less favored* than radical-site reactions.

Oxygen is intermediate in its ability to influence either α or i reactions. In an aliphatic ether *attraction* of an electron pair initiated by the localized positive charge on the oxygen can form the alkyl ion and the alkoxyl radical; in this case the charge site is moved (reaction 4.22; see Figure 3I). Note

$$\text{C}_2\text{H}_5\overset{\curvearrowleft}{\underset{}{-}}\overset{+}{\text{O}}\text{C}_2\text{H}_5 \xrightarrow{\quad i \quad} \text{C}_2\text{H}_5{}^+ + \cdot\text{OC}_2\text{H}_5 \tag{4.22}$$
$$\text{40\%}$$

$$\text{CH}_3\text{CH}_2\text{CH}_2{-}\text{CH}_2\overset{\overset{\longrightarrow}{\cdot\cdot+}}{-}\text{Cl} \xrightarrow{\quad i \quad} (\text{M} - \text{HCl})^{+\cdot}, \ \text{C}_4\text{H}_9{}^+, \ \text{C}_3\text{H}_7{}^+ \tag{4.23}$$
$$\qquad\qquad\qquad\qquad\qquad\quad \text{100\%} \qquad\quad \text{3\%} \qquad \text{30\%}$$

$$(\text{CH}_3)_2\text{CH}{-}\text{CH}_2\overset{\overset{\longrightarrow}{\cdot\cdot+}}{-}\text{Cl} \xrightarrow{\quad i \quad} (\text{M} - \text{HCl})^{+\cdot}, \ \text{C}_4\text{H}_9{}^+, \ \text{C}_3\text{H}_7{}^+ \tag{4.24}$$
$$\qquad\qquad\qquad\qquad\quad \text{7\%} \qquad\quad \text{4\%} \qquad \text{100\%}$$

that the bond cleaved ($\text{R}'{-}\text{CH}_2{\!+\!}\text{OR}$) is *not* that cleaved by the radical-site initiation ($\text{R}'{\!+\!}\text{CH}_2{-}\text{OR}$, reaction 4.13). The reaction is also competitive for many nitroalkanes, alkyl iodides, secondary and tertiary alkyl bromides, and tertiary alkyl chlorides, but not for $\text{RCH}_2{-}\text{Y}$ compounds of higher C—Y bond strengths, such as *n*-alkanols and *n*-alkyl chlorides; for these $[(\text{M} - \text{HY})^{+\cdot}] > [(\text{M} - \text{Y})^+]$ as in reaction 4.23 (see Section 4.9). As is expected for an inductive effect, the influence of the charge site can affect a bond that is farther away if it has more polarizable electrons; the loss of CH_2Cl is much larger in reaction 4.24 than in 4.23.

This reaction will not cleave a multiple bond to a heteroatom. For the carbonyl group (4.25), electron-pair attraction to the charge site yields the canonical resonance structure; inductive cleavage of this then gives the alkyl ion and acyl radical. However, these products are the *complements* of the alkyl radical and acylium ion formed by α-cleavage (reaction 4.14), in contrast to the α- and i-reactions (4.13 and 4.22) at a saturated heteroatom.

$$\text{R}\diagdown\text{C}\overset{\cdot\!+}{=}\ddot{\text{O}}\; \Big(\longleftrightarrow\; \text{R}\diagdown\overset{+}{\text{C}}-\text{O}\cdot\Big)\;\overset{i}{\longrightarrow}\; \text{R}^{+} + \text{R}'-\dot{\text{C}}=\text{O} \qquad (4.25)$$

Considering reactions 4.14 and 4.25, one would expect aliphatic ketones to show four major ions from cleavage of the two bonds to the carbonyl group; the favored acylium ion should be the one formed through loss of the larger alkyl group, and the favored carbonium ion should be the more stable one. Unknowns 4.9 and 4.10 are the spectra of 3-pentanone and 3-methyl-2-butanone; which is which?

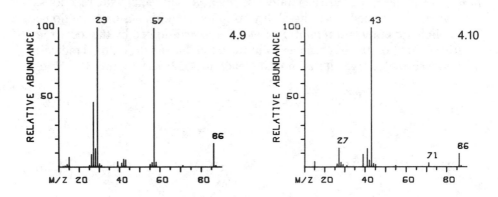

Even-electron ions. As discussed in Section 4.4, the favored decomposition of an EE^{+} ion yields another EE^{+} ion and a molecule as the neutral product. Examples are the second steps of reactions 4.26 and 4.27, which are two-step pathways for forming the same products produced by reactions 4.22 and 4.25, respectively. However, these are energetically less favorable by 0.2–0.8 eV (20–80 kJ/mol) than their one-step counterparts.

$$\text{R}'-\text{CH}_2-\overset{\cdot\!+}{\text{O}}-\text{R}\;\overset{-\text{R}'\cdot}{\underset{\alpha}{\longrightarrow}}\;\text{CH}_2=\overset{+}{\text{O}}-\text{R}\;\overset{i}{\longrightarrow}\;\text{CH}_2\text{O} + \text{R}^{+} \qquad (4.26)$$

$$\text{R}'\diagdown\text{C}\overset{+}{=}\ddot{\text{O}}\;\overset{-\text{R}'\cdot}{\underset{\alpha}{\longrightarrow}}\;\text{R}-\text{C}\overset{+}{\equiv}\ddot{\text{O}}\;\overset{i}{\longrightarrow}\;\text{R}^{+} + \text{CO} \qquad (4.27)$$

$$\text{R}-\text{OH}\;\overset{\text{H}^{+}}{\underset{\text{CI}}{\longrightarrow}}\;\text{R}-\overset{+}{\text{OH}}_2\;\overset{i}{\longrightarrow}\;\text{R}^{+} + \text{H}_2\text{O} \qquad (4.28)$$

The initial ions produced by chemical ionization (CI) are mainly EE^+ ions such as $(M + H)^+$ and $(M - H)^+$; also field ionization often produces EE^+ ions (Section 6.1). For decompositions of the $(M + H)^+$ ion (4.28), the loss of molecules of lower proton affinity (Table A.3) is generally more favorable (Field 1972). The EI mass spectrum of ephedrine (Figure 4E) shows no molecular ion; the spectrum is so dominated by the α-cleavage reaction leading to $CH_3N^+H{=}CHCH_3$ (*m/z* 58) that there are only weak peaks indicative of the rest of the molecule. However, the CI spectrum (Fales 1971; Figure 4E) has abundant ions at 166, $(M + H)^+$, indicating the molecular weight, and at 148, $(M + H - H_2O)^+$. The latter, formed by inductive cleavage (4.28), thus indicates the presence of the hydroxyl group.

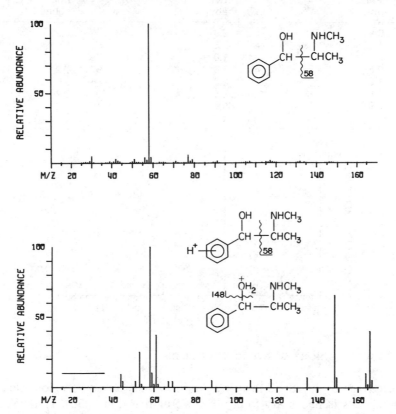

Figure 4E. Electron ionization and chemical ionization (CH_4 reagent) mass spectra of ephedrine.

Despite the lower tendency for charge-site reactions of the amino group, the small peak produced in the CI spectrum at *m/z* 135 (4.29) is very useful, since it indicates the N-methyl group.

$$C_6H_5-CHOH-CHCH_3-\overset{\curvearrowleft}{N}^+H_2CH_3 \xrightarrow{i}$$
$$C_6H_5-CHOH-\overset{+}{C}HCH_3 + H_2N-CH_3 \quad (4.29)$$

In Unknown 4.11 the m/z 130 peak has the elemental composition $C_8H_{18}O$. Utilize the above mechanisms in elucidating the structure of this molecule. (*Hint:* The molecular-ion information from Table A.4 is also valuable.)

Unknown 4.11

m/z	Rel. abund.	m/z	Rel. abund.	m/z	Rel. abund.
27	6.1	42	2.9	59	2.2
28	1.6	43	9.8	71	1.7
29	15.	44	1.0	73	1.1
30	0.5	45	3.6	87	7.9
31	1.2	55	2.3	88	0.5
39	5.6	56	4.8	101	0.7
40	0.9	57	100.	115	0.2
41	22.	58	4.5	130	2.1 $C_8H_{18}O$

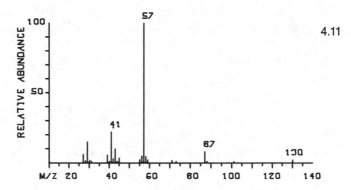

4.8 Decompositions of cyclic structures

The cleavage of one bond in a ring produces only an isomeric ion; cleavage of two bonds is necessary to produce a fragment ion (4.30). Thus (Section 4.4) this product must be an *odd-electron* ion. This is produced from the acyclic isomer by an α-cleavage reaction. The unpaired electron is donated to form a new bond to the adjacent carbon atom, with concomitant cleavage of another bond to that α-carbon. The mass spectrum of cyclohexane is shown in Figure 4F (see also Figure 3F). Note that the important OE$^{+\cdot}$ ion $C_4H_8^{+\cdot}$ (m/z 56) has the most hydrogens (Section 3.4) of any $C_4H_n^+$ ion.

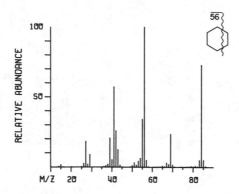

Figure 4F. Mass spectrum of cyclohexane.

In cyclohexene the π-electrons provide a favored site for the initial charge and radical (reaction 4.31). Donation of this unpaired electron produces an acyclic isomer by α-cleavage; a second such reaction eliminates neutral C_2H_4. Note that the other product is ionized 1,3-butadiene, so that this corresponds to a retro-Diels-Alder reaction. In reactions yielding an OE$^{+\cdot}$ product, the stabilization of the unpaired electron as well as the charge is an important driving force; note that both of these are much better stabilized in the ion product of 4.31 than in that of 4.30.

m/z 54: R = H, 80% ($C_5H_7^+ = 100\%$)
($I = 9.1$) R = C_6H_5, 0.4%

(charge retention) (4.31)

(charge migration) (4.32)

R = H ($I = 10.5$), < 5%
R = C_6H_5 ($I = 8.4$), 100%

As with other hydrocarbon fragmentation reactions, the retro-Diels-Alder reaction will be important only if no preferable fragmentations are possible. Thus the addition of a functional group can dramatically reduce the abundance of ions from this reaction. Unknowns 4.12 and 4.13 show the spectra of α- and β-ionone (Figure 4G). Pair the structures and spectra, and justify your choice.

α-Ionone   β-Ionone

Figure 4G.

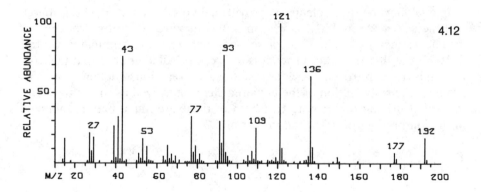

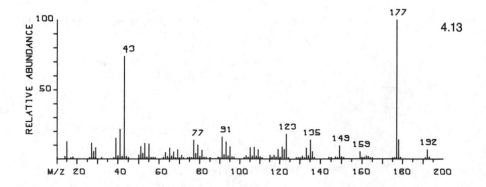

Charge migration. The final step of 4.31 is shown as a homolytic bond cleavage, with each of the separating fragments retaining one of the electrons. However, heterolytic cleavage (*i*) is also possible (4.32), moving the positive charge to produce the ionized alkene (Figure 4H). In the transition state the alkene and olefin are competing for an electron; according to "Stevenson's Rule" the product more likely to lose the electron is the one with the lowest ionization energy (*I*, Table A.3). Thus for R = H, the ionization of butadiene is favored over that of ethylene. However, for R = phenyl, the resulting styrene ion should be favored. If the *I* values are similar, both ions should be present. Look for both; note that the mass sum of these complementary OE‡ ions is equivalent to the molecular weight.

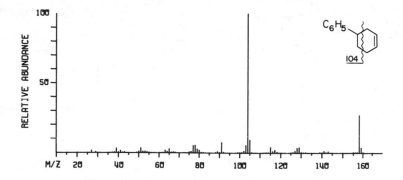

Figure 4H. Mass spectrum of 4-phenylcyclohexene.

Unknown 4.14. Ionization of a nonbonding electron can also trigger such ring fragmentations. Predict an abundant odd-electron fragment ion in the mass spectrum of *p*-dioxane,

$$\begin{array}{c} H_2C \underset{\underset{H_2C}{|}}{\overset{O}{\diagup}} \underset{\underset{O}{CH_2}}{\overset{CH_2}{\diagdown}} \end{array}$$

4.9 Radical-site rearrangements

There are also important mass-spectral reactions which produce ions whose atoms have not retained the connectivity relationships of the original molecule; part of the precursor ion has reacted with another part before or during decomposition. Sometimes such rearrangements are so extensive that they make the product ion useless (or misleading) for deducing structure; such "random" rearrangements are common where σ- (and sometimes π-) ionization is necessary. Fortunately, many rearrangements occur by means of "specific" mechanisms which are now well-understood, and these product ions are thus valuable for deducing structure. Hydrogen-atom rearrangements initiated at a radical site, the main subject of this section, are an important class of reactions exhibited by a wide variety of structures.

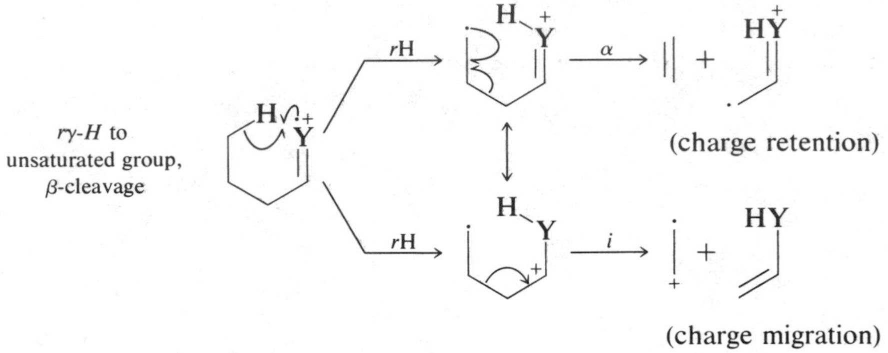

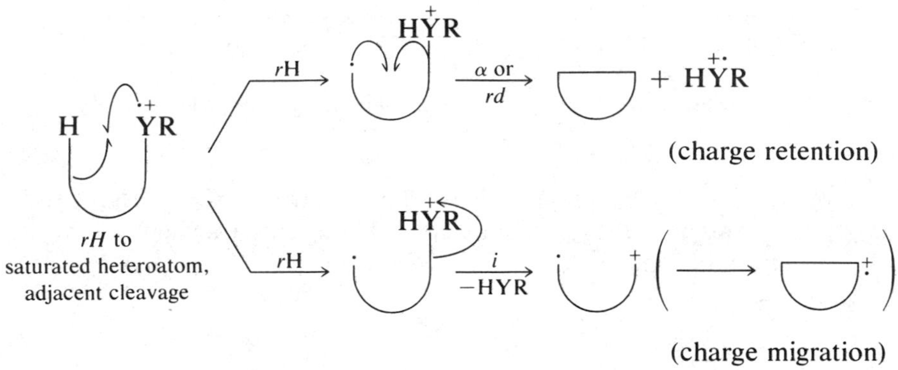

$r\gamma$-H to unsaturated group, β-cleavage

(charge retention)

(charge migration)

rH to saturated heteroatom, adjacent cleavage

(charge retention)

(charge migration)

γ-H rearrangement to an unsaturated group with β-cleavage. An unpaired electron can also be donated to form a new bond to an adjacent atom through space; as in reaction 4.14, the second electron of this pair is supplied by transfer from another bond to this adjacent atom, resulting in its cleavage (4.33). Flatteringly, this is usually referred to as the "McLafferty rearrangement" (McLafferty 1956; Kingston, Bursey, and Bursey 1974). For compounds containing an unsaturated functionality such as the carbonyl group, the γ-hydrogen atom is transferred by a sterically favorable six-membered ring transition state. However, in this process the initial cleavage does not result in the loss of part of the ion, only in a change in the position of the radical site. The new radical site can now initiate an α-cleavage reaction resulting in fragmentation of the carbon–carbon bond which is beta to the carbonyl group with loss of an olefin or other stable molecule to form the odd-electron ion; an OE^{+} is formed because two bonds are cleaved to effect fragmentation in M^{+} in this rearrangement reaction.

(charge retention)

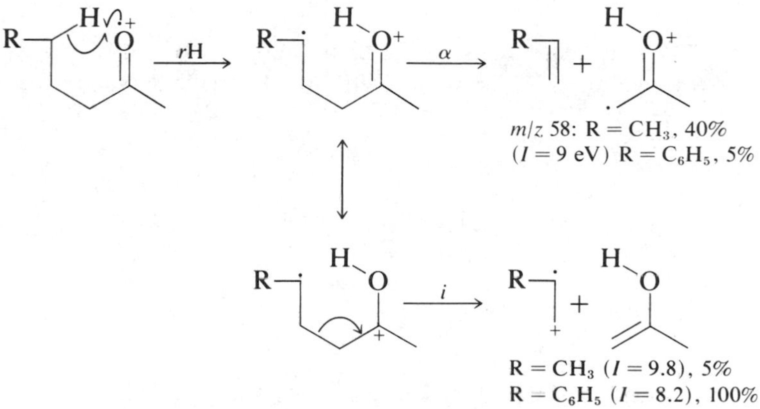

$$ (4.33) $$

m/z 58: R = CH$_3$, 40%
($I = 9$ eV) R = C$_6$H$_5$, 5%

$$ (4.34) $$

R = CH$_3$ ($I = 9.8$), 5%
R = C$_6$H$_5$ ($I = 8.2$), 100%

(charge migration)

Note the concomitant *formation* of two bonds, accounting for the characteristically low critical energy of such reactions. Part of the driving force for this rearrangement is the resonance stabilization of the radical site in the product ion, which is isoelectronic with the allyl radical; note that this requires β-bond cleavage in the second reaction step, and thus necessitates γ-H transfer to produce the reactive intermediate.

Such hydrogen rearrangement through a six-membered-ring intermediate yields *characteristic* OE^{+} *ions for a wide variety of unsaturated functional groups,* such as aldehydes, ketones (Figures 3J and 3K), esters (Figures 3M to 3O), acids (Figure 3L), amides (Figures 3S and 3T), carbonates, phosphates, sulfites, ketimines, oximes, hydrazones (reaction 4.35), alkenes, alkynes, and phenylalkanes (4.36; see Figure 3G).

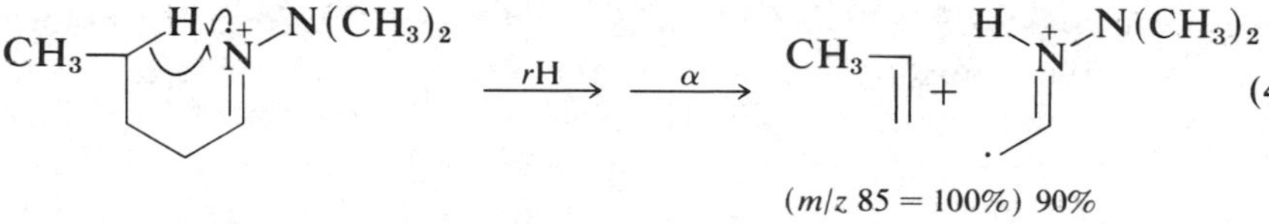

$$ (4.35) $$

(m/z 85 = 100%) 90%

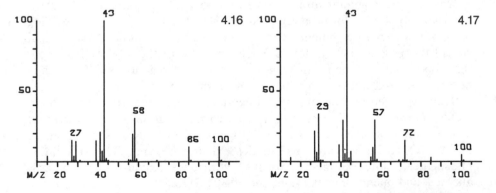

$$(m/z\ 91 = 100\%)\ m/z\ 92,\ 60\%$$

Unknown 4.15. What is the mass of the characteristic $OE^{\ddagger}$ ion (or ions) which will be formed by this six-membered ring H rearrangement from each of the following $M^{\ddagger}$:

$$CH_3CH_2CH_2CH(C_2H_5)\overset{\overset{\displaystyle O}{\|}}{C}NH_2^{\ddagger},\ CH_3CH_2CH_2C(CH_3){=}CH_2^{\ddagger},$$

$$CH_3CH_2CH_2\overset{\overset{\displaystyle O}{\|}}{C}OCH_2CH_3^{\ddagger},\ C_2H_5OCH_2C_6H_5^{\ddagger},\ CH_3CH_2O\overset{\overset{\displaystyle O}{\|}}{C}OCH_2CH_3^{\ddagger}?$$

Draw out the mechanism, complete with "fishhooks," for each.

Unknowns 4.16 and 4.17. These are the spectra of 3- and 4-methyl-2-pentanone; which is which?

Note that these isomers produce the same peaks by α-cleavage (m/z 43 and 85) and by i-cleavage (m/z 15 and 57), whereas the β-cleavage rearrangement makes it possible to distinguish between the isomers by characterizing the substituents on the α-position. The tendency for rearrangement is greater for hydrogen on branched, allylic, oxygen, or other labile sites.

Unknown 4.18. Does an important odd-electron ion aid in the solution of this unknown?

Unknown 4.18

m/z	Rel. abund.	m/z	Rel. abund.	m/z	Rel. abund.
15	0.3	52	1.9	92	55.
26	1.2	53	1.1	93	3.9
27	11.	55	0.5	103	2.0
28	1.1	63	3.3	104	1.4
29	3.9	64	1.0	105	8.5
38	1.2	65	10.	106	0.8
39	10.	66	0.7	115	1.0
40	0.9	76	0.6	116	0.3
41	5.3	77	5.9	117	0.7
42	0.4	78	6.2	119	0.8
43	2.7	79	2.7	134	24.
50	2.7	89	1.7	135	2.7
51	7.4	91	100.		

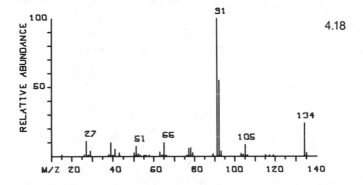

4.18

Charge migration can also occur (reaction 4.34). In the dissociation transition state, the separating products can compete for the ionic charge, paralleling the complementary OE^{+} ion formation in ring cleavages (Section 4.8). The favored ion product should arise from the fragment of lower ionization energy (I) (Stevenson's Rule, 1951). Although the spectrum of 4-methyl-2-pentanone (Unknown 4.16) shows 30 per cent of the rearrangement product $CH_2\text{==}CH(OH)CH_3^{+}$ (m/z 58, $I = 9.1$ eV), a far greater amount than that of the complementary $C_3H_6^{+}$ ($I = 9.7$ eV), substitution of a 5-phenyl (Figure 4I) reverses this, yielding 100 per cent $C_6H_5CH\text{==}CH_2^{+}$ (m/z 104, $I = 8.4$ eV). The spectrum of $C_3H_7COOC_2H_5$ shows $[C_3H_7COOH^{+}]$ ($I = 10.2$ eV) $\gg [C_2H_4^{+}]$ ($I = 10.5$ eV), whereas those of $C_3H_7COOC_3H_7$ (n- or iso-) show $[C_3H_6^{+}]$ ($I = 9.8$ eV) $> [C_3H_7COOH^{+}]$. Although the charge retention OE^{+} product has gained an H atom in the rearrangement, the charge migration OE^{+} product has *lost* a hydrogen; thus the former usually has the most H atoms in its peak group, but the latter may have one less. Again, the sum of these complementary masses equals that of M^{+}; in Figure 4I, $58 + 104 = 162$.

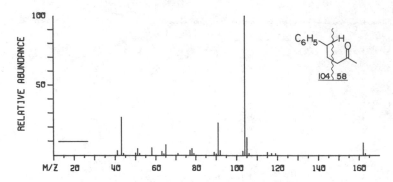

Figure 4I. Mass spectrum of 5-phenyl-2-pentanone.

Hydrogen rearrangement to a saturated heteroatom with adjacent cleavage. Another important mass-spectral reaction triggered by intramolecular H transfer involves an initial radical site on a saturated heteroatom such as reaction 4.38. Again, the unpaired electron is donated to form a new bond to an adjacent (in appropriate conformations) H atom, with concomitant cleavage of another bond to that hydrogen. However, a charge-site reaction

$$C_2H_5 \underset{OH}{\overset{H}{\frown}} \xrightarrow{rH} C_2H_5 \overset{H}{\underset{OH}{\frown}} \xrightarrow{rd}$$

$$C_2H_5 \square + HOH^+ \quad (4.37)$$
$$m/z \ 18, \ < 3\%$$

(charge retention)

$$\xrightarrow{rH} \quad C_2H_5 HOH \xrightarrow[-HOH]{i}$$

$$C_2H_5 \xrightarrow{i} C_2H_5 + \underset{H_2C}{\overset{CH_2}{\diagdown}} \quad (4.38)$$

$$m/z \ 84, \ 11\% \qquad m/z \ 56, \ 100\%$$

(charge migration)

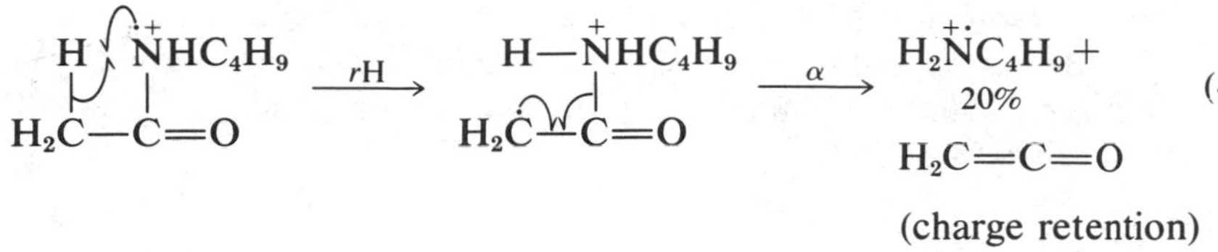

$$
\begin{array}{ccccc}
\underset{H_2C-C=O}{H \overset{+}{\underset{|}{NHC_4H_9}}} & \xrightarrow{rH} & \underset{H_2C-C=O}{H-\overset{+}{\underset{|}{NHC_4H_9}}} & \xrightarrow{\alpha} & \underset{20\%}{H_2\overset{+\cdot}{N}C_4H_9 +} \\
& & & & H_2C=C=O
\end{array}
$$ (4.39)

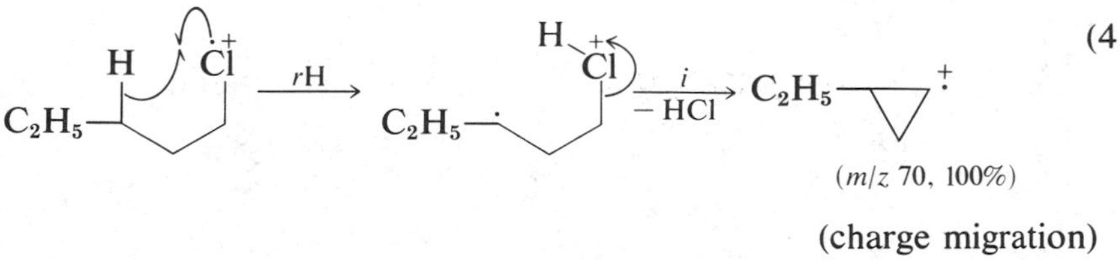

(charge retention)

(4.40)

$$\xrightarrow[-HCl]{i} \quad C_2H_5\!-\!\triangle^{\;\cdot}_{+}$$

(*m/z* 70, 100%)

(charge migration)

is now possible, forming in this case the $(M - H_2O)^{\ddagger}$ peak by cleavage of a bond to the heteroatom. Further charge-site reactions can give secondary products by cleavage of a bond that is two or more bonds further from the heteroatom (4.38). The products formed are not greatly affected by which hydrogen is transferred in the first step (a six-membered-ring intermediate is not required). Further, the resulting heteroatom-containing fragment is saturated, and so competes less well for the charge (Table A.3), making charge migration (4.38, 4.40) much more common. The charge-migration reaction favors electronegative species; small saturated molecules of higher ionization energy (Table A.3) such as H_2O, C_2H_4 (second step of 4.38), CH_3OH, H_2S, HCl, and HBr are commonly lost in this way (found in Figures 3H, 3T, 3U, 3W, and 3X, but not 3P). For charge retention the second step instead involves a radical site reaction (4.37, 4.39). The "*ortho*-effect" reaction (4.41) indicative of aromatic ring position results from a combination of a labile hydrogen available for rearrangement and a stable ion product from charge migration.

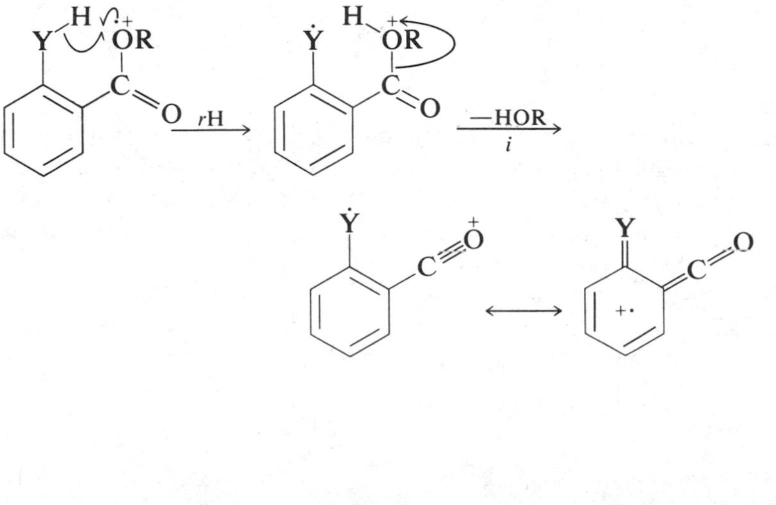

(4.41)

Unknown 4.19 involves some important points that have been covered previously. Be sure you understand its solution before proceeding. The utility of the *m/z* 96.5 peak formed by metastable ion decomposition will be discussed in Section 6.4.

Unknown 4.19

m/z	Rel. abund.	m/z	Rel. abund.	m/z	Rel. abund.
15	0.7	45	11.	84	0.8
18	1.1	46	0.2	85	1.7
19	0.9	55	86.	86	5.6
27	20.	56	6.6	87	57.
28	2.0	57	14.	88	3.4
29	12.	58	3.6	96.5	0.1
30	0.6	59	1.0	112[a]	7.4
31	11.	69	100.	113	1.0
39	6.4	70	8.4	127	0.3
40	1.0	71	3.7	128	0.5
41	34.	72	7.1	129[a]	0.9
42	4.0	73	61.	130	0.1
43	33.	74	2.8		
44	9.2	83	1.6		

[a]The elemental compositions of the *m/z* 112 and 129 ions are C_8H_{16} and $C_8H_{17}O$, respectively, by high-resolution mass spectrometry.

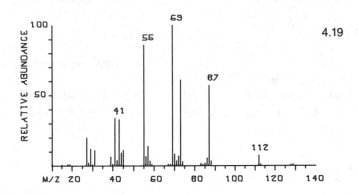

Displacement reactions, rd. Only brief mention will be made here of one of several types of nonhydrogen rearrangement (see Section 8.10). Two atoms or groups (one usually bearing the radical site) can react intramolecularly; formation of this new bond is accompanied by cleavage of another bond to one (or both) of these groups. Cyclization (reaction 4.42) to form a divalent chloronium ion with displacement of an alkyl group accounts for the largest peak in the spectra of appropriate 1-chloro- and 1-bromoalkanes (Figures 3W, 3X). This reaction is much less important in branched haloalkanes and for functionalities with a higher tendency for radical-site reactions. The mass spectrum of n-dodecyl mercaptan (Figure 3U) shows the following $C_nH_{2n+1}S^+$ ions: $n = 1$, 23 per cent ($CH_2=SH^+$ from α-cleavage); $n = 2$, 9 per cent; $n = 3$, 1 per cent; $n = 4$, 11 per cent (five-membered ring); and $n = 5$, 2 per cent. Note the similar "ion series" (Section 5.2) in n-dodecylamine (Figure 3P).

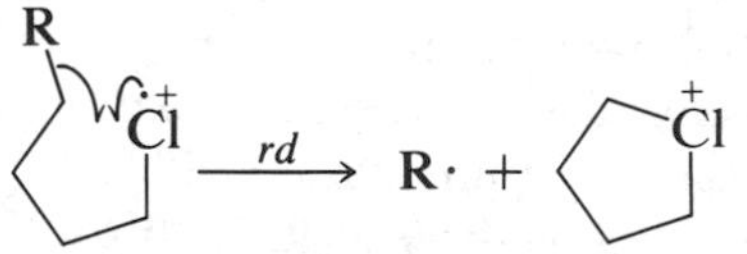

$$(4.42)$$

4.10 Charge-site rearrangements

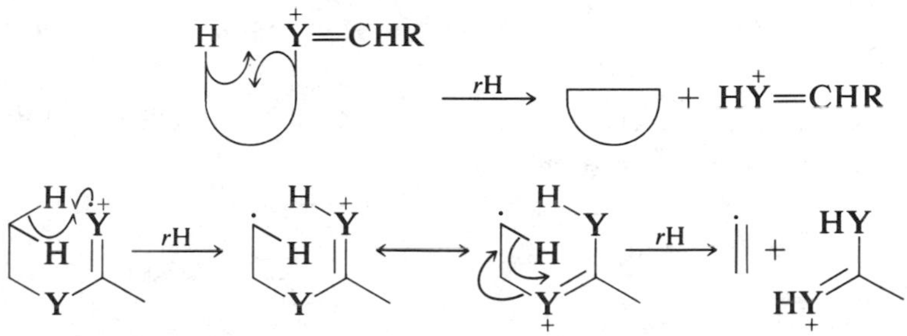

$$(4.43)$$

Even-electron acyclic ion decompositions in which the charge site retains its location and the electrons remain paired, both of which are usually desirable from an energetic standpoint, involve rearrangement (Section 4.4). For hydrogen rearrangement, these are favored if loss of a molecule of low proton affinity is involved. A variety of mechanisms are possible, and the resulting product ions can have their atoms scrambled to a confusing degree; thus the structural implications of these product ions must be interpreted with caution. However, the rearrangement of hydrogen to an unsaturated group containing a heteroatom, with elimination of an unsaturated neutral, can provide valuable information. This reaction gives the second largest peak in the spectrum of diethylamine, Figure 4D (reaction 4.44). The transition-state size requires an ethyl or larger group on the nitrogen for this rearrangement. Thus this reaction provides a spectral distinction between $(CH_3)_3CNH_2$ (Figure 4C), $(CH_3)_2CHNHCH_3$, and $CH_3CH_2N(CH_3)_2$ (Unknowns 4.5 and 4.6). Although these three spectra also show a base peak of $(M - CH_3)^+$ by α-cleavage, their m/z 30 peaks are much lower than that of diethylamine (note that these are still appreciable, so that such data must be utilized with caution).

$$CH_3CH_2\overset{+\cdot}{N}H \overset{}{\diagdown}CH_2 \overset{}{\diagdown} CH_3 \xrightarrow{-CH_3\cdot} \quad \overset{H}{\underset{H_2C \overset{}{\diagdown} CH_2}{\diagup}} \overset{+}{NH}=CH_2 \xrightarrow{rH}$$

$$H_2C=CH_2 + H_2\overset{+}{N}=CH_2 \qquad (4.44)$$

Unknown 4.20. Predict the principal peaks in the mass spectrum of N-methyl-N-isopropyl-N-*n*-butylamine.

Even-electron ions can also undergo H rearrangements which do *not* involve the charge site (nor an unsaturated heteroatom), such as reaction 4.45. The resulting peaks should be used with caution for assigning func-

$$\overset{CH_3}{\underset{\overset{+\cdot}{Cl}-CHCH_2CH_2Cl}{\mid}} \xrightarrow[i]{-Cl\cdot} CH_3\overset{+}{C}H-\overset{H}{\underset{\mid}{C}}H\overset{\overset{\cdot}{Cl}}{\underset{\mid}{\diagdown}}CH_2 \xrightarrow{rH}$$

$$CH_3\overset{+}{C}HCH=CH_2 + HCl \qquad (4.45)$$

tional group locations, however; note that *m/z* 55 is the base peak in both 1,3- and 1,1-dichlorobutane (Figure 4J).

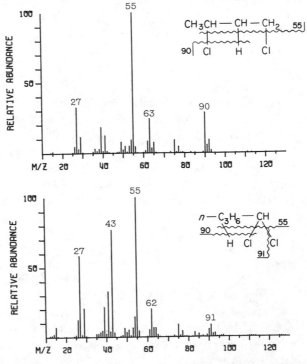

Figure 4J. Mass spectra of 1,3- and 1,1-dichlorobutane.

Rearrangement of two hydrogen atoms (sometimes called the "McLafferty + 1" rearrangement) is a characteristic decomposition of esters and similar functional groups that can be rationalized (reaction 4.46) using a transfer of the second hydrogen by a mechanism analogous to reaction 4.43. An added driving force is the resonance stabilization of the EE$^+$ product ion (three bonds are cleaved in the reaction). Although this ion is often formed in relatively low abundance, you should be able to recognize it from the unusual masses 27, 41, 55, etc. lost (Table A.5) in its formation. This rearrangement provides a useful characteristic of esters (Figures 3N and 3O), thioesters, amides (Figure 3T), and phosphates.

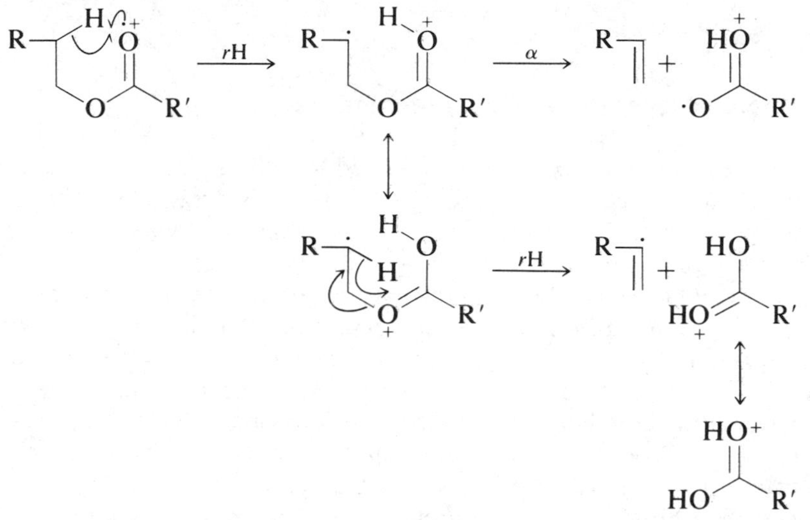

(4.46)

One of the above mechanisms should help in the elucidation of Unknown 4.21.

Unknown 4.21

m/z	Rel. abund.	m/z	Rel. abund.	m/z	Rel. abund.
27	3.6	56	19.	121	0.3
28	2.5	57	1.5	122	17.
29	5.1	65	0.4	123	68.
39	2.4	76	2.0	124	5.3
40	0.3	77	37.	125	0.5
41	6.0	78	3.0	135	1.3
42	0.3	79	5.1	149	0.3
43	0.9	80	0.3	163	0.3
50	3.0	104	0.7	178	2.0
51	1.1	105	100.	179	0.3
52	0.8	106	7.8		
55	2.7	107	0.5		

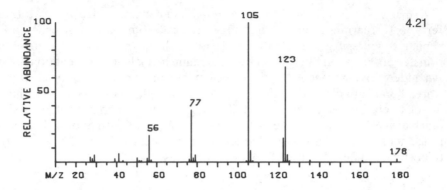

4.11 Summary of types of reaction mechanisms

The summary inside the back cover is intended to underscore the relationships of the common types of reactions which ions can undergo, using examples from this chapter. You should use this as a checklist when you are attempting to predict the spectrum of a molecule; to use it effectively, however, you will have to understand the material of the preceding sections.

The following conventions have been employed. R indicates an alkyl group, but this could also contain another functionality; if a species contains more than one R group, these are not necessarily identical. Y indicates a heteroatom functionality, but it is not limited to the valence indicated (thus "YR" might actually be a chlorine atom). ⌣ indicates a saturated connecting chain of two or more atoms; ▽ indicates a stable cyclic or unsaturated molecule. In reactions in which an alkyl group is lost, the loss of the largest is generally favored.

5

POSTULATION OF MOLECULAR
STRUCTURES

There are several major kinds of general structural information available in the mass spectrum. These are particularly helpful if you have no information from other sources concerning the unknown molecule. The over-all appearance of the spectrum and the low mass peaks can give a general indication of the type of compound. Furthermore, the neutral fragments lost in forming the high mass peaks can provide information on specific structural features. The fragmentation behavior of the indicated type of molecule is then used to postulate structures for the unknown. Be sure to examine the unknown spectrum for each of the types of information described here, as outlined in the Standard Interpretation Procedure inside the back cover.

5.1 General appearance of the spectrum

A brief inspection of the bar-graph spectral presentation can tell the experienced interpreter a substantial amount about the unknown molecule. You have already seen how the mass and relative abundance of the molecular ion indicate the size and general stability of the molecule (Section 3.6). In addition, the number of abundant ions in the spectrum and their distribution in the mass scale are indicative of the type of molecule and the functional groups present. For example, a glance at the spectrum of Unknown 5.1 should tell you that this is a highly stable molecule. Deduce its elemental composition; this should correspond to a value for rings plus double bonds which is consistent with this stability. Not only are there no weak bonds in the molecule, but apparently the main fragmentation pathways are of approximately the same low probability. What is a possible structure?

Unknown 5.1

m/z	Rel. abund.	m/z	Rel. abund.	m/z	Rel. abund.
38	1.8	64	10.	102	7.1
39	3.9	64.5	1.1	103	0.6
40	0.2	74	4.7	125	0.8
50	6.4	75	4.9	126	6.1
51	12.	76	3.3	127	9.8
52	1.6	77	4.1	128	100.
61	1.4	78	2.7	129	11.
62	2.7	87	1.4	130	0.5
63	7.4	89	0.7		
63.5	0.9	101	2.7		

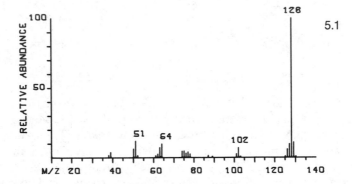

A molecule composed of stable substructures connected by relatively weak bonds should give a spectrum with a few prominent fragment ions, such as that of nicotine, Figure 5A.

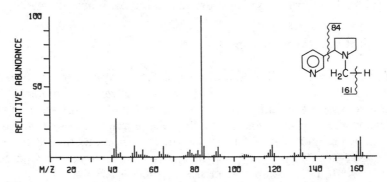

Figure 5A. Mass spectrum of nicotine.

In the spectrum of Unknown 5.2 the m/z 43 peak is by far the most prominent in the spectrum. The bond that is cleaved in the formation of this ion would be expected to be the weakest in the molecule from its known chemical reactivity.

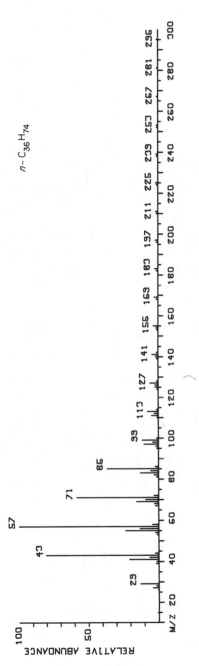

Figure 5B. Mass spectrum of n-hexatriacontane. The relative abundances of the m/z 506 and 507 peaks are 0.46 per cent and 0.18 per cent respectively. The relative abundances of the $C_nH_{2n+1}{}^+$ peaks decrease regularly from m/z 309 = 0.7 per cent to m/z 505 = 0.1 per cent.

Unknown 5.2

m/z	Rel. abund.
15	34.
16	0.5
27	1.4
28	1.0
29	2.0
41	1.8
42	7.2
43	100.
44	2.1
45	0.2
86	11.
87	0.4

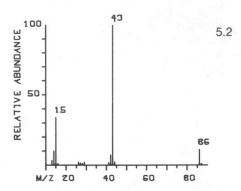

Unknown 5.3 is the spectrum of a larger molecule producing only a few prominent peaks. In evaluating the *m/z* 51 peak, don't forget that the importance of a peak decreases with decreasing mass, as well as with decreasing abundance.

Unknown 5.3

m/z	Rel. abund.	m/z	Rel. abund.	m/z	Rel. abund.
38	0.4	75	1.7	127	0.4
39	1.1	76	4.3	151	1.1
50	6.2	77	62.	152	3.4
51	19.	78	4.2	153	1.8
52	1.4	104	0.4	154	1.4
53	0.3	105	100.	181	7.4
63	1.3	106	7.8	182	55.
64	0.6	107	0.5	183	8.3
74	2.0	126	0.6	184	0.6

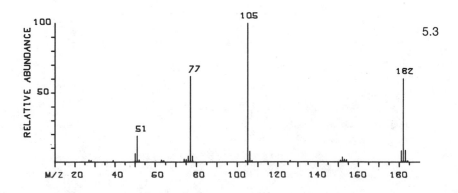

In *n*-alkanes, the most easily cleaved bonds (the σ-bonds between secondary carbons) are nearly equivalent in bond strength. The resulting mass spectra (Figures 3B and 5B) have many peaks of regularly varying abundances. Study these typical "picket fence" spectra so that you can recognize them on sight. Note the striking similarity of the spectra of Figures 3B and 5B, despite their difference in molecular weight. All the "important" peaks except M^{+} are even-electron ions. The rates of the initial decomposition reactions of the molecular ion involving cleavages of the different carbon-carbon bonds are comparable to each other, as are the secondary decompositions of the primary product ions. This accounts for the regular increase in abundances with decreased sizes of the alkyl ions. The possibility of rearranged products of greater stability becomes higher with the secondary reactions, so that the structures of the smaller ions, such as $C_3H_7^+$ and $C_4H_9^+$ are largely the more stable branched carbonium ion structures. Thus the distribution of ions is maximized in the C_3 and C_4 region of higher *n*-alkanes.

Further substitution of a carbon atom increases the probability of cleaving its bonds; compare Figures 3B, 3C, and 4A. After you reread Section 4.5, try Unknowns 5.4 to 5.6, which are $C_{16}H_{34}$ isomers. Don't forget that σ-bond cleavages give EE^+ $C_nH_{2n+1}^+$ ions, *not* OE^{+} $C_nH_{2n}^{+}$ ions; ignore the latter in these unknowns.

5.2 Low-mass ion series

Study the mass spectra in Figures 3B to 3Y and this chapter for such "picket-fence" peak series, especially at lower masses. You should find many spectra in which several groups of peaks are separated by 14 mass units; others have a slightly smaller separation (12 or 13 mass units). Such low-mass *ion series* are important for providing general information (heteroatoms, rings-plus-double-bonds) on particular substructures. The individual peaks at higher masses representing primary fragmentation products are more important as indicators of *specific* structural features. Although secondary decompositions commonly involve rearrangements (EE^+, Section 4.10), the low mass $C_nH_{2n+1}^+$ ion series so produced in Unknowns 5.4 to 5.6 is a reliable indicator of the presence of an alkyl chain. Unknown 5.5 actually has CH_3 and C_4H_9 (but not C_3H_7) groups attached to a tertiary carbon, although $[C_3H_7^+] > [C_4H_9^+] > [CH_3^+]$; it is the *losses* of $CH_3 \cdot$ and $C_4H_9 \cdot$ that produce the "important" peaks at m/z 211 and 169, respectively.

Try to identify any low-mass ion series before postulating specific structures for higher mass peaks. At lower masses there are fewer possible ion assignments for a particular mass value (and these are usually EE^+); a sizable m/z 29 peak is usually $C_2H_5^+$ or CHO^+, whereas m/z 129 has literally hundreds of isobaric possibilities. Thus the information from low-mass ion series can limit the possible assignments for high mass ions.

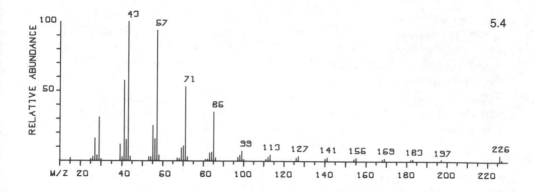

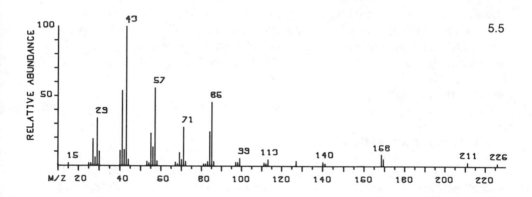

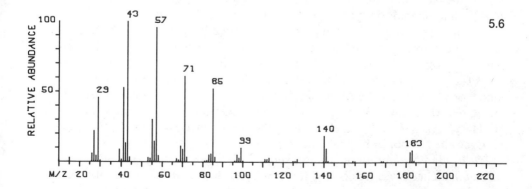

Series separated by CH₂ groups. Many structural features give rise to a significant homologous series of ions starting at the low-mass end of the spectrum, such as the continuous series of alkyl ions, $C_nH_{2n+1}^+$, in alkane spectra. The high probability of rearrangement on forming the lowest-mass ions produces a much more complete homologous series with much less variation in ion abundances than at higher masses. Thus a quick inspection of the spectrum of Unknown 5.7, which is known to contain *no* oxygen, starting at the low mass end indicates the presence of an alkyl moiety. The fact that the $C_nH_{2n+1}^+$ ions can be traced from CH_3 to $C_5H_{11}^+$ (to $C_7H_{15}^+$ at higher sensitivity) is a valuable indication of the maximum size of the alkyl moiety.

Unknown 5.7

m/z	Rel. abund.	m/z	Rel. abund.	m/z	Rel. abund.
15	3.2	64	0.8	94	0.2
27	5.4	65	10.	98	0.3
29	7.1	66	0.7	103	2.6
39	13.	67	0.6	104	5.9
40	1.7	69	1.4	105	12.
41	21.	70	0.9	106	1.1
42	3.0	71	0.5	117	1.9
43	17.	75	0.4	118	0.6
44	0.6	76	0.5	119	2.2
50	1.5	77	6.1	120	0.2
51	5.1	78	6.5	133	5.1
52	1.6	79	3.8	134	0.6
53	1.9	83	0.3	147	1.1
55	4.7	89	1.8	161	0.3
56	2.0	90	1.4	175	0.1
57	17.	91	100.	190	26.
58	0.8	92	96.	191	4.1
63	2.4	93	7.5	192	0.3

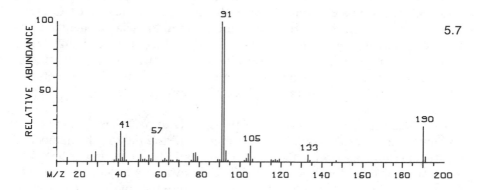

As shown in Table 5.1, the $C_nH_{2n+1}^+$ alkyl series gives peaks at m/z 15, 29, 43, etc. Insertion of an oxygen atom (such as substitution of H by OH)

Table 5.1. *Common Ion Series* (see Tables A.6 and A.7)

Function[a]	Formula	Index, Δ[b]	m/z values					
Alkyl	$C_nH_{2n+1}^+$	+2	15	29	43	57	71	85
Aldehydes, ketones	$C_nH_{2n-1}O^+$	+2		29	43	57	71	85
Amines	$C_nH_{2n+2}N^+$	+3		30	44	58	72	86·
Alcohols, ethers	$C_nH_{2n+1}O^+$	+4		31	45	59	73	87
Acids, esters	$C_nH_{2n-1}O^+$	+3			45	59	73	87
Thiols, sulfides	$C_nH_{2n+1}S^+$	+5		33[c]	47	61	75	89
Chloroalkyl	$C_nH_{2n}Cl^+$	−6		35	49	63	77	91
Nitriles	$C_nH_{2n-2}N^+$	−1		40	54	68	82	96
Alkenyl, cycloalkyl	$C_nH_{2n-1}^+$	0	27	41	55	69	83	97
Alkenes, cycloalkanes, alkyl-Y[d]	$C_nH_{2n}^{+·}$	+1	28	42	56	70	84	98
Aromatic	$C_nH_{\leqslant n}$	−2 to −8	38, 39, 50–52, 63–65, 75–78[e]					

[a]Connected to a saturated aliphatic substructure
[b]Δ = mass − $14n$ + 1 (Dromey 1976)
[c]H_3S^+ ion
[d]For which HY is a molecule of high ionization energy
[e]All of these peaks may not be of significant abundance in a particular spectrum, and neighboring peaks are sometimes observable.

gives the m/z 31, 45, 59, etc., $C_nH_{2n+1}O^+$ series which is characteristic of saturated alcohols and ethers; see Figures 3H and 3I. Saturated amines yield the $C_nH_{2n+2}N^+$ series m/z 30, 44, 58, etc., only now these EE^+ ions are at odd mass numbers (Figures 3P to 3R). Increasing the number of rings-plus-double-bonds by one decreases the mass by two units; aliphatic aldehydes and ketones give the $C_nH_{2n-1}O^+$ series at m/z 29, 43, 57, etc. (Figures 3J and 3K). Unfortunately, this overlaps the $C_nH_{2n+1}^+$ series, since CO corresponds to C_2H_4 in mass. Be sure you have tried to assign elemental compositions to these peaks; if $[44^+]/[43^+]$ is only 2.2 per cent, the 43^+ peak cannot be part of a $C_nH_{2n+1}^+$ series. Table 5.1 shows a similar overlap in the 31, 45, 95, etc., series between the $C_nH_{2n+1}O^+$ ions (alcohols, ethers) and the $C_nH_{2n-1}O_2^+$ ions (acids, esters); a more complete list is given in Table A.6.

In evaluating such ion series, use the criteria of peak importance; don't forget that in a "peak group" (compositions differing only in the number of hydrogens) the peak with the most H atoms is the most important. The spectra of alkanes (Figures 3B and 3C, Unknowns 5.4 to 5.6) contain both $C_nH_{2n-1}^+$ and $C_nH_{2n+1}^+$ series; obviously, the presence of the former should not be interpreted as indicating an alkenyl or cycloalkyl group (Table 5.1). Both of these series also appear in the spectrum of 1-dodecene (Figure 3E); however, now the $C_nH_{2n-1}^+$ series is more important than $C_nH_{2n+1}^+$, since the former can be followed up to much higher masses. Note that the OE$^+$ series $C_nH_{2n}^+$ is also quite important in this spectrum.

The spectra of Unknowns 5.8 and 5.9 are very similar, with m/z 224 shown to be $C_{16}H_{32}^+$ in both. However, one is unique in containing an important ion series (note that these are best identified by starting at the low-mass end of the spectrum). What possible kinds of molecules are these?

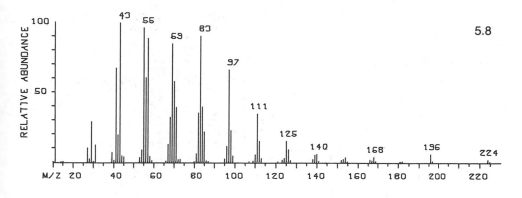

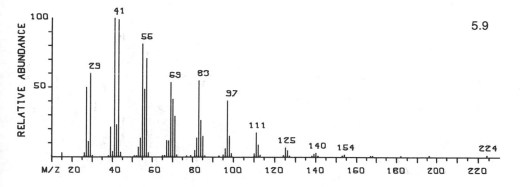

Ion series separated by CH and C. Compounds with a hydrogen-to-carbon ratio much less than two cannot show a significant series of ions spaced at CH_2 intervals (see Unknown 5.1, naphthalene, $C_{10}H_8$). The low-mass ions of such compounds can still show characteristic series. A very important series is that shown by aromatic hydrocarbons at m/z 38–9, 50–2, 63–5, and 75–8 ($C_nH_{0.5n}$ to C_nH_n). Heterocyclic compounds containing oxygen and nitrogen atoms and compounds containing these elements near an aromatic ring show peaks at similar m/z values plus additional peaks at m/z 40, 53, 66, and 79 (the "high" aromatic series of Table A.6) because of replacement of CH by N, or CH_2 by O. Although the aromatic series is often of relatively low abundance, the uniqueness of its mass values (Table 5.1) usually makes it quite recognizable; note its absence in Unknowns 5.8 and 5.9 compared to 5.7. Using Figures 3B to 3Y as unknowns, see if you can deduce which contain aromatic moieties using this series.

As can be seen in Figures 3W and 3X, the replacement of hydrogen atoms by halogen atoms, X, changes the homologous series spacings by CHX or CX_2, causing a marked change in the appearance of the spectrum. Additionally, the electronegative halogen atoms are much more easily lost than hydrogen atoms, so that ions separated by X and HX in mass are also found. For chlorine and bromine, these are easily identified because of the "isotope clusters." Fluorine and iodine have only one natural isotope each—^{19}F and ^{127}I—but these can usually be recognized by the very unusual mass differences they cause.

In Unknowns 5.10 and 5.11 notice in particular the general appearance of the spectra and their ion series.

Unknown 5.10

m/z	Rel. abund.	m/z	Rel. abund.	m/z	Rel. abund.
38	3.7	62	3.2	88	0.3
39	4.8	63	5.8	98	1.2
40	0.6	64	1.5	99	0.9
43	0.4	64.5	3.9	100	0.9
44	1.6	65	0.8	101	5.6
49	2.0	74	7.1	102	24.
49.5	0.6	75	9.9	103	7.6
50	12.	76	9.0	104	0.6
50.5	0.4	77	3.8	127	1.8
51	19.	78	2.5	128	16.
51.5	1.0	79	0.5	129	100.
52	4.2	81	0.4	130	10.
53	0.4	87	0.9	131	0.5

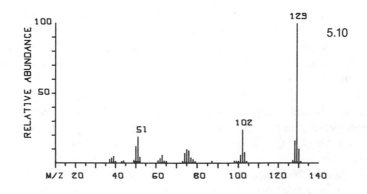

Unknown 5.11

m/z	Rel. abund.
12	13.
14	2.1
19	2.0
24	2.7
26	11.
31	22.
32	0.3
38	6.2
50	25.
51	0.3
69	100.
70	1.1
76	46.
77	1.2
95	2.4

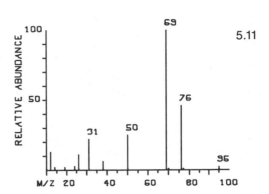

5.3 Small neutral losses

Perhaps the most simple and specific assignments that can be made in the spectrum are for the small neutral species lost in the formation of the fragment ions of highest mass in the spectrum, especially those formed directly from the molecular ion. For example, important ions at masses $(M - 1)^+$, $(M - 15)^+$, $(M - 18)^+$, and $(M - 20)^+$ almost always represent the losses of H, CH_3, H_2O, and HF, respectively, from the molecular ion. Because formation of such large primary ions involves the lowest probability of a randomizing rearrangement, such "small-neutral loss" peaks are of major significance in deducing the molecular structure. Thus an abundant peak corresponding to $(M - 1)^+$ indicates a labile hydrogen atom (and the absence of other labile substituents). A list of common small neutral fragments lost in the formation of important spectral peaks is set forth in Table A.5; their use as tests for the molecular ion was discussed in Section 3.5. As discussed in Section 3.4, these high-mass peaks are of high relative importance; even those of < 1 per cent abundance can be useful for structure determination.

In Unknown 5.12, chemical ionization and exact mass measurements show that m/z 119 is formed by the loss of C_2H_5 from the molecular ion; such small neutral losses are indicative of the structure. Did you identify *both* important ion series?

Unknown 5.12

m/z	Rel. abund.	m/z	Rel. abund.	m/z	Rel. abund.
27	7.0	55	19.	84	0.8
28	1.5	56	6.5	91	0.6
29	10.	57	100.	93	0.2
30	0.2	58	4.4	98	8.4
36	0.4	63	1.1	99	22.
39	5.4	65	0.4	100	1.6
40	0.7	69	3.3	105	0.3
41	21.	70	3.9	112	0.2
42	3.6	71	1.7	119	4.7 $(M - C_2H_5)^+$
43	21.	77	1.0	120	0.3
44	0.7	79	0.3	121	1.5
49	0.2	83	4.2	148	0.1

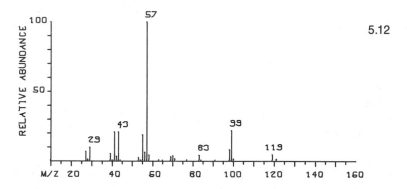

5.12

5.4 Characteristic ions

Deduction of the structural information from a larger ion is more difficult because of the larger number of theoretically possible assignments for it. Despite this, for many specific mass (and, especially, elemental composition) values, there are only a few characteristic structural groupings that commonly give rise to the corresponding peak in mass spectra. Such "characteristic ions" are surprisingly helpful in suggesting possible fragments of the molecule. You should already be familiar with some of these, like the m/z 30 from amines, m/z 77 from phenyl, and m/z 105 from benzoyl (Unknown 5.3). If Figure 5C were an unknown, the experienced interpreter would immediately *suspect* a phthalate from the presence of the base peak at m/z 149. In a data base representing 29,000 different compounds (Stenhagen et al, 1974), over half the spectra giving a m/z 149 base peak are phthalates. Phthalates are ubiquitous impurities found in mass spectra, since they are common components of plasticizers (tubing, cap liners, gaskets) and chromatographic column packings.

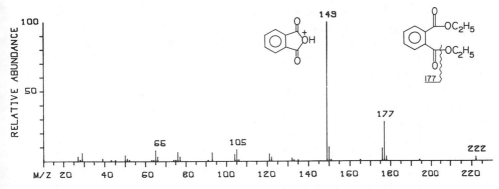

Figure 5C. Mass spectrum of diethyl phthalate.

Thus the next step in the interpretation of the spectrum is to note the possible structural significance of all important peaks. Where more than one possible interpretation exists, be sure to note all of them; then if other evidence in the spectrum eliminates one or more of these possibilities, the others will be much more meaningful. Possible assignments for even-electron ions are given as part of the ion series of Table A.6; Table A.7 lists common odd-electron fragment ions. *Mass Spectral Correlations* (McLafferty and Venkataraghavan 1982) shows what elemental compositions and structures are possible for a peak, and what the *relative* probability is for each of these postulations. The *Eight-Peak Index of Mass Spectra* (Mass Spectrometry Data Centre 1970) provides a listing of compounds which give a particular m/z value as the largest, second largest, etc., peak in their spectra.

5.5 Assignment of the most probable structure

All the information and postulations gathered above must now be used to deduce the most logical structure. It is difficult to outline a generally applicable pathway from these to such a deduction, for each case is usually dependent on the particular spectrum and the other available information. Although practice and experience are invaluable in finding the most logical and efficient pathways, some particulars may be helpful.

Deducing a molecular structure by most spectroscopic techniques involves postulating a particular structure, predicting its spectrum, and then comparing this with the unknown spectrum. A main reason for the first eight steps of the "Standard Interpretation Procedure" (inside back cover) is that these will generally suggest structures for examination. Thus at this point you should review the information and postulations from these steps and the possible molecular structures which they suggest. Before each of these structures is examined critically, a real effort should be made to list *all* that appear possible, although there is usually sufficient information in the mass spectrum to narrow the structural possibilities rapidly. The elimination of all but one possibility for the molecular structure of course does not prove that this one is correct, unless a reference spectrum of the compound can be obtained.

Comparison with reference spectra should be made for all probable assignments, if possible. Bar-graph spectra of 19,000 different compounds are assembled in the *Registry of Mass Spectral Data* (Stenhagen *et al.* 1974); these are arranged according to the exact molecular weight and structure of the compound. The EPA/NIH *Mass Spectral Data Base* (Heller and Milne 1978) contains bar-graph spectra of 25,600 different compounds with structural sketches. If the reference spectrum is not available, try to predict it by using the rules in Chapters 4 and 7 and the correlations of Chapter 8. Special techniques are discussed in Chapter 6.

Follow the "Standard Interpretation Procedure" in attempting to assign structures to Unknowns 5.13 and 5.14.

Unknown 5.13

m/z	Rel. abund.	m/z	Rel. abund.	m/z	Rel. abund.
27	14.	64	2.0	92	0.5
28	1.1	65	15.	93	1.6
38	2.2	66	12.	94	100.
39	14.	67	0.7	95	6.6
40	2.1	74	0.9	96	0.4
41	0.3	75	0.6	107	29.
50	3.0	76	0.6	108	2.2
51	10.	77	22.	121	1.6
52	0.9	78	1.8	156	41.
53	0.5	79	5.0	157	3.7
55	1.5	80	0.5	158	13.
63	19.	91	1.5	159	1.2

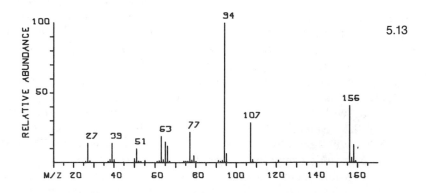

Unknown 5.14

m/z	Rel. abund.	m/z	Rel. abund.	m/z	Rel. abund.
15	9.2	54	5.7	82	11.
26	1.3	55	74.	83	46.
27	24.	56	100.	84	51.
28	6.1	57	38.	85	4.9
29	32.	58	2.9	87	1.0
30	0.8	59	3.0	88	0.7
33	0.4	60	4.9	89	11.
34	1.1	61	18.	90	0.6
39	19.	62	1.6	91	0.5
40	4.0	63	0.9	97	4.3
41	76.	67	4.9	98	0.3
42	47.	68	18.	103	0.9
43	74.	69	59.	112	23.
44	3.3	70	70.	113	2.0
45	8.2	71	15.	145	1.7
46	3.6	72	0.8	146	37.
47	29.	73	0.9	147	3.7
48	1.6	74	0.6	148	1.8
53	4.7	75	1.1		

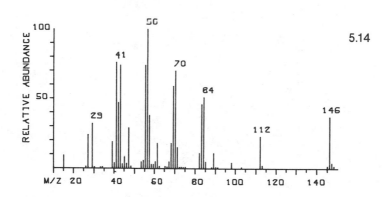

5.14

6

AUXILIARY TECHNIQUES

The limitations of the structural information in the normal mass spectrum can be partly offset by special mass-spectral techniques. Although a complete description of these is beyond the scope of this book, you should have a basic familiarity with their capabilities.

6.1 Alternative ionization methods

Low-energy electrons. Lowering the energy of the bombarding electrons will also lower the average (but *not the minimum*) internal energy of the molecular ions. For structure elucidation it is sometimes useful to take a second mass spectrum at a lower electron energy (*ca.* 15 eV), since this will eliminate higher-energy reactions giving secondary product ions which are much less representative of the original structure (although these do provide "ion series" information; Section 5.2). However, lower electron energies increase the relative abundance of primary rearrangement reactions, in the same manner that these increase in metastable ion decomposition (Section 6.4). Further, lowering the electron energy decreases the *absolute* abundance of all ions; although the *relative* abundance of the molecular ion increases, it nevertheless is more difficult to detect at lower electron energies (unless it has been obscured by fragment ions from a higher-molecular-weight impurity).

Chemical ionization (CI). Probably the most commonly used alternative ionization technique is CI (Munson and Field 1966; Field 1968, 1972; Richter and Schwartz 1978; Arsenault 1980) in which the sample molecules undergo ion-molecule reactions with "reagent" ions in a high pressure ($\sim$100 Pa) ion source. Abundant product ions are commonly formed from the sample molecules by the gain or loss of hydrogen; these even-electron ions are usually of relatively low internal energy, so that most (although not all) molecules that do not yield molecular ions by EI produce CI ions indicative of the molecular weight. Analogous ion-molecule reactions useful for measuring molecular weight can even be caused by high sample pressures under EI conditions (McLafferty 1957).

For CI the reagent gas RH is introduced into the ion source at a concentration in large excess ($\sim 10^4:1$) to that of the sample, and is ionized by electron bombardment (usually more vigorous conditions than for EI, such as 500 V) or electric discharge (Hunt 1975). The reagent molecular ions $R^{+\cdot}$ can react further with other reagent molecules to form reactive ion species, which can then react with the sample molecule M:

$$R + e^- \longrightarrow R^{+\cdot} + 2e^- \tag{6.1}$$

$$R^{+\cdot} + R \longrightarrow RH^+ + (R-H)\cdot, \text{ or } (R-H)^+ + RH\cdot \tag{6.2}$$

$$RH^+ + M \longrightarrow R + (M+H)^+ \text{ (protonation)} \tag{6.3}$$

$$(R-H)^+ + M \longrightarrow (R-2H) + (M+H)^+ \text{ (protonation)} \tag{6.4}$$

$$(R-H)^+ + M \longrightarrow R + (M-H)^+ \text{ (hydride abstraction)} \tag{6.5}$$

$$R^{+\cdot} + M \longrightarrow R + M^{+\cdot} \text{ (charge exchange)} \tag{6.6}$$

Although a wide variety of ionization reactions are possible, the most common are: protonation (6.3 and 6.4), which is favored for sample molecules of proton affinity higher than that of the reagent; hydride abstraction (6.5), which is common for lower proton affinity molecules, such as alkanes; and charge exchange (6.6), which is favored for reagents of high ionization potential, such as helium. Fragmentations of these even-electron (formed by reactions 6.3 to 6.5) and odd-electron (by 6.6) ions in general obey the mechanisms outlined in Chapters 4 and 8, with the amount of internal energy determined by the exothermicity of the ion-molecule reaction, the ion source temperature and pressure, and other instrumental factors. Thus CI spectra can provide structural as well as molecular weight information.

Field ionization (FI). Placing a high potential (~ 10 KV) between the MS ion-entrance slit and a "field emitter," usually a collection of microneedles grown on a fine wire or edge such as a razor blade, can produce electric fields at the emitter as high as 1 V/Å (Beckey 1977; Schulten 1979). Removal of an electron from a molecule in such a high field occurs by a "tunneling" process requiring less energy than normal electron ionization, giving a higher abundance of molecular ions of low internal energy. Thus $M^{+\cdot}$ are observable in the FI spectra of molecules for which the normal EI spectra show negligible $M^{+\cdot}$. Ion-molecule reactions can also occur at or near the surface of the field emitter. In general, measurements of the molecular weight of volatile molecules are more often done by CI than by FI; the most advantageous application of FI has been in "field desorption (FD)" of less volatile samples (Section 6.2).

Ion cyclotron resonance (ICR). A low-kinetic-energy ion in a magnetic field exhibits a cycloidal motion whose frequency is dependent on the ion's mass. The ICR spectrometer measures the absorption of RF energy as a function of frequency to produce a mass spectrum. The relatively long ion lifetimes (milliseconds) result in high cross sections for ion-molecule reactions at relatively low pressures ($\sim 10^{-2}$ Pa), and a double-resonance feature makes it possible to identify both the precursor and the products of such reactions. This technique is especially useful for measuring ion energetics, such as proton affinities, and studying gaseous-ion reactivities.

Negative ions. EI generation of negative ions generally gives poor sensitivity, although the resulting spectra can give useful structural information for particular electronegative species (Fales 1971; Bowie and Janposri 1976; Jennings 1979b; Dillard 1980). However, CI methods (Hunt *et al.* 1976) can yield even lower detection limits for many electronegative compounds than are achievable by EI-MS of any compounds. Fragmentation of such ions, particularly by CA (Section 6.3; Bowie 1973; Bursey *et al.* 1979) appears to be a promising source of molecular structure information.

6.2 Low volatility samples

For conventional direct-probe introduction, a sample must have sufficient vapor pressure and thermal stability to be vaporized in the ion source. This severely limits the application of mass spectrometry to important samples such as biological macromolecules, industrial polymers, and petroleum residues. Intensive recent research activity has resulted in promising methods such as electrospray (Dole *et al.* 1971; Friedman 1979), rapid heating (Gaffney, Pierce, and Friedman 1977), Californium-252 plasma desorption (Macfarlane and Torgerson 1976; Macfarlane 1980), direct laser ionization (Posthumus *et al.* 1978), secondary ion bombardment (Benninghoven and Sichtermann 1978), and flash vaporization of dilute solution droplets (Lory and McLafferty 1980). Two techniques which are more generally available and reliable, Direct Ionization and Field Desorption, will be described here.

Direct ionization. Exposure of the solid sample to the ion plasma in a CI source lowers the sample volatility requirements (Baldwin and McLafferty 1973a). Such "Direct Chemical Ionization" is especially effective for dilute solutions introduced directly in the source, using the solvent as the CI reagent gas (see LC/MS, below). For "Desorption Chemical Ionization"

(Hunt *et al.* 1978), a thin layer of sample on a filament is heated electrically in the CI plasma. Lower-volatility solid samples can also be ionized by direct "in-beam" electron bombardment (Dell *et al.* 1975).

Field desorption. Low-volatility samples can be field-ionized after being deposited directly from solution onto the FI emitter, a "bottle brush" of microneedles grown out from a fine wire (Beckey 1977; Schulten 1979). Although FI is basically a "soft ionization" technique, substantial ion fragmentation can be induced by heating the emitter electrically above the minimum temperature required for field desorption.

Pyrolysis. Useful structural information about complex mixtures of high molecular weight, such as bacteria, plant specimens, coal, high polymers, and sewage sludge, can be obtained from the mass spectra of their thermal decomposition products (Meuzelaar *et al.* 1980). To minimize secondary reactions, small samples (10^{-4} to 10^{-7} g) are preferable, as are reproducible pyrolysis conditions, such as rapid RF heating of a wire to its Curie temperature. Laser, spark, and other flash-heating methods have also been used. Of particular interest is an automated system for Curie-point pyrolysis in the MS vacuum chamber with rapid multiple scanning of the product mass spectra. This system gives highly reproducible spectra for ~ 30 samples/hour, making it possible to distinguish between different strains of bacteria.

6.3 Exact mass measurements (high resolution)

Measurement of the mass of an ion with sufficient accuracy provides an unequivocal identification of its elemental (and isotopic) composition. This technique is often referred to as "high resolution" mass spectrometry, since double-focusing instruments (Figure 6A) are usually used for such measurements (McDowell 1963). However, if interfering masses requiring high resolution are absent, sufficiently accurate mass measurements (~ 10 ppm) can be achieved with single-focusing or quadrupole instruments of 1000–2000 resolution. Ion cyclotron resonance spectrometers are promising for high-resolution mass measurements (Comisarow 1979; Ledford *et al.* 1980).

Elements can be identified in this way because monoisotopic atomic weights are not exact whole numbers. When we take as standard the mass of ^{12}C as 12.00000000 (see Table A.1), then m for $^1H = 1.00782506$, for $^{14}N = 14.00307407$, and for $^{16}O = 15.99491475$ (the least significant figure, 10^{-8} daltons, represents 9.3 eV of energy). Every isotope has a unique, characteristic "mass defect"; so the mass of the ion, which shows the total mass defect, identifies its isotopic and elemental composition. For example, an ion with a mass of 43.0184 must be $C_2H_3O^+$; it cannot be $C_3H_7^+$ ($m = 43.0547$), $C_2H_5N^+$ ($m = 43.0421$), $CH_3N_2^+$ ($m = 43.0296$), $CHNO^+$ ($m = 43.0058$), or

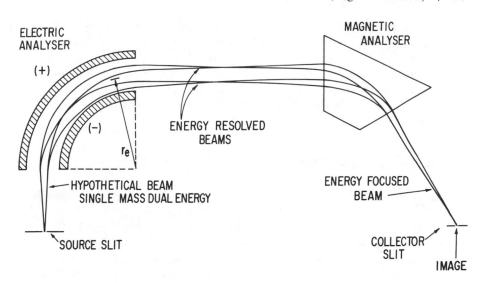

Figure 6A. Nier-Johnson double-focusing mass spectrometer. The two sets of ion paths illustrate focusing for dispersion and for ion kinetic energy (shown in exaggerated form; Ligon 1979).

C_2F^+ ($m = 42.9984$). To distinguish these compositions requires an accuracy in measuring mass of 130 ppm.

The usefulness of elemental composition information increases exponentially with increasing mass, as does the required mass-measuring accuracy. Figure 6B shows the exact mass of a variety of possible ions of molecular weight 310 containing carbon, hydrogen, not more than three nitrogen atoms, and not more than four oxygen atoms. Note, however, that identifying a number of these requires a mass-measuring accuracy of *2 ppm,* which is near the limit attainable with all but the most sophisticated instruments (Kilburn *et al.* 1979).

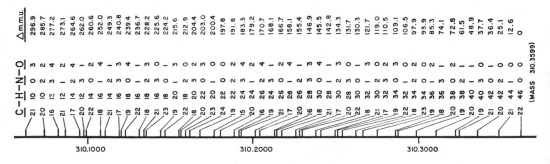

Figure 6B. Exact masses of possible ions of m/z 310 containing carbon, hydrogen, not more than 3 nitrogen atoms, and not more than 4 oxygen atoms.

Such measurements are made by comparing the focus position of the unknown peak with that of a reference peak whose composition, and thus exact mass, is known. Manual "peak-matching" can give ppm accuracy, but is time-consuming (2 to 10 minutes per peak). However, with modern on-line computer systems, such measurements and elemental-composition calculations can be made for all peaks of a few per cent abundance or greater for scan times as short as 10 sec/decade.

6.4 Collisional activation and metastable ion spectra

"Double resonance" techniques are useful in a number of spectroscopic methods. In mass spectrometry there are analogous techniques for studying the decompositions of individual fragment ions, i.e., measuring mass spectra of the peaks in a mass spectrum. Just as normal mass spectra can characterize a molecule's structure, these secondary mass spectra can provide structural information about fragment ions. For example, in the mass spectrum of the hypothetical molecule A_2B_2C, the presence of ions corresponding in mass to AB^+ and ABC^+ could indicate either of the molecules A—B—C—B—A or A—B—B—C—A. However, the presence of AB^+ in the secondary mass spectrum of the ABC^+ fragment ion would be possible, barring rearrangements, only for the structure A—B—C—B—A. In the mass spectrum of N-methyl-N-isopropyl-N-n-butylamine (Figure 9R) α-cleavages give the prominent ions at m/z 86 and 114, and their further rearrangement decompositions yield, respectively, other abundant ions at m/z 44 and 58 (6.7). If this were the spectrum of an unknown, the compound N-methyl-N-ethyl-N-2-pentylamine would also have to be considered, since it should also give rise to these peaks (6.8). However, the observation of m/z 58 in the secondary spectrum of m/z 114 would eliminate the latter compound from consideration; its spectrum should show m/z 114 $\rightarrow$ m/z 44.

$$(CH_3)_2-CH-\overset{+\cdot}{\underset{|}{N}}CH_2-C_3H_7$$
$$CH_3$$

$$\xrightarrow[\alpha]{-CH_3\cdot} CH_3CH=\overset{+}{\underset{\underset{CH_3}{|}}{N}}-CH_2C_3H_7 \xrightarrow[rH]{-C_4H_8} CH_3CH=\overset{+}{\underset{\underset{CH_3}{|}}{N}}H$$
$$m/z\ 114 \qquad\qquad m/z\ 58 \qquad (6.7)$$

$$\xrightarrow[\alpha]{-C_3H_7\cdot} (CH_3)_2-CH-\overset{+}{\underset{\underset{CH_3}{|}}{N}}=CH_2 \xrightarrow[rH]{-C_3H_6} H\overset{+}{\underset{\underset{CH_3}{|}}{N}}=CH_2$$
$$m/z\ 86 \qquad\qquad m/z\ 44$$

$$CH_3-CH_2\overset{+\cdot}{\underset{\underset{CH_3}{|}}{\overset{\overset{CH_3}{|}}{N}}}CH-C_3H_7$$

$$\xrightarrow[\alpha]{-CH_3\cdot} CH_2=\overset{+}{\underset{\underset{CH_3}{|}}{\overset{\overset{CH_3}{|}}{N}}}CH-C_3H_7 \xrightarrow[rH]{-C_5H_{10}} CH_2=\overset{+}{\underset{\underset{CH_3}{|}}{N}}H \qquad (6.8)$$

$$\underset{m/z\ 114}{} \qquad \underset{m/z\ 44}{}$$

$$\xrightarrow[\alpha]{-C_3H_7\cdot} CH_3CH_2-\overset{+}{\underset{\underset{CH_3}{|}}{\overset{\overset{CH_3}{|}}{N}}}=CH \xrightarrow[rH]{-C_2H_4} H\overset{+}{\underset{\underset{CH_3}{|}}{\overset{\overset{CH_3}{|}}{N}}}=CH$$

$$\underset{m/z\ 86}{} \qquad \underset{m/z\ 58}{}$$

Instrumentation. The simplest method for observing further decompositions of fragment ions uses the single-focusing magnetic mass spectrometer (Figure 1.5). Ions of mass m_1 which decompose to yield m_2 in the field-free drift region ahead of the magnetic field will be found in the spectrum at a mass m^* determined by both m_1 and m_2. Ion acceleration involves m_1 and magnetic deflection involves m_2, which yields $m^* = m_2^2/m_1$. This equation is best solved by trial and error, using the major normal ions of the spectrum as possible assignments for m_1 and m_2, and logical mass differences (Table A.5) for the neutrals lost. Note that more than one answer may be possible ($m^* = 10.0$ can arise from m/z 40 → 20, 90 → 30, etc.) Such ambiguities may be avoided with double-focusing mass spectrometers of both normal and reverse geometries by using several combinations of the ion-accelerating and electrostatic analyzer potentials and magnetic field (Cooks *et al.* 1973; Boyd and Beynon 1977). Secondary-ion mass spectra cannot be obtained using the quadrupole mass spectrometer in normal operation, but a system with tandem quadrupoles (Yost and Enke 1979) is very promising.

Types of decompositions. The most useful secondary mass spectra are those from spontaneous decompositions, yielding "metastable ion" (MI) mass spectra (Cooks *et al.* 1973), and those caused by collision with molecules, yielding "collisional activation" (CA) mass spectra (McLafferty 1979; McLafferty *et al.* 1973a, 1973b; Cooks 1978). Metastable ions with appropriate lifetimes ($\sim 10^{-5}$ sec) will decompose in the field-free region; for example, isopropyl alcohol molecular ions (mass 60) losing CH_4 thus yield the

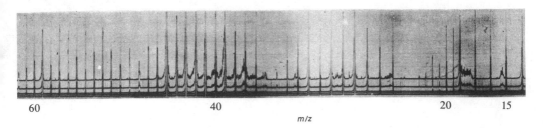

60   40   20   15

                   m/z

Figure 6C. Mass spectrum of isopropyl alcohol.

m/z 32.3 (44²/60) "metastable peak" in Figure 6C. The "flat-topped metastable" centered at *m/z* 18.7 results from a decomposition with a substantial release of kinetic energy.

Unknown 6.1. Assign appropriate ion dissociations to *m/z* 18.7, 8.0 (not shown), and any other metastable peaks of Figure 6C. Also try Unknown 4.19 again, using the metastable ion data.

If a gas such as helium at $\sim 10^{-2}$ Pa is in the path of the ion beam in a field-free drift region, collisions occur which convert part of the ions' translational energy into internal energy over a broad range of values (0–20 eV), causing decomposition in the drift region. Thus the CA spectrum resembles a normal EI mass spectrum in showing peaks from high-energy reactions as well as the threshold-energy decompositions of long-lived ions shown by MI spectra. CA spectra can be produced by gas leaks (the "Aston bands" of early poorly pumped instruments), such as one caused by the careful loosening of a drift-region vacuum flange which has been encased in a helium-filled bag. A differentially pumped collision chamber or He molecular beam (McLafferty *et al.* 1980a) placed near an ion focal point is much more efficient.

The CA spectrum as a "fingerprint" of ion structure. When run under the same experimental conditions, the normal EI mass spectrum of a particular molecule, such as toluene, is independent of the sample's source, and can thus serve as a useful "fingerprint" for identification. In the same way, the CA spectrum of a particular ion is virtually independent (± 5 per cent for peaks from higher energy decompositions) of its source and internal energy: $CH_3CH{=}OH^+$ ions from $(CH_3)_2CHOH$, $(CH_3)_2CHOCH(CH_3)_2$, and 5α-pregnan-20β-ol-3-one (which has a CH_3CHOH- side chain) yield CA spectra which are identical within experimental error and substantially different from that of $CH_3O{=}CH_2^+$ (McLafferty *et al.* 1973b). The relative abundances of MI spectra and of CA peaks produced by lower-energy processes can be affected by precursor-ion internal energy; thus for CA ion structure identification, the peaks found in the ion's MI spectrum are omitted from its CA spectrum.

Unknown 6.2. Use the secondary decompositions indicated by the diffuse peaks to eliminate all but one isomeric possibility.

Unknown 6.2

m/z	Rel. abund.
15.5	diffuse peak
30	25.
31	0.4
33.6	diffuse peak
39.1	diffuse peak
41	8.2
42	10.
43	7.3
44	5.7
56	4.2
57	1.3
58	19.
59	0.7
70	1.2
71	1.1
72	3.6
84	1.1
85	0.5
86	100.
87	5.9
98	1.3
100	32.
101	2.2
114	1.4
115	8.3
116	0.6

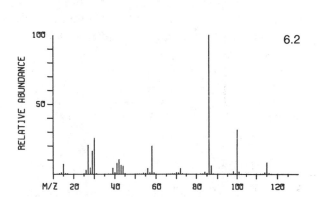

6.5 Mixture analysis

The mass spectrum of a mixture should represent a linear superposition of the spectra of its individual components run under the same conditions using the same amounts. Thus if you can identify correctly one component of an unknown mixture spectrum, subtract out the reference spectrum of that component ("spectrum stripping"), and attempt to interpret the residual spectrum.

A main application of the analytical mass spectrometer in the period ~1940–1960 (before gas chromatography) was for routine quantitative analysis of mixtures of light hydrocarbons and other common gases. Abundance reproducibilities of ±0.5 per cent could be achieved, making possible the routine analysis of mixtures of more than 20 components. Recently there has been a dramatic resurgence of interest in quantitative mass spectrometry (Millard 1978).

Separation-identification systems. Instrument systems combining a gas chromatograph and mass spectrometer (GC/MS) have proven to be of high value for analysis of complex mixtures, such as those from human biological fluids, plant extracts, pollutants, industrial processes, and samples for forensic study. With an on-line computer, useful mass spectra of the GC effluent can be obtained and stored as rapidly as one per second. The high-performance liquid chromatograph (LC) can separate mixtures not amenable to GC separation; LC/MS is particularly valuable because the vapor-pressure requirements for MS are not as stringent as for GC (Dawkins and McLafferty 1978; Arpino and Guiochon 1979). A further separation/identification system for which a wide variety of unique applications have been reported recently is "MS/MS" (McLafferty 1980a). Ions are formed from the mixture components and separated by mass analysis; a secondary mass spectrum (CA or MI) is then used to identify the individual components. At present there are serious drawbacks to the use of MS/MS to identify components of unknown mixtures; it is difficult to find "soft-ionization" conditions applicable to all components, and few reference CA and MI spectra are available. However, MS/MS often has unique advantages in analyzing for a *specific* component (or components) in a complex mixture, since ionization conditions (EI at high or low energies, CI, FI/FD, negative or doubly charged ions) can be optimized for that component, and calibration spectra can be prepared from known mixtures. A key advantage for routine or continuous analyses is the virtually instantaneous separation by MS versus GC or LC.

6.6 The shift technique

In many cases the addition of a small functional group to a large molecule such as an alkaloid changes the spectrum by merely increasing the mass of the specific ion fragments which contain this functional group, but without changing the relative abundances of these ions to a great extent. An obvious case is isotopic substitution; deuterium labeling of a compound increases the mass of each ion by one for each deuterium atom incorporated, but isotope effects on ion relative abundances are generally small. Replacement of a hydrogen by a group of low influence (such as methyl, methoxyl, or chlorine) at an unreactive position (for example, a ring, especially an aromatic ring) often gives ion abundances similar to those of the original spectrum. Using this to discover the nature and position of a substituent added to a molecule has been termed the "shift technique" by Biemann (1962), who demonstrated its great utility for the indole alkaloids. These stable ring structures contain functional groups that strongly influence the decomposition, minimizing the effect on the spectrum of molecular substitution. The method is less useful for molecules containing less influential groups, such as terpenes or steroidal alcohols.

Unknown 6.3. The spectrum of this unknown indole alkaloid (middle) is compared to the spectra of two known structures. What functional groups does the unknown contain? Where are they most probably located on the molecular skeleton?

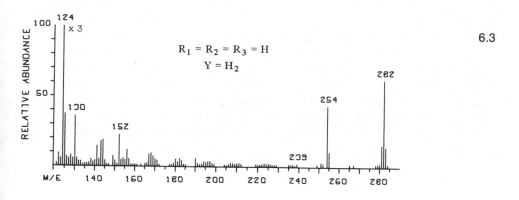

$R_1 = R_2 = R_3 = H$

$Y = H_2$

6.3

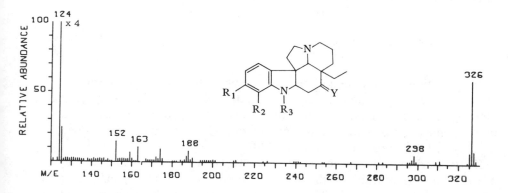

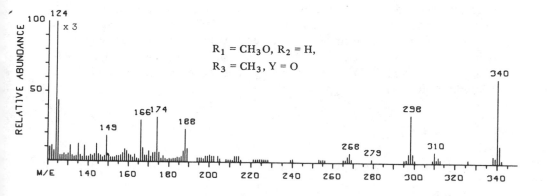

$R_1 = CH_3O, R_2 = H,$

$R_3 = CH_3, Y = O$

6.7 Chemical derivatives

Chemical conversion of a compound to an appropriate derivative can sometimes improve the resulting mass-spectral data, either by increasing the
compound's vapor pressure or by making its spectrum more easily inter-
pretable. A derivative that decreases the polarity of ionic groups in the
molecule generally increases the vapor pressure; for example, oligopeptides
of relatively high molecular weight (containing ten amino acids) can be made
sufficiently volatile by subjecting them to acetylation of the terminal amino
groups followed by permethylation of the carboxyl and amide functions (6.9).
Trimethylsilylation is probably the most widely used technique for increasing
sample volatility; formation of the trimethylsiyl ether of an alcohol also has
a beneficial effect on its fragmentation (for example, the large $(M - 15)^+$
peak is indicative of the molecular weight). Condensation of a 1,2-diol with
acetone to yield the acetonide can provide a more useful volatile derivative.

$$H_2N \dashv CHR_i - CO - N \overset{\underset{|}{H}}{\dashv}_n CHR_j - COOH \xrightarrow{\text{acetylation}} \xrightarrow{\text{permethylation}}$$

$$CH_3CON \dashv CHR_i - CO - N \overset{\underset{|}{CH_3}}{\dashv}_n CHR_j - CO - OCH_3 \qquad (6.9)$$

Introduction of a functional group which will strongly direct the fragmentation (for example, amino, ethylene ketal) will substantially change the
mass-spectral information. This could be used, for example, to cause fragmentation in part of a steroid molecule about which the original mass spectrum gave little information (Section 9.6).

Derivatization can be used as a chemical test of specific structural features
of the unknown molecule, with the mass spectrum of the subsequent product
used to measure the results. Thus the number of enolizable hydrogen atoms
in a molecule can be readily calculated from the mass spectrum after
exchange with deuterium; the exchange need not be complete, since it is
only the mass (not the relative abundance) of the product formed with the
largest number of deuterium atoms that needs to be observed. Such exchange
using CI ionization can be particularly convenient, for example, with
H_2O/D_2O or NH_3/ND_3 as the ionizing reagent gases (Hunt *et al.* 1972;
Lin and Smith 1979).

7

THEORY OF UNIMOLECULAR ION DECOMPOSITIONS

To understand thoroughly the capabilities, and especially the limitations, of mass spectrometry for structure elucidation, one must be familiar with the basic theoretical aspects of unimolecular ion decompositions. Sections 7.1 and 7.2 are appropriate for introductory courses designed to teach fundamental interpretive skills. (Lifshitz 1977 and Levsen 1978 are recommended for further details.)

7.1 Energy deposition and rate functions, $P(E)$ and $k(E)$

Ionization of the sample molecules with 70 eV electrons produces molecular ions whose internal energy values (E) cover a broad range (0 to >20 eV), averaging a few eV (1 eV = 23 kcal = 96.5 kJ). In the upper part of the Wahrhaftig diagram (Figure 7A), this distribution is described by the probability function $P(E)$. This range of energies is dramatically different from the <1 eV thermal-energy distribution usually found in condensed-phase reactions. The initially formed M^{+} ions range from cold to "red hot"; not surprisingly, these thus show a wide range of decomposition behavior.

Only *unimolecular* reactions are possible for the gaseous ions formed under the usual MS operating conditions. Without collisions, it is the energy originally deposited in the ion from the ionization process (plus the small amount of thermal energy originally in the precursor molecule) that causes the ion to decompose or isomerize. The probability of any such reaction is expressed as the rate constant k, which obviously must change with the precursor ion's internal energy, E. Hypothetical $k(E)$ functions showing the change of log k with E are also given in Figure 7A (lower half), which shows that ions of the lowest internal-energy values will not have appreciable decomposition rates, and so will appear as M^{+} in the spectrum. A rate con-

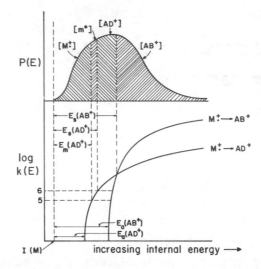

Figure 7A. The Wahrhaftig diagram: relationship of $P(E)$ and $k(E)$ for unimolecular ion decompositions of $ABCD^{\ddagger}$. See text for definitions. (Wahrhaftig 1962.)

stant of $\sim 10^6$ sec^{-1} or greater is necessary for ion source decomposition (lifetimes $\sim 10^{-6}$ sec or less). Log $k = 6$ on the curve (Figure 7A, lower half) for the reaction $M^{\ddagger} \rightarrow AD^+$ thus defines the minimum $M^{\ddagger}$ internal energy for AD^+ formation, which is somewhat higher than the reaction critical energy, E_0 (AD^+); this terminology is preferable to "activation energy," and is defined as the difference between the zero-point energy of $M^{\ddagger}$ and that of the activated complex for $M^{\ddagger} \rightarrow AD^+$ (Robinson and Holbrook 1972; Maccoll 1980). At higher $M^{\ddagger}$ energies $k(M^{\ddagger} \rightarrow AB^+)$ becomes greater than $k(M^{\ddagger} \rightarrow AD^+)$, so that the formation of AB^+ is favored. (Metastable ion decompositions, m^*, are discussed later.)

The actual ion abundances are determined by both the $P(E)$ and the $k(E)$ relationships. The relative proportion of $M^{\ddagger}$ ions that remain undissociated is determined by the lowest energy required for ion source dissociation *and* the relative proportion of $M^{\ddagger}$ ions formed with internal energies below this energy. These relationships are shown by the vertical dashed lines in Figure 7A, with the shaded areas of the $P(E)$ curve representing the relative proportions formed. If the ionizing electron energy is lowered, this will reduce the number of $M^{\ddagger}$ ions formed with higher energies more than it will reduce the number of those formed with lower (note that *all* are lowered), so that the *relative* proportion of $M^{\ddagger}$ ions that do not dissociate will increase.

Note that the shaded areas of the $P(E)$ curve represent the *initial* abundances of AD^+ and AB^+. Those ions formed with sufficient internal energy will decompose further; the abundances of the resulting product ions (and of the undissociated AD^+ and AB^+) will depend in turn on the $P(E)$ and $k(E)$ functions of AD^+ and AB^+.

7.2 Thermodynamic versus kinetic effects

The hypothetical system of Figure 7A in which $[AB^+] > [AD^+]$ illustrates another important aspect of unimolecular decompositions: the reaction of lowest critical energy does not necessarily produce the most abundant ion. Similarly, the temperature of a condensed-phase synthesis reaction is raised to optimize the yield of the desired product through "kinetic control." In Figure 7A the reaction $ABCD^{+} \rightarrow AD^+$ has the more favorable enthalpy, but $ABCD^{+} \rightarrow AB^+$ has the more favorable entropy. The transition states depicted in Figure 7B for these reactions rationalize this behavior. In the rearrangement forming AD^{+} two new bonds are formed, offsetting the energy required to cleave A—B and C—D. The critical energy for AB^+ formation is higher, since this requires B—C cleavage. However, the steric requirements for AD^{+} formation are much more strict than those for AB^+ formation; in the terminology often used for such transition states, these are, respectively, "tight" and "loose" activated complexes. For higher-energy $ABCD^{+}$ ions, the B—C bond dissociation can take place whenever sufficient energy accumulates in this bond. However, for AD^{+} formation, the energy requirements must be met at the same time that A and D are within bonding distance, which is true for only a small proportion of all possible conformations of $ABCD^{+}$.

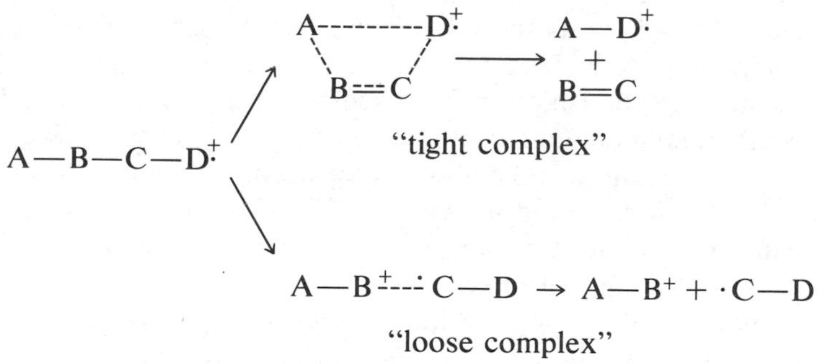

$$A—B—C—D^{+} \nearrow \begin{array}{c} A\text{--------}D^{+} \\ B\!=\!C \end{array} \longrightarrow \begin{array}{c} A—D^{+} \\ + \\ B\!=\!C \end{array}$$

"tight complex"

$$A—B^{+}\text{---}C—D \rightarrow A—B^+ + \cdot C—D$$

"loose complex"

Figure 7B

These offsetting enthalpy and entropy effects in general lead to a substantial number of *competing* primary reactions as well as the *consecutive* secondary and further reactions; thus the mass spectrum of a larger molecule can have hundreds of peaks. The competitive nature of these reactions can mean that relatively small changes in molecular structure result in large differences in peak abundances. For this reason, in interpreting an unknown spectrum it is helpful to study the spectra of closely related compounds.

7.3 Quasi-equilibrium theory

The quasi-equilibrium theory (QET), originally formulated by Rosenstock *et al.* (1952), provides a physical description of mass-spectral behavior which is now generally accepted. Ionization of the molecule, which takes place in approximately 10^{-16} sec, initially yields the excited molecular ion without change in bond length (a Franck-Condon process). Except for the smallest molecules, transitions between all the possible energy states of this ion are sufficiently rapid so that a "quasi-equilibrium" among these energy states is established before ion decomposition takes place. Thus the probabilities of the various possible decompositions of an ion depend only on its structure and internal energy, and not on the method used for the initial ionization, or on the structure of the precursor for, or formation mechanism of, the ion undergoing decomposition. Thus an ion's decomposition depends only on its internal energy and its $k(E)$ functions. When we show mechanistically the molecular ion with a localized charge and radical site, it is not necessary that the ionization actually remove an electron at that site; this is just an attempt to depict the ground electronic state before decomposition. Few exceptions have been found to the QET. These generally involve small molecules (fewer, more widely separated states), excited electronic states that are similarly "isolated," and very fast decompositions ($<10^{-11}$ sec).

The thermochemical appearance energy, $A_t(\text{AD}^+)$, is the minimum energy necessary to produce AD^+ from the ground-state neutral molecule. The critical energy, $E_o(\text{AD}^+)$, is the minimum internal energy of $\text{M}^{+\cdot}$ required for the decomposition to yield AD^+, and thus $A_t(\text{AD}^+) = I(\text{M}) + E_o(\text{AD}^+)$. (These and other relationships to be discussed later are shown in Figure 7C.) Molecular ions containing internal energy $< E_o$ cannot decompose, regardless of the amount of time allowed for decomposition. However, the probability that $\text{M}^{+\cdot}$ ions of internal energy $> E_o$ will produce AD^+ in the ion source depends on the rate constant, k; $\ln [\text{M}^{+\cdot}]_0/[\text{M}^{+\cdot}] = kt$. We shall define $E_s(\text{AD}^+)$ (Figure 7A) as the internal energy of precursor ions which have an equal probability of leaving the ion source as $\text{M}^{+\cdot}$ or AD^+. This causes the measured appearance energy to be higher than the thermochemical value by approximately $E_s - E_o$, which is known as the "kinetic shift"; although this is often <0.01 eV, it can be as large as 2 eV. The kinetic-shift error in such E_o measurements for low-energy reactions can be minimized by observing the ions produced by slow metastable decompositions (Baldwin 1979) or by detecting the product ions with very high sensitivity, such as in photo-ionization studies (McLoughlin and Traeger 1979).

For fragment ions other than the one of lowest appearance potential, such as AB^+ in Figure 7A, an additional kinetic factor, the "competitive shift," also increases the measured appearance energy with reference to $E_o(\text{AB}^+)$.

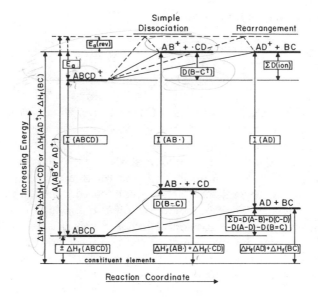

Figure 7C. Thermochemical energy relationships for unimolecular ion decompos -
tions in the mass spectrometer.

Most M^{+} ions whose energy corresponds to that required for $k = 10^6$ sec^{-1}
for $M^{+} \rightarrow AB^+$ will decompose instead to produce AD^+, since k for this
reaction is much larger. The internal energy of M^{+} must be $E_s(AB^+)$ as
shown in Figure 7A to yield equal rates of formation for AB^+ and AD^+.

MI spectra. The decompositions of metastable ions in a field-free drift region
are reactions of rate-constant values just below the minimum required for
ion source decomposition. The reactions yielding MI spectra thus represent
a narrow range of k values, $\sim 10^5$–10^6 mol/sec. In Figure 7A AD$^+$ ions in the
MI spectrum of M^{+}, designated as m^* ($M^{+} \rightarrow AD^+$), thus arise with highest
probability from M^{+} ions of energies between $E_m(AD^+)$ and $E_s(AD^+)$,
and their abundance is represented by the area of the corresponding narrow
window in the $P(E)$ curve. Tight complex reactions, which cause a larger
kinetic shift, will thus tend to give more abundant metastable ion decompositions (Hickling and Jennings 1970). In the same manner the metastable
ions $m^*(M^{+} \rightarrow AB^+)$ must arise from M^{+} of energies corresponding to k
between 10^5 and 10^6, but the probability of such ions decomposing instead
in the ion source by $M^{+} \rightarrow AD^+$ is much higher (see Figure 7A), so that
$[m^* (M^{+} \rightarrow AB^+)]$ will be only a small fraction of the abundance indicated
by the relative area of the corresponding window in the $P(E)$ curve of M^{+}.

7.4 Derivation of $P(E)$ functions

Most methods for approximating $P(E)$ of a molecular ion involve measurement of the energy transferred in producing M^+ from M, to which must be added the internal energy of M before ionization. Because ionization involves a vertical transition, the energy necessary to produce M^+ depends on the energy level in the neutral molecule of the electron which is expelled. Thus the effect of a structural feature on the energy of a molecular state can be reflected in the internal energy of M^+ produced by ionization from this state. The representation of the $P(E)$ function of M^+, such as in Figure 7A, results from plotting E for each state against the relative transition probability (Franck-Condon factor) for the state, convoluted with the internal energy of the neutral molecule. The photoelectron spectrum, which represents such transition probabilities for photoionization, provides a useful approximation of $P(E)$. These can be made more accurate by a correction procedure (Meisels *et al.* 1972); the resulting $P(E)$ curves for 1,2-diphenylethane and its *p*-amino derivative are shown in Figure 7D (McLafferty *et al.* 1970).

The lower energy limit of the $P(E)$ curve is defined by the ionization energy (I). The change in I caused by the addition of the substituent to a molecule such as ABCD in Figure 7C is due to the difference in stabilization energy conferred by the substituent on $ABCD^+$ and ABCD (neutral). The ionization energies of a variety of substituted aromatic compounds show a good correlation with σ^+ constants for para substituents; note Figure 7D and the I data of Table A.3. The shapes of $P(E)$ functions above I can also be related to molecular structure. For example, ionization of the lone pair or π-electrons of added substituents can introduce characteristic new low-energy states, and thus enhance the corresponding low-energy portions of the $P(E)$ curve (Figure 7D).

The $P(E)$ function for a primary product ion A^+, *immediately after formation* from molecular ion M^+ (see Figure 7E), will be determined by: (1) $P(E)$ of M^+; (2) $k(E)$ for the reaction forming A^+ from M^+; (3) $k(E)$ for other reactions competitive with A^+ formation ($M^+ \rightarrow B^+$, $M^+ \rightarrow C^+$); and (4) partitioning of the excess energy in M^+ between the A^+ ion and the neutral product N^o. Figure 7F shows the $P(E)$ of $C_3H_6O^+$ formed from 2-hexanone and from 2-undecanone. The much lower average energy of the latter $P(E)$ is due mainly to the partitioning of the excess energy of M^+, which is divided in proportion to the number of vibrational degrees of freedom in the ion and neutral products. This is seen in the so-called "degrees-of-freedom" effect (Bente *et al.* 1975), which relates the metastable ion abundance, $[m^*]$, for the further decomposition of a particular fragment ion to the size of the molecular ion from which it is formed. In general for a homologous series of molecular ions a linear relation is found for log ($[m^*]/[AB^+]$) versus the reciprocal of the number of vibrational degrees of freedom in the molecular ion yielding AB^+.

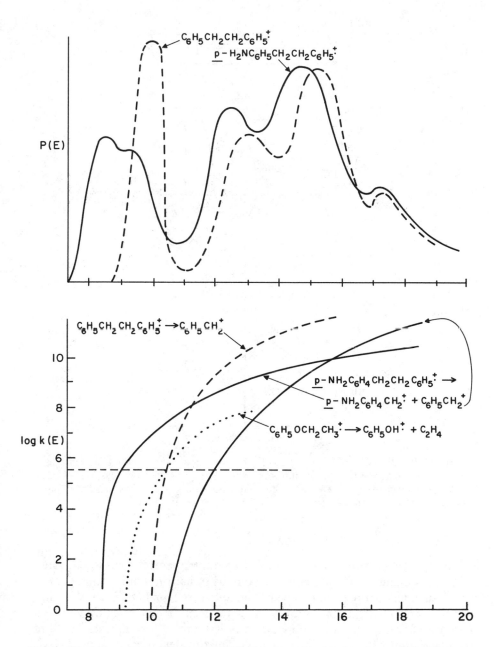

Figure 7D. Approximate $P(E)$ and $k(E)$ functions for the reactions $C_6H_5CH_2CH_2$-$C_6H_5^+$ $\rightarrow$ $C_6H_5CH_2^+$ and p-$NH_2C_6H_4CH_2CH_2C_6H_5^+$ $\rightarrow$ $NH_2C_6H_4CH_2^+$ + $C_6H_5CH_2^+$, and $k(E)$ for $C_6H_5OC_2H_5^+$ $\rightarrow$ $C_6H_5OH^+$. Note that the energy scale is modified from that of Figure 7A to show the effect of adding the p-NH_2 substituent.

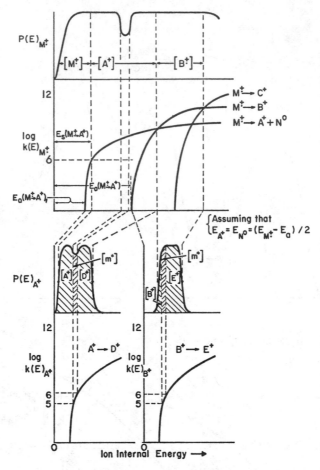

Figure 7E. Relationship of $P(E)$ and $k(E)$ for molecular and product ions in mass-spectral relations; see text for definitions. It is assumed for the process $M^{\ddot{+}} \rightarrow A^+ + N^0$ that $E_0(\text{rev}) = 0$, and that the excess internal energy of $M^{\ddot{+}}$ is divided equally between A^+ and N^0. (McAdoo *et al.* 1974; Bente *et al.* 1975.)

7.5 Calculation of *k(E)* functions

Some of the accessible energy states of an ion can correspond to activated complexes capable of undergoing decompositions; the minimum energy at which such an energy state can be populated corresponds to E_o for the minimum energy reaction. The rate constant for a particular reaction will be a function of the energy state population of its activated complex relative to the population of all other energy states of the decomposing ion. This is described by the Rice-Ramsperger-Kassel-Marcus (RRKM) theory (Robinson and Holbrook 1972):

$$k(E) = \frac{1}{h} \frac{Z^{\ddagger}}{Z^*} \frac{\Sigma P^{\ddagger}(E - E_o)}{\rho^*(E)}, \qquad (7.1)$$

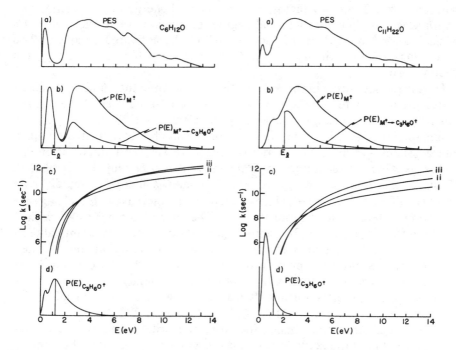

Figure 7F. Data and results of calculations for 2-hexanone and 2-undecanone: (a) photoelectron spectrum of the molecule (experimental data); (b) calculated functions for $P(E)_{M^+}$ and $P(E)_{M^+\to C_3H_6O^+}$; (c) calculated $k(E)$ functions for (i) $M^{\ddagger} \to C_3H_6O^+$; (ii) $M^{\ddagger} \to C_2H_3O^+$; (iii) $M^{\ddagger} \to C_n H_m^+$; (d) calculated value of $P(E)_{C_3H_6O^+}$ functions. The vertical line at 1.2 eV represents the "metastable window" for $C_3H_6O^{\ddagger} \to C_2H_3O^+$.

where h is Planck's constant, Z is the partition function for the adiabatic degrees of freedom, $\ddagger$ refers to the activated complex, * refers to the active molecule (which in this case is an ionic species), $P^{\ddagger}(E - E_o)$ is the number of states in the energy range $E - E_o$, and $\rho(E)$ is the density of states. Adiabatic, in contrast to active, degrees of freedom cannot contribute their energy freely to the dissociating bond. In the active molecule the nonfixed internal energy, E, is randomly distributed over all degrees of freedom. In the activated complex one degree of freedom has been transformed into the translational coordinate requiring E_o, so that only $E - E_o$ is available for distribution. The partition function includes the symmetry factor, which is the number of identical pathways by which the reaction takes place.

The RRKM form of the theory is preferred because it uses an exact enumeration of states. However, some discussions of mass-spectral reactions continue to use the approximate form of the theory,

$$k(E) = \nu \left(\frac{E - E_o}{E}\right)^{n-1}, \tag{7.2}$$

where v is the frequency factor, E is the internal energy of the reacting ion, and n is the number of vibrational degrees of freedom. (The magnitude of v is qualitatively related to the "looseness" of the activated complex.) The predictions of this theory are poorer, especially for lower-energy ion decompositions.

In general, the terms in the RRKM theory reflecting the number of energy states are those affected the most by the structural features of the decomposing ion. The probability that energy sufficient to cause reaction, E_o, will reside in the activated complex depends on the number of activated-complex energy states available for energy residence versus the number of energy states elsewhere in the ion to which the internal energy could be distributed. If the excess internal energy, $E - E_o$, is small, k increases rapidly with increasing E (see Figure 7A), since an incremental increase in E will increase $P^{\ddagger}$ proportionately much more than ρ^*. The rate of increase of k with E should decrease as $E - E_o$ becomes larger, approaching a constant value as E_o becomes very small compared to E. The maximum reaction rate is limited by the time required to separate the two atoms of the dissociating bond ($\sim 10^{-13}$ sec), which depends on its vibrational frequency.

Equation 7.1 predicts further that a change in structure can affect the $k(E)$ function by effects on (1) E_o, (2) $P^{\ddagger}/\rho^*$, and (3) $Z^{\ddagger}/Z^*$.

(1) E_o. If a structural change increases the critical energy, the excess energy, $E - E_o$, of the activated complex must decrease, so that the number of states found between E_o and E should also decrease. Thus k will exhibit a lower value, since the probability of accumulating the energy E_o in the reaction coordinate will be decreased. The energy states are quantized, so that at $E = E_o$ only the ground state of the activated complex is available; the equation demands that $k = 0$ for $E < E_o$. The minimum rate,

$$k_{min} = \frac{1}{h} \frac{Z^{\ddagger}}{Z^*} \frac{1}{\rho^*(E_o)}, \qquad (7.3)$$

will also be reduced by an increase in E_o because of the increase in $\rho^*(E_o)$. Note that an increase in E_o will increase both E_s and the kinetic shift, $E_s - E_o$, since the slope of $k(E)$ will decrease.

(2) $P^{\ddagger}/\rho^*$. A modification of the molecular ion structure which adds or subtracts states, or changes the energy values of states, will change $P^{\ddagger}/\rho^*$, and will thus also change the $k(E)$ function.

If, while other parameters are constant, the number of vibrational degrees of freedom (n) in the reactant ion is increased, $\rho^*(E)$ will increase. This will lower k_{min}, and will thus usually increase the kinetic shift, which is the amount of excess internal energy necessary to reach $k = 10^6$. A major factor that determines $P^{\ddagger}/\rho^*$ is the nature (energy levels and degeneracies) of the vibrational degrees of freedom, all of which are assumed to be active. It is also possible that the number of *active* internal rotational states, the so-called "free rotors," can change in going from the active molecule to the activated complex. A reaction is said to have a "loose complex" if the number of internal rotational states have increased at the expense of vibrational states; for example, stretching of a C—C bond in the activated complex would allow increased rotation of the attached groups while decreasing the vibrational frequency of the bond. A reaction is said to have a "tight complex" if rotational degrees of freedom are frozen out in the transition state. Rearrangements are a common example of this, since the juxtaposition of atoms demanded by the transition state effectively stops rotation about their adjacent bonds. A change in the number of free rotors can have an important effect on the rate constant, since the density of rotational states is much larger than that of vibrational states. This predicts that reactions with tight complexes will have a much slower increase of k with increase in E (lower slope of $k(E)$ in Figure 7A) and will exhibit a much larger kinetic shift.

(3) $Z^{\ddagger}/Z^*$. The partition function includes the symmetry factor, which takes account of the fact that the total rate will be proportional to the number of identical reaction pathways. $Z^{\ddagger}/Z^*$ also reflects the moments of inertia of the activated complex relative to the active molecule; such values usually fall between 1 and 10.

7.6 Thermochemical relationships

Equation 7.4 is a simple cleavage reaction to form an even-electron ion and a neutral radical. The thermochemical relationships for this reaction are

$$ABCD \rightarrow ABCD^{\ddagger} \rightarrow AB^+ + \cdot CD \qquad (7.4)$$

shown in Figure 7C. Let $\Delta H_f(ABCD)$ signify the heat of formation of ABCD from the constituent elements, and $D(B—C)$ be the dissociation energy of the B—C bond. $A_t(AB^+)$ is the *thermochemical* appearance energy of AB^+, so that $A_t(AB^+) - I(ABCD) = E_o(ABCD^{\ddagger} \rightarrow AB^+)$. The activated complex may be of higher energy than the sum of the energies of the products in their ground states; this excess energy is a reflection of the critical energy for the reverse reaction, $E_o(\text{rev})$. For simple cleavage reactions in neutral systems it is usually assumed that $E_o(\text{rev})$ is negligible, since most gas-phase radical-recombination reactions proceed without an activation energy. Similar conclusions have been reached for simple cleavage reactions of ions. However, for rearrangement reactions such as Equation 7.5, in which two bonds are cleaved and two are formed to yield an odd-electron ion and a molecule, $E_o(\text{rev})$ often has an appreciable value.

$$ABCD^{\ddagger} \longrightarrow AD^{\ddagger} + B{=}C \qquad (7.5)$$

If $E_o(\text{rev})$ is negligible, E_o is equal to the bond-dissociation energy of the ion and is predicted directly from several thermochemical relationships (Figure 7C). For reaction 7.4,

$$E_o(\text{AB}^+) = \Delta H_f(\text{AB}^+) + \Delta H_f(\cdot \text{CD}) - \Delta H_f(\text{ABCD}^{\ddagger}) \quad (7.6)$$

$$= \Delta H_f(\text{AB}\cdot) + I(\text{AB}\cdot) + \Delta H_f(\cdot \text{CD}) \\ - \Delta H_f(\text{ABCD}) - I(\text{ABCD}) \quad (7.7)$$

$$= I(\text{AB}\cdot) + D(\text{AB}\!-\!\text{CD}) - I(\text{ABCD}) \quad (7.8)$$

A substantial collection of ionic thermochemical data has been tabulated, most of which is for smaller molecules (Rosenstock et al. 1977). Heats of formation can be estimated for other species (Bowen and Williams 1977).

Note that Equation 7.8 uses bond-dissociation energies of *neutral* species. to which the organic chemist's intuition and knowledge of mechanistic principles should be directly applicable. For two competing reactions of the same molecular ion (and thus the same $I(\text{M})$ values), Equation 7.8 predicts that ΔE_o is determined by ΔI of the respective radicals and ΔD of the respective bonds; this substantiates the conclusions of correlation studies, that the relative stabilization of the product ion and the relative bond strengths are important driving forces for mass-spectral reactions. The term $I(\text{AB}\cdot)$ reflects the electron affinity of the product ion, AB^+. This term, and thus E_o, will decrease with increasing ability of the product ion to stabilize the positive charge. A great deal is known from organic chemistry about the factors governing such stabilization, especially in aromatic systems. The substituted benzyl radicals illustrate this well; the experimental values of $I(\text{YC}_7\text{H}_6\cdot)$ correlate closely with σ^+ constants. For the compounds $\text{YC}_7\text{H}_6\text{R}$ the remaining term of Equation 7.8, $I(\text{M})$, is correlated by σ_p^+ with a smaller positive ρ value.

7.7 Examples

The applicability of Equation 7.1, and especially the usefulness of the concept of the "looseness" or "tightness" of the activated complex, are illustrated by a few model calculations which have been carried out for larger molecules (reactions 7.9 to 7.11). The best agreement between the experimental data and the calculated $k(E)$ functions (Figure 7D) were obtained by assuming that the activated complex undergoes an increase of one or two free rotors over the ion for Equation 7.9, a *reduction* of one free rotor in Equation 7.10, and an increase of three to four free rotors in Equation 7.11. In Equation 7.9 the importance of product ion stability (see Chapter 4) suggests that the activated complex resembles the products more than the precursor. In such a model the central C—C bond will be substantially dissociated, thus lowering the barrier to rotation about this bond, and increasing the number of free rotors in the activated complex.

$$C_6H_5CH_2CH_2C_6H_5^{+\cdot} \longrightarrow C_7H_7^+$$
$$(E_o = 1.3, E_s = 1.8 \text{ eV}) \qquad (7.9)$$

$$p\text{-}H_2NC_6H_4CH_2CH_2C_6H_5^{+\cdot} \longrightarrow p\text{-}H_2NC_7H_6^+$$
$$(E_o = 0.7, E_s = 1.4 \text{ eV}) \qquad (7.10)$$

$$\longrightarrow C_7H_7^+$$
$$(E_o = 2.3, E_s = 7 \text{ eV}) \qquad (7.11)$$

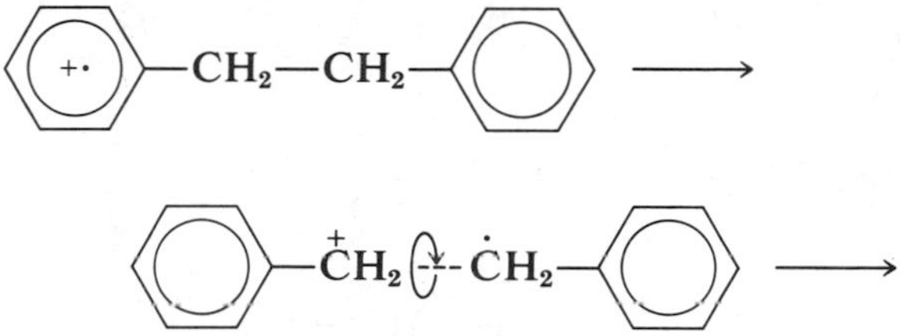

Quite a different situation is indicated for the decomposition of p-amino-1,2-diphenylethane. The presence of the p-amino group should lower the ionization potential of the benzyl radical, lowering E_o, and thus lower the kinetic shift. This is not observed, however, suggesting a *substantial reduction in the "looseness" of the activated complex.* The presence of the amino

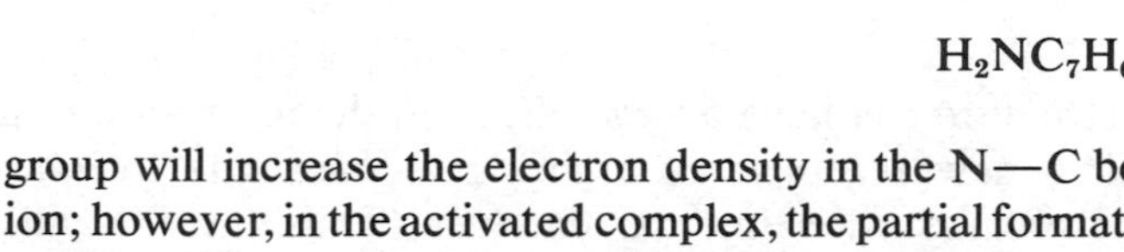

$$H_2NC_7H_6^+ + \cdot C_7H_7 \qquad (7.13)$$

group will increase the electron density in the N—C bond of the molecular ion; however, in the activated complex, the partial formation of the resonance-stabilized p-aminobenzyl ion should also reduce the free rotation about the C(aryl)—CH$_2$ bond, offsetting the increased rotation about the central C—C bond. This indicates that the configuration of the activated complex for Equation 7.10 is substantially "tighter" than that for the active molecule, in direct contrast to the situation in Equation 7.9.

The critical energy for Equation 7.11 is even higher than that for Equation 7.9 because of the lowered value of I, and there is an even greater increase in the value of the combined kinetic and competitive shifts. The significant abundance of this product ion in the 70 eV spectrum indicates that $k(E)$ for Equation 7.11 must rise much more rapidly than that for Equation 7.10. The transition-state Equation 7.11 must involve an even greater *increase* in free rotors than for Equation 7.9, based on the fact that the partial charge on the amino group and the adjacent aromatic ring must now *decrease* in the transition state, lowering the double-bond character of the adjacent bonds.

$$H_2\overset{+}{N} = \langle \cdots \rangle \text{---} CH_2 \text{---} CH_2 \text{---} \langle \bigcirc \rangle \longrightarrow$$

$$H_2N \text{---} \langle \bigcirc \rangle \text{---} \overset{\cdot}{C}H_2 \text{---} \overset{+}{C}H_2 \text{---} \langle \bigcirc \rangle \longrightarrow$$

$$H_2NC_7H_6\cdot + C_7H_7^+ \qquad (7.14)$$

RRKM calculations have also been made for the major decomposition of the phenyl ethyl ether molecular ion, which yields the phenol ion through loss of C_2H_4. The reaction proceeds through a four-membered-ring transition state supporting this mechanism; calculations assuming the loss of three free rotors in the activated complex give results which are in agreement with the observed product-ion abundances. Note in Figure 7.4 that the critical energy, the tightness of the activated complex, and the size of the ion are of major importance in determining the shape of the $k(E)$ curve.

$$\langle \bigcirc \rangle \overset{H}{\underset{\overset{+}{O}}{}} \longrightarrow \| + \langle \bigcirc \rangle \overset{\cdot+}{HO} \qquad (E_o = 1.0, E_s = 2.5 \text{ eV}) \qquad (7.15)$$

Small changes in structure can have a large effect on the mass spectrum if these make possible new reactions of higher rate constants over an appreciable range of precursor-ion internal energies. Thus the dramatic differences between the spectra of *n*-dodecane and 1-amino-*n*-dodecane (Figures 3B and 3P) are due primarily to the high rate constant for the new reaction forming the $CH_2{=}NH_2^+$ ion. Increasing the critical energy of a reaction decreases the slope of the $k(E)$ curve, so that a higher-energy reaction must have a sufficiently looser activated complex in order to be competitive.

Structural effects on the $P(E)$ function of the molecular ion can also affect the mass spectrum, particularly in the abundances of the molecular and metastable ions. Addition of a m-amino group to 1,2-diphenylethane only lowers the critical energy for benzylic cleavage by 0.2 eV, but lowers the ionization energy by 1.1 eV; this produces a doubling of the molecular ion abundance. Adding a p-nitro group to 1,2-diphenylethane make no appreciable change in either the I value of the molecule or the A value of the benzyl ion (10.5 $\pm$0.2 eV). However, the nitro group greatly increases the population of the $P(E)$ curve of $M^{+\cdot}$ around 10.5 eV, presumably through ionization at the nitro group, resulting in a several-fold increase in the ion abundance arising from the metastable decomposition $M^{+\cdot} \rightarrow C_7H_7^+$; in Figure 7D note the low population of $C_6H_5CH_2CH_2C_6H_5^{+\cdot}$ ions in the energy region required for this metastable decomposition.

8

DETAILED MECHANISMS OF
ION FRAGMENTATION

There are now many excellent reviews that give detailed discussions of mechanisms of unimolecular ion reactions: Howe and Williams 1973; Bursey *et al.* 1973; Kingston *et al.* 1974, 1975; Johnstone 1975, 1977, 1979; Levsen 1978; Bowen *et al.* 1979; Williams 1979; Nibbering 1979; McLafferty 1980b. The reader should consult these for further study. A thorough discussion is not possible here; in this chapter, as in Chapter 4, the main emphasis will be on mechanisms of particular use for interpretation of EI, CI, FI, CA, and MI mass spectra.

8.1 Unimolecular ion decompositions

To review and reemphasize the positive ion reactions important for deducing structure, the mass spectrum shows *competing* and *consecutive* unimolecular reactions (Figure 8A). Primary decomposition products such as ABC+ and

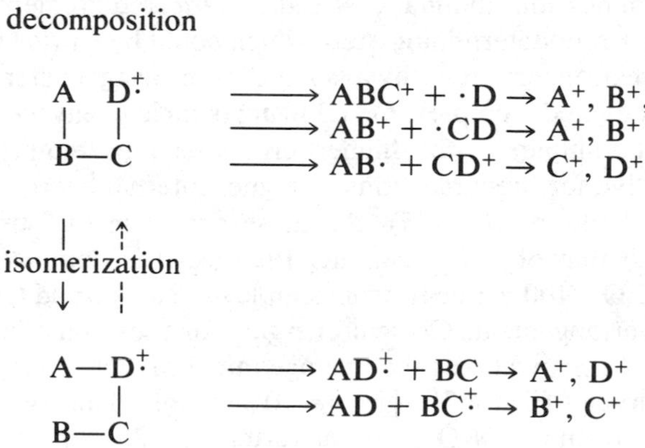

decomposition

$$
\begin{array}{ccl}
A & D^{+} & \\
| & | & \longrightarrow ABC^+ + \cdot D \to A^+, B^+, C^+ \\
| & | & \longrightarrow AB^+ + \cdot CD \to A^+, B^+ \\
B & \!\!-C & \longrightarrow AB\cdot + CD^+ \to C^+, D^+
\end{array}
$$

$$
\begin{array}{c}
| \quad \uparrow \\
isomerization \\
\downarrow \quad \vdots
\end{array}
$$

$$
\begin{array}{ccl}
A & \!\!-D^{+} & \longrightarrow AD^+ + BC \to A^+, D^+ \\
| & & \longrightarrow AD + BC^+ \to B^+, C^+ \\
B & \!\!-C &
\end{array}
$$

Figure 8A.

119

CD^+ represent pieces of the original molecule, and thus provide direct evidence of molecular structure. However, at lower energies isomerization can compete with primary decomposition; secondary ion products formed after (or accompanying) isomerization can be rearranged ($AD^{+\cdot}$, $BC^{+\cdot}$). *Specific rearrangements* giving structurally indicative ions commonly involve direct isomerization to a structure with a competitively favorable decomposition pathway. Ions undergoing *random rearrangements* usually require higher critical energies for dissociation, so that several isomerization steps (or isomeric equilibrium) are competitive. In *consecutive* decompositions the chances are greatly increased that one or more of the intermediate ion decompositions will involve isomerization and thus yield products of lower structural integrity. However, such products are useful in low-mass "ion series" to indicate heteroatoms and rings-plus-double-bonds values in particular substructures.

8.2 Basic factors affecting ion abundance

In interpreting mass spectra it is important to understand the influence of product stabilities, steric effects, and bond labilities on ion-decomposition pathways. Product stability is of primary importance; for competing reactions the relative critical energies are determined by the respective heats of formation, and thus stabilities, of the corresponding ionic and neutral product pairs (Figure 7C; this assumes that the critical energies of the corresponding reverse reactions are negligible or similar). Increased stability is of additional importance for increasing the abundance of product ions by reducing their secondary dissociation. However, product stability cannot be the only important factor, for then a molecular ion would produce very few peaks; these would be formed by whatever combination of isomerization and decomposition pathways yields the most stable products. A second important factor that determines ion abundances is the *steric* requirement(s) of the reaction pathway's rate-determining step, which could be a prior isomerization. In general, rearrangement pathways have more stringent steric requirements (tighter activated complexes), allowing simple cleavage reactions (looser activated complexes) of higher critical-energy requirements to be more competitive for precursor ions of higher internal energy.

For example (Bente *et al.* 1975), the mass spectrum of 2-undecanone (closely similar to that of 2-dodecanone, Figure 3J) is dominated by the primary ions $C_2H_3O^+$ (100 per cent) from simple α-cleavage and $C_3H_6O^{+\cdot}$ (50 per cent) from rearrangement. Critical energies for these reactions are 2.5 and 1.0 eV, respectively, whereas the average initial $M^{+\cdot}$ ion energy is 3.8 eV from the $P(E)$ curve of Figure 7F ($[M^{+\cdot}] = 10$ per cent). There is an average energy of 2.3 eV in the $C_3H_6O^{+\cdot}$ ions at formation; 25 per cent of these decompose before leaving the ion source (Figure 7E).

Bond lability. As discussed in Section 7.4, the critical energy for $ABCD^{+\cdot}$ $\rightarrow AB^+ + \cdot CD$ is equivalent to the dissociation energy of the AB—CD bond in the *neutral* molecule plus the ionization energy of the radical $AB\cdot$, if the reverse critical energy is negligible. Bond-dissociation energies for competitive reactions are generally 2 to 4 eV, compared to radical ionization energies of 6 to 10 eV (Table A.3); so variations in the latter are generally more important in determining the relative activation energies of possible competing processes. Further, product-ion stability is often increased by factors increasing bond lability, such as the effect of the degree of branching on carbon.

Bond lability can be the deciding factor, however, for competitive pathways that yield ions of nearly equal stability. Consider the aromatic ion $Br—CH_2C_6H_4CH_2—I^{+\cdot}$; the loss of Br or I will form benzyl ions, for which the corresponding radicals should have nearly equal ionization energies. The C—I bond is weaker than the C—Br bond, and so I loss will be favored. The base peak in the mass spectrum of $C_6H_5CH_2COOCH_2C_6H_5$ is $C_7H_7{}^+$, as expected. Isotopic labeling (Hoffman and Wallace 1973) shows that 65 per cent of this ion arises by cleavage of the C—O bond, which in the neutral molecule should be weaker than the benzylic C—C bond.

Charge competition in $OE^{+\cdot}$ decompositions. Stevenson's Rule states the *energy* considerations that govern charge retention vs. charge migration in $OE^{+\cdot}$ decompositions. During dissociation the separating fragments compete for the unpaired electron, with the loser becoming ionized. If in $M^{+\cdot}$ one bond is cleaved (8.1), two incipient radical species compete for the electron, whereas with two bonds cleaved (8.2) two incipient molecules compete. Energetically, the fragment of highest ionization energy (I) is favored to retain the unpaired electron and to become the neutral product. To make such predictions, Table A.3 includes I values of molecules, radicals, and other less-stable neutrals (e.g., enols). The fragmentation of methyl *n*-butyl ether ions (Equation 8.3) at the heteroatom favors radical retention on oxygen, which is borne out by $[C_4H_9{}^+] > [OCH_3{}^+]$ in the spectrum. However, for the α-cleavage reaction, formation of $CH_2{=}OCH_3{}^+$ is favored by 1.2 eV, and is the base peak of the spectrum.

$$C_2H_5{}^{+\cdot}H \quad \begin{array}{l} \longrightarrow C_2H_5{}^+ + \cdot H \\ \longrightarrow C_2H_5\cdot + H^+ \end{array} \qquad (8.1)$$

$$C_2H_4{}^{+\cdot}\begin{array}{c} H \\ \vdots \\ \vdots \\ H \end{array} \quad \begin{array}{l} \longrightarrow C_2H_4{}^{+\cdot} + H_2 \\ \longrightarrow C_2H_4 + H_2{}^{+\cdot} \end{array} \qquad (8.2)$$

$$C_3H_7CH_2OCH_3^+ \begin{cases} \longrightarrow C_3H_7CH_2^+\hspace{-0.3em}\cdot OCH_3 \longrightarrow C_4H_9^+ \text{ or } OCH_3^+ \\ \qquad\quad 8.0 \text{ eV} \quad 9.8 \text{ eV} \qquad\qquad 25\% \qquad 1\% \\ \\ \longrightarrow C_3H_7^+\hspace{-0.3em}\cdot CH_2OCH_3 \longrightarrow \\ \qquad\quad 8.1 \text{ eV} \quad 6.9 \text{ eV} \end{cases} \qquad (8.3)$$

$$C_3H_7^+ \text{ or } CH_2{=}OCH_3^+$$
$$4\% \qquad\qquad 100\%$$

Stevenson's Rule may also be unique in that steric effects appear to be of little importance. If simple cleavage of the same bond is the rate-determining step in the two competing reactions, steric effects should be similar, as well as small, for both reactions. For example, the retro-Diels-Alder reaction of 4-vinylcyclohexene (butadiene dimer) molecular ions yields the same products, C_4H_6 and $C_4H_6^+\hspace{-0.3em}\cdot$, with either charge retention or migration (8.4). Isotopic labeling shows that each path forms nearly equal amounts ($[b]/[a] = 1.15$) of the $C_4H_6^+\hspace{-0.3em}\cdot$ ions; the difference is consistent with conformational effects on the ionization energies of the incipient a and b products in the activated complex. Other diene dimers show similarly small or negligible steric effects (Turecek and Hanus 1980).

$$(8.4)$$

The original formulation of Stevenson's Rule (1951) was slightly different than that used by mass spectrometrists for many years (Harrison *et al.* 1971, p 1316*n*).

Charge competition in EE$^+$ decompositions. The closest parallels to Stevenson's Rule which are applicable to EE$^+$ ions are proposals by Field (1972) and by Bowen *et al.* (1978) pointing out the *energetic favorability of forming neutrals with low proton affinities* (PA). Consider first the one-bond cleavage of an EE$^+$ ion, $AY^+ \rightarrow A^+ + Y$. According to Field (1972), the tendency for the reverse reaction $A^+ + Y \rightarrow AY^+$ (for which the critical energy should be negligible) should depend on the affinity of the molecule Y for the A ion. The relative affinities of Y molecules for a particular A ion should be similar to those for the proton, so that the tendency for different AY^+ ions to lose the same A ion should be approximately in the reverse order of the proton affinities (Table A.3) of the corresponding Y molecules. This would predict the loss of NH_3 (PA = 9.1 eV) from a protonated amine to be much smaller than the loss of H_2O (PA = 7.7) from the corresponding protonated alcohol, and this smaller than the loss of HCl (PA = 6.4) from the corresponding protonated chloride, as is observed. For precursor EE$^+$ ions of equal internal energies in 8.5 and 8.6, there should be a higher tendency for loss of formaldehyde (PA = 7.9 eV) than of acetaldehyde (PA = 8.3). However, in evaluating competing molecular losses from a particular EE$^+$ ion (formation of different A$^+$ ions), one must take into account that the stabilities of the complementary product ions also affect the relative energy requirements, and the product abundance can also be affected by further dissociation (Harrison and Onuska 1978; Respondek et al. 1978).

$$C_2H_5\overset{+}{O}=CH_2 \xrightarrow{\quad i \quad} C_2H_5^+ + O=CH_2 \qquad (8.5)$$

$$C_2H_5\overset{+}{O}=CHCH_3 \xrightarrow{\quad i \quad} C_2H_5^+ + O=CHCH_3 \qquad (8.6)$$

For EE$^+$ rearrangement reactions (Equation 4.43) taking place at low threshold energies, Bowen et al. (1978) propose that proton affinity values are also useful for predicting relative product abundances because these reactions involve a proton-bound intermediate (Equation 8.7). There is no $C_2H_5^+$ in the MI spectrum of $C_2H_5O=CH_2^+$ (8.5), but $HO=CH_2^+$ represents a major peak; C_2H_4 (PA = 7.3 eV) competes unsuccessfully with CH_2O (PA = 7.9) for the proton. However, for higher-energy ions, steric effects will play a larger role because of the tight activated complex and the possibility of reverse activation energy. From higher energy precursors the formation of $C_2H_5^+$ (8.5) is increased; in the CA spectrum of $C_2H_5O=CH_2^+$, 9 per cent of the product ions are $C_2H_5^+$, and 58 per cent are $CH_2=OH^+$ (8.7). The latter should also be formed by direct H rearrangement to the oxygen (see reaction 4.43), for which the activated complex should be less tight than that of reaction 8.7.

$$C_2H_5\overset{+}{O}=CH_2 \xrightarrow{\quad rH \quad} C_2H_4 \cdots \overset{+}{H} \cdots O=CH_2 \xrightarrow{\quad rH \quad}$$

$$C_2H_4 + H\overset{+}{O}=CH_2 \qquad (8.7)$$

Benoit and Harrison (1977) point out that ionization energy values correlate with PA values within a particular compound class, such as alcohols and ethers or aldehydes and ketones; similar relationships with bases have been found (DeKock and Barbachyn 1979). Thus there is a qualitative relationship between Stevenson's Rule governing OE$^{\dot{+}}$ decompositions and these for EE$^+$ decompositions. However, to reiterate, these rules can be misleading for reactions which are primarily under kinetic, not thermodynamic, control.

Loss of the largest alkyl. Zahorsky (1979) has made extensive quantitative studies recently on the relative abundances of ions produced by competitive loss of alkyl radicals from $M^{\ddot{+}}$ of a variety of compound types. In many cases the *abundance ratio of the resulting product ions is inversely proportional to their masses* for the loss of ethyl or larger radicals. If the primary product ions decompose further, the abundance of any secondary products must be added to that of their primary precursor. The α-cleavage of ethyl propyl butyl carbinol at 13 eV (to eliminate secondary decompositions) produces peaks at 129, 115, and 101 with relative abundances of 78, 85, and 100 per cent, respectively; predicted abundances are $101/129 = 78$ per cent and $115/101 = 88$ per cent. In methyl dialkyl carbinols (Equation 4.17) methyl-loss ion abundances are 25 to 45 per cent of those predicted. Deviations are also found at low electron energies if the appearance energies for the competing processes are unequal. In the formation of highly stable ions, such as the immonium ion α-cleavage products of alkylamines, the increased radical stabilization of the larger alkyl product results in a lower appearance energy; thus the loss of the largest alkyl group is not favored at threshold ionization energies.

The impressive quantitative correlation between product abundance and the mass of the alkyl lost (Zahorsky 1979) may be another manifestation of the "degree-of-freedom" effect in which the abundance of a metastable ion product from homologous precursors is found to be inversely related to the number of vibrational degrees of freedom (which for a homologous series is closely approximated by the mass) of the decomposing ion (McAdoo *et al.* 1974; Bente *et al.* 1975; Section 7.4).

Steric factors. Of the several ways that the competitiveness of unimolecular reaction pathways can be affected by steric relationships, the most significant involves the "looseness" or "tightness' of the activated complexes involved (Sections 7.5, 7.7). The gain or loss of free rotors in the reaction transition state is a major factor (along with the activation energy) in determining the $k(E)$ function. The structural factors important for reactions of favorable critical energy usually have unfavorable steric factors, such as (Section 7.7) the competitive formation of $p\text{-}NH_2C_6H_4CH_2^+$ and $C_6H_5CH_2^+$ from $p\text{-}NH_2C_6H_4CH_2\text{—}CH_2C_6H_5^{\ddot{+}}$. Although the first product is favored by a critical energy difference of 1.6 eV because of its higher resonance stabilization, this also leads to an increased bond order in the reaction activated complex which lowers the number of free rotors.

The much tighter activated complexes of rearrangement reactions cause their competitiveness to be reduced even more dramatically with increasing precursor energy. In many cases dominant rearrangement products in MI spectra are hardly observable in normal EI spectra. Rearrangements show a reduced probability of forming the activated complex; for example, in the

γ-H/β-cleavage rearrangement (Equation 4.33), the γ-H will be within bonding distance of the radical site in only a small proportion of the total possible conformations of the molecular ion. The steric restrictions of the molecular ion may reduce such reactive conformations, or even make them impossible. The γ-H/β-cleavage rearrangement takes place in 16-ketosteroids, in which the distance of closest approach of the γ-H and carbonyl oxygen is 1.5 Å. However, the rearrangement does not take place in 11- or 15-ketosteroids, whose corresponding distances are 1.8 and 2.3 Å, respectively. Such steric unfavorability in the molecular ion can be changed by its isomerization, however; in 12-ketopregnanes the 20-H is rearranged despite an initial 3 Å distance, apparently through initial cleavage of the 13–17 C—C bond (Djerassi and Tokes 1966).

The ring size of the rearrangement transition state can depend more heavily on steric factors if there are minimal differences in the stabilities of the competing products. Thus in hydrogen rearrangement to a saturated heteroatom (Section 4.9, reactions 4.37 and 4.40) the covalent radius of the heteroatom affects the preferred ring size. 1,3-Elimination (five-membered ring) predominates for HCl and HBr eliminations from alkyl chlorides and bromides (Equation 4.40); ~40 per cent 1,3- and 60 per cent 1,4-elimination of H_2S is observed from thiols; and mainly 1,4-elimination of H_2O is found from alcohols (Equation 4.38, Budzikiewicz et al. 1967). Kuster and Seibl (1976) find that 1,3- and 1,4-elimination are equally important in H_2O loss from the EE$^+$ ion n-C_5H_{11}—CH=OH$^+$ formed from $(n$-$C_5H_{11})_2$CHOH. Similarly, the displacement reaction 4.42 of R—$(CH_2)_n$—Y compounds yields cyclic $(M - R)^+$ ions in which the five-membered-ring species predominates for Y = Cl, Br, and SH, and the six-membered ring for Y = NH_2.

Even conformational effects have been observed; the 1,3-elimination of HCl from 2-chloropentane (8.56; that of H_2O from 2-hydroxypentane gives similar results) shows $[(M - DCl)^{+}]/\{[(M - DCl)^{+}] + [(M - HCl)^{+}]\} = 0.26$ for the trans-4-D- and 0.30 for the cis-4-D-isomer (a and b, Figure 8B; Green 1976). Despite a large isotope effect, the data show an experimentally significant difference favoring abstraction of the cis-hydrogen through the five-membered-ring transition state. The analogous cis-epimer of 1,3-dimethylcyclopentane is more stable than the trans-.

Steric factors can also influence product stability; for example, α-cleavage loss of methyl gives the most abundant fragment ion from Figure 8Bc but of only 1 per cent abundance for d, whose product ion would be significantly strained. The protonated cis-1,4-cyclohexanediol e is substantially more stable than the corresponding trans-ion for which the bridging H-bond would not be possible (Winkler and McLafferty 1974; VanGaever et al. 1977).

"Ortho effect" rearrangements (Schwarz 1978), such as Equation 4.41, involve neighboring groups on an aromatic ring; the corresponding reaction

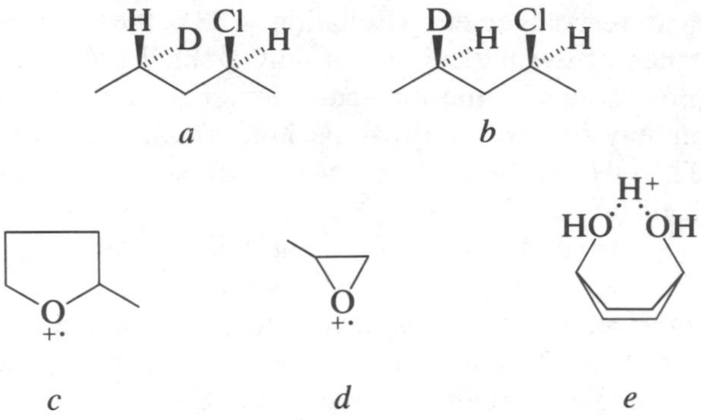

Figure 8B.

of substituents located *meta*- or *para*- to each other is of much lower probability because the necessary reactive conformations are improbable. Rearrangement of a γ-H to an aromatic ring (reactions 4.36 and 8.8) is unfavorable when both *ortho* positions are substituted. Steric crowding can increase the rate of a reaction. The appearance energy of $(M - CH_3)^+$ from *o*-di-*tert*-butylbenzene is 0.9 eV lower than that of the *m*- and *p*-isomers, which has been attributed to the difference in strain energies of the corresponding molecular ions (Arnett *et al.* 1967).

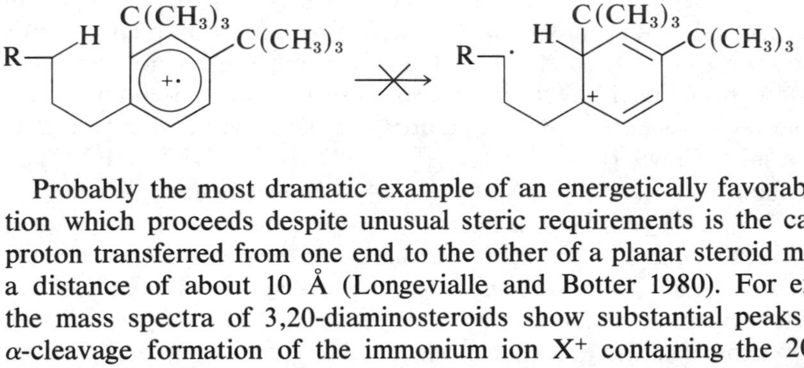

$$\tag{8.8}$$

Probably the most dramatic example of an energetically favorable reaction which proceeds despite unusual steric requirements is the case of a proton transferred from one end to the other of a planar steroid molecule, a distance of about 10 Å (Longevialle and Botter 1980). For example, the mass spectra of 3,20-diaminosteroids show substantial peaks for the α-cleavage formation of the immonium ion X^+ containing the 20-amino group and an $(M - X + 1)^+$ ion. Metastable decomposition of the latter involves the loss of the 3-amino group plus a hydrogen atom; labeling studies show that this hydrogen was originally on nitrogen-20. Theoretical calculations are consistent with the following explanation; during the initial α-cleavage, the immonium ion X^+ separates so slowly from the D-ring that there is time ($\sim 10^{-11}$ sec) for the steroid residue to rotate. The A-ring 3-amino substituent can thus approach X^+ and abstract a proton, in effect transferring a hydrogen from the 20- to the 3-position. Note the similarities of this mechanism to reaction 8.7.

Unknown 8.1. What is unusual about this ion series?

m/z	Rel. abund.	m/z	Rel. abund.	m/z	Rel. abund.
15	0.6	56	17.	106	0.6
27	18.	57	21.	107	5.0
28	3.5	58	0.9	108	0.7
29	18.	69	6.0	109	4.7
30	0.4	70	0.5	121	0.5
39	11.	71	0.5	123	0.4
40	2.1	83	0.7	135	50.
41	37.	84	0.6	136	2.2
42	14.	85	49.	137	49.
43	100.	86	3.2	138	2.1
44	3.4	93	1.2	164	2.2
55	34.	95	1.2	166	2.2

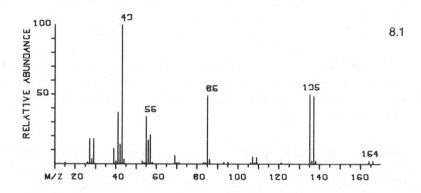

8.1

Deducing stereochemistry. Extensive recent investigations have shown a variety of cases in which mass-spectral information can be used to distinguish between possible stereoisomers (Brion and Hall 1966; Eadon 1977); the reviews of Green (1976) and Mandelbaum (1977) are particularly recommended. Caution in interpreting the data is necessary, however. Reliability is greatly increased by comparing the unknown's mass-spectral behavior with that of closely related molecules or of a contrasting isomer. An elegant example reported by Green *et al.* (1970) involves the losses of H_2O and HCl from cyclohexanol and cyclohexyl chloride, respectively. Deuterium labeling shows that 1,4-elimination is stereospecific in both (8.9), but 1,3-elimination is stereospecific only in the chloride. This indicates that ring cleavage precedes the 1,3-elimination of water, possibly by reaction 8.10. The distance of closest approach for the 1,-4-cyclohexyl substituents is 1.7 Å, but 2.3 Å for 1,3-isomers. Stereospecific 1,3-elimination of HCl is probably made possible by the ~0.4 Å greater bond length of C—Cl vs. C—OH.

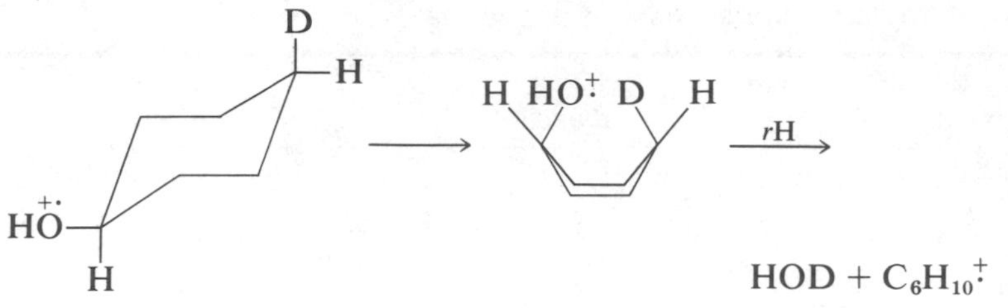

$$HOD + C_6H_{10}^{+\cdot} \qquad (8.9)$$

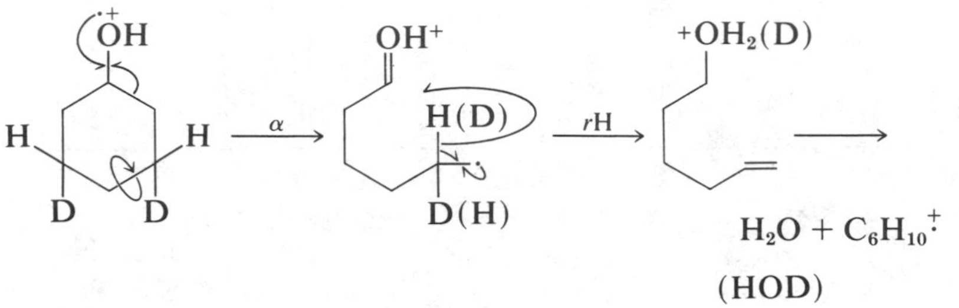

$$H_2O + C_6H_{10}^{+\cdot} \qquad (8.10)$$

$$(HOD)$$

Stability differences in ions produced by CI, such as Figure 8B*e*, are also valuable for determining stereochemistry. Such internal H bonds in cyclic diols can be detected by their effect on gas-phase basicities and acidities (Winkler and Stahl 1979).

Thermodynamic vs. kinetic effects. Thus a basic dilemma in predicting mass spectra is how to judge the degree to which favorable energy requirements are offset by steric effects among possible competing fragmentation pathways. To interpret an unknown mass spectrum, one must judge to what extent the presence of an important peak is due to favorable energetic or steric factors. Many rearrangement reactions go through stepwise pathways; the favorability of each step must be predicted to assess the competitiveness of the rate-determining step.

Yet another problem in predicting abundances is the effect of structure on $P(E)$, the energy deposition function. Higher-molecular-weight 2-alkanones show an unusual $C_3H_7O^+$ (m/z 59) peak, which apparently arises from a low-energy double-H rearrangement similar to reaction 4.46. Although the abundance of this peak is 20 per cent of that of the base peak in 2-undecanone, it is <1 per cent in 2-hexanone. The $P(E)$ functions of Figure 7F offer an explanation; there is a "hole" in the $P(E)$ for 2-hexanone, indicating that relatively few molecular ions are formed with internal energies in the range 1 to 2 eV. To reiterate, for either interpretation or prediction of mass spectra, it is important to study the spectra of closely related molecules.

8.3 Reaction initiation at radical or charge sites

Use of this mechanistic approach for a more complex molecule containing several possible sites for the localized charge and unpaired electron requires weighting of the relative importance of each site. Further, at each site the importance relative to each other of the radical and charge as reaction initiators must also be predicted. These factors will be discussed in this section, as will cases of decompositions similar to radical or charge-site reactions which do not formally have that initiating site.

Relative importance of possible sites. For a molecule containing noninteracting functional groups, the ionization energy (I) usually corresponds to the value characteristic of the functional group of lowest I value. This is illustrated for the amine, sulfide, and selenide groups in Figure 8C; apparently the electronegative carboxyl group does influence the amine I value in glycine (Svec and Junk 1967). The most abundant ions of these polysubstituted molecules correspond to α-cleavage products from initiation at the radical site of lowest ionization energy; the base peaks in the spectra of the lower rows of compounds are $H_2N=CH_2^+$, $CH_3S=CH_2^+$, and $CH_3Se=CH_2^+$, respectively. For the latter two this is true despite the fact that the appearance energies are ~ 2.5 eV higher than that for their $(M-COOH)^+$ peaks; note that secondary decomposition of higher energy $CH_3YCH_2CH_2CH=NH_2^+$ should yield additional $CH_3Y=CH_2^+$ ions.

$$CH_3COOH \qquad H_2NCH_2CH_3 \qquad CH_3SCH_2CH_2CH_3 \qquad CH_3SeCH_2CH_2CH_3$$

   10.4 eV   8.9 eV   8.7 eV   8.2 eV

$$H_2NCH_2COOH \qquad CH_3SCH_2CH_2CH(NH_2)COOH$$

   9.2 eV   8.6 eV

$$CH_3SeCH_2CH_2CH(NH_2)COOH$$

   8.3 eV

Figure 8C.

Unknown 8.2. A derivative of a substance essential for nutrition.

m/z	Rel. abund.	m/z	Rel. abund.	m/z	Rel. abund.
29	4.6	62	1.9	93	0.5
30	15.	63	5.3	102	1.7
31	3.6	64	2.6	103	12.
39	8.9	65	12.	104	1.8
40	1.4	66	0.6	117	2.2
41	4.2	76	1.0	118	7.3
42	6.9	77	7.1	119	4.0
43	5.0	78	2.0	120	77.
44	1.9	79	2.6	121	7.1
45	1.8	85	1.2	122	0.3
50	1.7	86	7.9	131	0.9
51	7.9	87	1.8	147	0.5
52	1.5	88	100.	148	0.6
53	0.8	89	3.9	179	1.3
54	0.8	90	1.4	180	1.4[a]
60	14.	91	30.		
61	2.0	92	5.9		

[a]Relative abundance increases with increasing sample pressure.

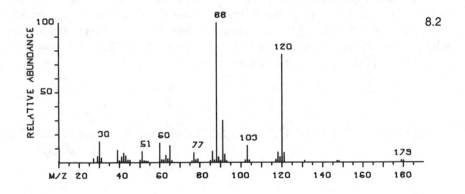

Charge vs. radical stabilization. For product stability as a reaction-path determinant, it is generally more important that the product provide a stable environment for the charge than for the unpaired electron. For example, isomerization of the *n*-butyl to the *tert*-butyl structure reduces the heat-of-formation value of the radical by 0.4 eV, but of the ion by 1.5 eV. Electron-sharing stabilization provided by the nonbonding electrons of an adjacent heteroatom (e.g., H_2NCH_2, $HOCH_2$) is effective mainly for ions. Although radical stabilization at an electronegative heteroatom (e.g., $CH_3O\cdot$) is often more important than charge stabilization, the difference is less pronounced. For example, the heat of formation of CH_3O^+ (triplet) is 2.9 eV higher (less stabilization) than that of $CH_2{=}OH^+$, but that of $CH_3O\cdot$ is only 0.1 eV lower than that of $\cdot CH_2OH$ (Dill *et al.* 1979). Production of $CH_2{=}OH^+$ $\leftrightarrow\,{}^+CH_2{-}OH$ by α-cleavage involves formation of a partial new bond, compensating in part for the energy required for the bond broken. Production of $CH_3O\cdot$ and R^+ by inductive cleavage of CH_3OR involves no such compensating bond formation, however.

Radical vs. charge-site initiation. The initial charge site in the molecular ion has, in general, the most favorable environment for stabilization for the positive charge. In many cases this is also the site of optimum charge stabilization in the product ion, favoring reactions in which the positive charge does not move. For $OE^{\underset{\cdot}{+}}$ ions such charge-retention reactions involve *initiation at the radical site* (Table 4.1). Thus such reactions are generally more common than charge-site reactions; reasons for this based on Stevenson's Rule were advanced in Section 8.2. The types of functional groups favoring such reactions were discussed in Chapter 4; if only electronegative substituents are present, charge-site reactions are often favored.

Initiation at incipient radical/charge sites. The radical/charge site concept used in Chapter 4 is basically a means to force consideration of possible decomposition pathways which indicates their resulting product stabilities and steric requirements. Some of its simplifying assumptions, however, limit its use as a mechanistic tool for mass-spectral interpretation. In some cases even-electron ions undergo reactions thought to be characteristic of a radical site, although ostensibly all electrons are paired in such ions. The first reported example, reaction 8.11, was shown by Kraft and Spiteller (1967) to transfer hydrogen specifically through a six-membered-ring intermediate, and so is a "γ-H rearrangement to an unsaturated group with β-cleavage" closely similar to reactions 4.33, 4.35, and 4.36. In common with these, a major driving force for reaction 8.11 is the formation of stable products; the importance of this driving force for dissociation of this relatively stable EE^+ ion could more than offset the energy necessary to unpair electrons in the postulated intermediate.

$$(8.11)$$

Unsaturated molecules exhibiting such γ-H/β-cleavage rearrangements in their EI spectra often show counterpart rearrangements of the EE$^+$ proton-ated molecular ion (MH$^+$) in their CI spectra. Aliphatic and aromatic esters and phenylalkanes have been studied with the aid of isotopic labeling. Although here the transferred hydrogen can originate from several positions, in some cases the position distribution is quite similar to the corresponding rearrangement of the OE$^+$ molecular ion (Benoit and Harrison 1976; Leung and Harrison 1977). In other cases the distribution has been shown to be dependent on the proton affinity of the ionizing reagent gas, and thus pre-sumably on the internal energy of the fragmenting MH$^+$ (Audier *et al.* 1977).

Reactions usually visualized as being initiated at charge sites can also take place in EE$^+$ ions in which little of the charge density can be at the site. The loss of HCl from $CH_3CH^+CH_2CH_2Cl$ has already been given as an example (reaction 4.45). Most exceptions to the mechanistic usefulness of radical/charge-site initiation are decompositions of stable EE$^+$ ions. If decomposi-tion pathways using their preferred charge site are not particularly favorable, reactions should be considered at other sites which would be favorable if they did hold an unpaired electron or charge.

In applying the radical/charge concept mechanistically, you should not be concerned that in some cases either radical or charge-site initiation produces the same products. The cyclic ion decompositions of reaction 4.30 can also be written (8.12) as a charge-site reaction. The substantial m/z 56 peak in the spectrum of 2-methylpiperidine could arise (Budzikiewicz *et al.* 1967, p. 316) from a retro-Diels-Alder decomposition of the α-cleavage product ion (reac-tion 8.13). These products could also be formed by initial α-cleavage in the ring, followed by losses of C_2H_4 and $CH_3\cdot$. In contrast, the retro-Diels-Alder reaction 4.31 is written as involving radical-site initiation.

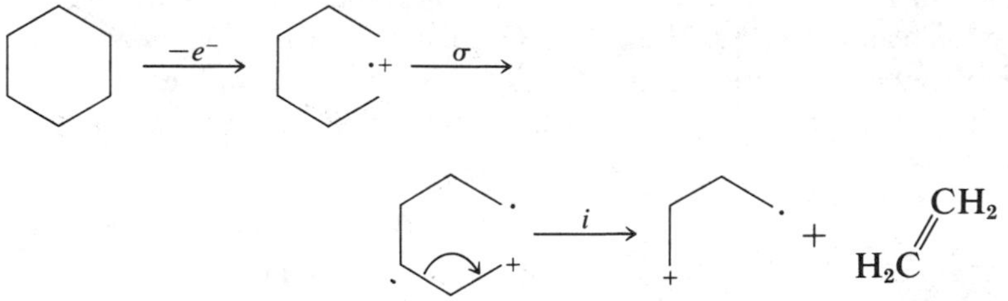

$$(8.12)$$

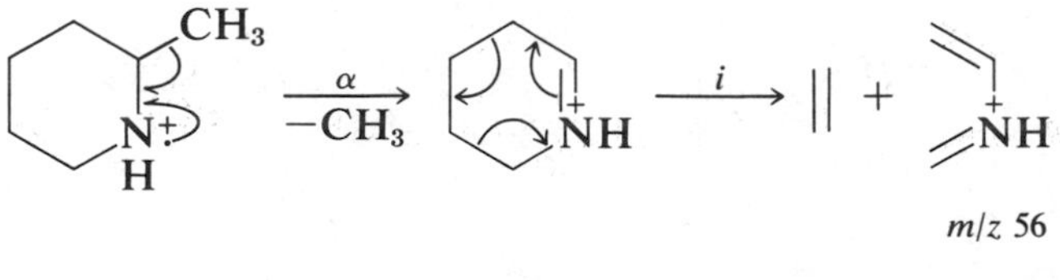

$$(8.13)$$

$$m/z\ 56$$

Unknown 8.3 contains a hydroxyl group.

m/z	Rel. abund.
15	2.9
28	12.
29	3.8
30	100.
31	3.4
39	4.6
41	13.
42	14.
43	18.
44	20.
45	5.7
56	4.5
57	1.3
58	3.1
70	21.
71	2.2
72	16.
73	0.7
88	24.
89	1.1
102	2.5
103	4.1
104	0.2

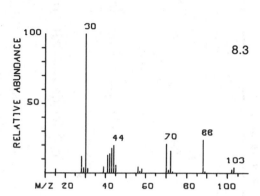

8.4 Reaction classifications

In Section 4.4, unimolecular ion reactions were classified by the number of bonds broken between the separating ion and neutral products. Here we will look in more detail at common reaction types within each class, as illustrated in Table 8.1 by the hypothetical ions $ABCD^{+\cdot}$ ($OE^{+\cdot}$) and $ABCD^+$ (EE^+), in which these letters can represent an atom, a functional group, or even several groups, such as an alkyl chain. Thus a second bond formed between A and D could form a ring incorporating their several groups as well as an $A{=}D$ double bond. The designations "α" and "i" include all types of reactions formulated as involving initiation at a radical or charge site, respectively. As discussed in the previous section, to be consistent with one of these types, some reactions must be considered as involving initiation at "incipient radical or charge sites."

Table 8.1 *Types of Ion Decompositions*

Ion Type	Reactant	Products from Charge retention	Charge migration	Example
One bond cleaved				
OE⁺·	AB·⁺CD	—σ→ AB· + CD⁺	(AB⁺ + ·CD)	$C_2H_5-CH(CH_3)_2 \longrightarrow C_2H_5^+,\ ^+CH(CH_3)_2$
OE⁺·	AB—⁺·CD	—σ→ AB· + CD⁺		$C_2H_5-\overset{+\cdot}{S}CH_3 \longrightarrow C_2H_5\cdot + {^+}SCH_3$
OE⁺·	AB—⁺·CD ; AB—⁺·C=D —i→		AB⁺ + ·CD	$C_2H_5-\overset{+\cdot}{O}CH_3 \longrightarrow C_2H_5^+ + \cdot OCH_3$; $C_2H_5-CR\overset{+\cdot}{=}O \longrightarrow C_2H_5^+ + \cdot CR=O$
OE⁺·	AB—C⁺·D —α→ AB· + C=⁺D			$C_2H_5-CH_2\overset{+\cdot}{O}R \longrightarrow C_2H_5\cdot + CH_2=\overset{+}{O}R$
EE⁺	AB—C⁺—D ; AB—⁺C=D —i→		AB⁺ + C=D	$(CH_3)_3C-CR_2\overset{+}{C}H_2 \longrightarrow (CH_3)_3C^+ + CR_2=CH_2$; $C_2H_5-\overset{+}{O}=CH_2 \longrightarrow C_2H_5^+ + O=CH_2$
OE⁺·	(A⌢·⁺C—D⌢B) —rd→ A—C⁺ + ·D (with B)			$D_2C\overset{\cdot}{S}\cdot H \longrightarrow D_2C=\overset{+}{S}D + \cdot H$
OE⁺·	⁺·A—C—D (B) —rd→ ⁺A—C + ·D (B)			$\overset{+}{Cl}(CH_2)_3-CH_2R \longrightarrow Cl\!\!-\!\!CH_2 + \cdot R\ (CH_2)_3$
EE⁺	⁺A—C—D (B) —rd→ A—C + D⁺ (B)			$H_2\overset{+}{C}(CH_2)_3-CH_2R \longrightarrow H_2C\!\!-\!\!CH_2 + R^+\ H_2C$

134

Table 8.1 *Types of Ion Decompositions (continued)*

Ion Type	Reactant	Products from Charge retention	Charge migration	Example

Two bonds cleaved

OE⁺˙

$$\begin{array}{c} A\overset{\wedge}{-}\overset{+\cdot}{D} \\ | \quad | \\ B\underset{\smile}{-}C \end{array} \xrightarrow{\alpha\alpha} \begin{array}{c} A\!=\!\overset{+}{D} \\ + \\ B\!-\!C \end{array} \quad or \quad \xrightarrow{i\alpha} \begin{array}{c} A\!=\!D \\ + \\ B\!-\!\overset{+\cdot}{C} \end{array}$$

$$\begin{array}{c} H_2C\!\!\overset{\wedge}{-}\!\!\overset{+\cdot}{CHOH} \\ | \quad | \\ H_2C\!-\!CH_2 \end{array} \longrightarrow \begin{array}{c} H_2C\!=\!\overset{++}{CHOH} \\ + \\ H_2C\!=\!CH_2 \end{array}$$

EE⁺

$$\begin{array}{c} A\overset{\curvearrowleft}{-}\overset{+}{D} \\ | \quad | \\ B\underset{\smile}{-}C \end{array} \xrightarrow{ii} \begin{array}{c} A\!=\!\overset{+\cdot}{D} \\ + \\ B\!-\!C \end{array}$$

$$\begin{array}{c} \overset{+}{H_2C}\!-\!CH \\ | \qquad \diagdown CH_2 \\ H_2C\!\!\!\curvearrowright\!\! CH_2 \end{array} \longrightarrow \begin{array}{c} \overset{+\cdot}{H_2C}\!-\!CH \\ | \qquad | \\ H_2C\!-\!CH_2 \\ + \\ H_2C\!=\!CH_2 \end{array}$$

OE⁺˙

$$\begin{array}{c} A\overset{\curvearrowleft}{\cdots}\overset{+\cdot}{D} \\ \overset{\curvearrowleft}{B}\underset{\smile}{-}C \end{array} \xrightarrow[(re)]{\alpha\alpha} \begin{array}{c} A\!-\!\overset{+\cdot}{D} \\ + \\ B\!=\!C \end{array} \quad or \quad \xrightarrow{ai} \begin{array}{c} A\!-\!D \\ + \\ B\!=\!\overset{+\cdot}{C} \end{array}$$

$$\begin{array}{c} H\!\!\smallfrown\!\!\overset{+\cdot}{O}\!\!=\!\!CR' \\ RHC\!\!\curvearrowright\!\!\overset{+\cdot}{CH_2}\!-\!CH_2 \end{array} \longrightarrow \begin{array}{c} HO\!=\!CR' \\ + \\ \cdot CH_2 \end{array} \quad or \quad RHC\!=\!CH_2^{\cdot}$$

EE⁺

$$\begin{array}{c} A\overset{\curvearrowleft}{\cdots}\overset{+}{D} \\ \overset{\curvearrowleft}{B}\underset{\smile}{-}\overset{+}{C} \end{array} \xrightarrow[(re)]{ii} \begin{array}{c} A\!-\!\overset{+}{D} \\ + \\ B\!=\!C^{(+)} \end{array}$$

$$\begin{array}{c} H\!\!\smallfrown\!\!\overset{\cdot\cdot}{O}\!\!=\!\!CHR \\ C_nH_{2n}\!-\!CH_2 \end{array} \longrightarrow \begin{array}{c} \overset{+}{HO}\!=\!CHR \\ + \\ C_nH_{2n}\!=\!CH_2 \end{array}$$

135

Table 8.1 Types of Ion Decompositions (continued)

Ion Type	Reactant	Products from	
		Charge retention	Charge migration

Three bonds cleaved

OE⁺	ααα	charge retention	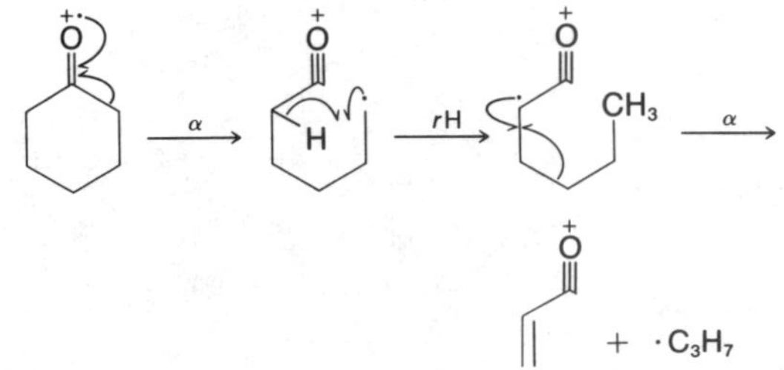
OE⁺	αii	charge retention	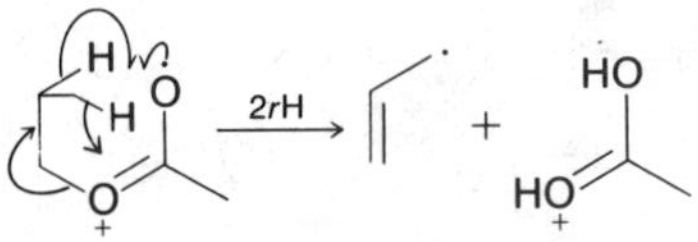
OE⁺	ααi (iii)		charge migration — less common
EE⁺	iii		charge migration — less common

All but the first three reactions involve formation of a new bond (or partial bond) for each bond cleaved, improving the energetic favorability of the reaction. This includes rearrangement reactions involving displacement (*rd*) and elimination (*re*). Displacement rearrangements (Sections 4.9 and 8.10) can involve cyclization to the radical site with displacement, $\overset{+\cdot}{A}-B-C-D \rightarrow {}^{+}\overline{A-B-C} + \cdot D$ (reactions 4.42 and 4.43), or an anchimerically assisted displacement at the radical site, $A-B-\overset{+\cdot}{C}-D \rightarrow B=\overset{+}{C}-A + \cdot D$. Elimination rearrangements involve extrusion of a small molecule or other small stable neutral species from the interior of the ion. These thus are analogous to hydrogen rearrangements, except that a larger group has replaced hydrogen as A in the reaction $A-B-C-\overset{+\cdot}{D} \rightarrow A-\overset{+}{D} + B=C$.

These types of decompositions will be discussed below, under the general headings of sigma-bond, radical-site, and charge-site initiation.

8.5 Sigma bond dissociation (σ)

$$OE^{\overset{+}{\cdot}} \quad AB\cdot + CD \xrightarrow{\ \sigma\ } AB\cdot + CD^{+} \ or \ AB^{+} + \cdot CD$$
$$OE^{\overset{+}{\cdot}} \quad AB-\overset{+\cdot}{C}D \xrightarrow{\ \sigma\ } AB\cdot + CD^{+} \tag{8.14}$$

If the localized ionization site is in the $C-C$ σ-bond to be dissociated, either of the carbon atoms will bear the charge, and the other the radical (Reaction 8.15). In terms of the bond polarization ($AB\cdot + CD$ *vs.* $AB + \cdot CD$) in the molecular ion, the more favored reaction will thus involve both charge and radical retention (neither the charge nor the unpaired electron migrates a full bond length). Thus for the σ-bond dissociation of $C_2H_5-CH(CH_3)_2{}^{+}$ the polarization $C_2H_5\cdot + CH(CH_3)_2$ should be favored, and the product abundance of ${}^{+}CH(CH_3)_2$ should be greater than $C_2H_5{}^{+}$. Loss of $CH_3\cdot$ from $C_2H_5CH(CH_3)_2{}^{\overset{+}{\cdot}}$ gives a product ($C_4H_9{}^{+}$) whose abundance is only half that of $C_3H_7{}^{+}$ formed by C_2H_5 loss, as expected from the "loss of the largest alkyl" rule.)

$$R_3C^{\overset{+}{\cdot}}CR'_3 \xrightarrow{\hspace{2cm}} R_3C^{+} + CR'_3 \ or \ R_3C\cdot + {}^{+}CR'_3 \tag{8.15}$$

If another functional group in the molecular ion is the nominal ionization site, however, σ-bond cleavage can be visualized as the charge-migration counterpart of, for example, α-cleavage at a saturated functionality (Table 8.1). For the spectrum of $(CH_3)_3C-CH_2OH$, $[C_4H_9{}^{+}] > [CH_2=\overset{+}{O}H]$

(Reaction 8.16); these abundances are consistent with the predictions of Stevenson's Rule. σ-Bond cleavages at primary and secondary carbon atoms are relatively high-energy reactions and thus poorly competitive. Alkyl-ion abundance from such reactions can be an ambiguous indicator of structure because competing isomerizations are common (Borchers *et al.* 1977; Levsen 1978; Lavanchy *et al.* 1979).

$$(CH_3)_3C—CH_2\overset{+\cdot}{O}H \xrightarrow{\alpha} (CH_3)_3C\cdot + CH_2=\overset{+}{O}H$$

$$I = 6.7 \text{ eV} \qquad 10\%$$

$$\xrightarrow{\sigma} (CH_3)_3C^+ + \cdot CH_2OH \qquad\qquad (8.16)$$

$$100\% \qquad I = 7.4 \text{ eV}$$

In classifying reactions which undergo "σ-ionization" initiation, we have somewhat arbitrarily chosen to include such reactions as $C_2H_5—\overset{+\cdot}{S}CH_3$ $\rightarrow C_2H_5\cdot + \overset{+}{S}CH_3$ (Table 8.1), although the charge in the molecular ion must reside much more in sulfur nonbonding orbitals than in C—S bonding orbitals. Electronically it should be classified with the α-cleavage reactions, since both involve charge retention; for $AB—\overset{+\cdot}{C}D$ this is the complementary reaction to the inductive reaction forming $AB^+ + \cdot CD$ by charge migration. However, "the" α-cleavage reaction of $C_2H_5—\overset{+\cdot}{S}CH_3$ produces $CH_3\cdot +$ $CH_2=\overset{+}{S}CH_3$, so that using the same classification for the production of $^+SCH_3$ is confusing. On the other hand, there should be substantial sigma ionization along the reaction coordinate, such as 8.17, so at least the later

$$C_2H_5—\overset{+}{S}CH_3 \longrightarrow C_2H_5\cdot{}^+SCH_3 \xrightarrow{\sigma} C_2H_5\cdot + \overset{+}{S}CH_3 \qquad (8.17)$$

stages of the reaction resemble alkane "σ-ionization" reactions. For interpretive purposes it should be recognized that ionization of *n*- or π-electrons on an atom generally does not lead to cleavage of a bond to that atom; for π-electrons this would be a vinylic cleavage, and for *n*-electrons this would result in a destabilizing *reduction* of the valence on the charged atom (*e.g.*, $\overset{+\cdot}{R}OR \rightarrow RO^+ + \cdot R$). For elements with *d*- (or higher) shell electrons, such as sulfur, this effect is greatly reduced. The heat of formation of CH_3S^+ (triplet) is ≈ 0.4 eV greater than that of $CH_2=SH^+$, but the heat of formation of CH_3O^+ (triplet) is ~ 3 eV above that of $CH_2=OH^+$ (the value for the C_{2v} complex $H_2\cdot\cdots CHO^+$ is 1.3 eV above that of $CH_2=OH^+$; Dill *et al.* 1979).

8.6 Radical-site initiation (one bond cleaved without rearrangement)

$$OE^{+\cdot} \qquad AB\!-\!C\!\overset{\curvearrowleft}{\underset{\curvearrowright}{=}}\!\overset{+\cdot}{D} \xrightarrow{\;\;\alpha\;\;} AB\cdot + C\!=\!\overset{+}{D} \qquad (8.18)$$

The mass-spectral reaction most commonly used in spectral interpretation is α-cleavage of an $OE^{+\cdot}$ ion. This involves donation of the unpaired electron to form a new bond to an adjacent (α) atom (for rearrangements, Section 8.9, adjacent through space), with concomitant cleavage of another bond to the α-atom (Equations 4.13 to 4.15). The reaction is competitive because the new bond compensates energetically for the one cleaved, in which process an electron has been donated to help stabilize the positive charge (8.19). By Stevenson's Rule (Harrison *et al.* 1971) the cleavage could involve charge migration if the ionization energy (I) of $R\cdot$ is less than that of $\cdot CH_2Y$ (Reaction 8.16).

$$R\!-\!CH_2\!-\!\overset{+\cdot}{Y}\begin{cases} \xrightarrow{\;\alpha\;} R\cdot + CH_2\!=\!\overset{+}{Y} \\ \\ \xrightarrow{\;\sigma\;} R^+ + \cdot CH_2Y \end{cases} \qquad (8.19)$$

For a *particular* radical site, usually several competitive α-cleavage reactions are possible, since each of the one or more (three in a tertiary amine) α-atoms can have as many as three substituents to be lost by α-cleavage. Among these possibilities the loss of the radical of highest ionization energy is energetically preferable, *except* that the loss of the largest alkyl group is favored (Section 8.2). Note that an abundant peak from the loss of H, the smallest alkyl group, thus indicates that no other α-cleavage is possible. Unknown 5.12 illustrates these rules; try it again. If several radical sites are possible in the $OE^{+\cdot}$ precursor ion, the favored α-cleavage product is usually the one for which the corresponding radical has the lowest ionization energy (this is true for the reactions $AYB^{+\cdot} \rightarrow A^+ + \cdot YB$ versus $AY\cdot + B^+$ if the summed heats of formation of $A\cdot$ and $\cdot YB$ equals that of $AY\cdot$ and $\cdot B$.) Note that the radical-site preference (Section 4.4) $N > S, O, \pi, R\cdot > Cl, Br > H$ is generally the inverse of the I values of the corresponding radicals; for example (Table A.3), $H_2N\!-\!CH_2\cdot$, 6.2 eV; $HS\!-\!CH_2\cdot$, 7.3; $HO\!-\!CH_2\cdot$, 7.4; $C_6H_5\!-\!CH_2\cdot$, 7.3; $CH_2\!=\!CH\!-\!CH_2\cdot$, 8.1; $Cl\!-\!CH_2\cdot$, 8.7; $H\!-\!CH_2\cdot$, 9.8 eV. Remberg and Spiteller (1970) have measured the competitive effect of such functional groups on product-ion abundances from α-cleavage in the same molecule. For the compounds $C_2H_5CH(OCH_3)CH_2CH_2CH_2CH_2\!-\!CY\!-\!C_2H_5$, the abundance of the $C_2H_5C\!=\!Y^+$ ion relative to that of the $C_2H_5CH\!=\!OCH_3{}^+$ ion in the same spectrum yields the following values for various $-CY-$ groups: $-CHCl-$, $-CHBr-$, $-CHI-$, <1;

—CHOH—, 5; —CHSH—, 5; —CO—, 50; —CH(SCH₃)—, 100; —CH(OCH₃)—, 100; —C(—OCH₂CH₂O—)— (ethylene ketal), 500; —CH(NH₂)—, 1,000; and —CHN(CH₃)₂—, 2,000. I values estimated by the method of Harrison *et al.* (1971) have been included in Table A.3 to reflect these relative rates.

Unknown 8.4 is difficult because it does not contain a molecular ion. If you have substantial difficulty, use the elemental compositions given at the beginning of the answer.

Unknown 8.4

m/z	Rel. abund.
13	2.0
14	6.5
15	41.
28	2.8
29	46.
30	3.4
31	13.
32	0.8
43	0.9
44	2.3
45	100.
46	2.3
47	3.5
75	44.
76	1.4

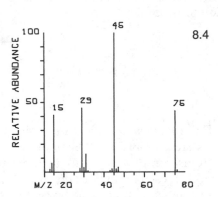

Unknown 8.5. The principal constituent of a defensive secretion of the willow-feeding larva Coleoptera chrysomelidae.

m/z	Rel. abund.	m/z	Rel. abund.	m/z	Rel. abund.
27	2.0	52	0.7	75	1.4
28	0.6	53	4.9	76	19.
29	3.6	54	0.5	77	2.3
31	1.0	55	1.6	92	1.8
37	3.3	61	2.9	93	18.
38	7.2	62	2.8	94	5.8
39	27.	63	5.6	95	0.4
40	6.6	64	2.6	104	13.
46	1.4	65	27.	105	1.0
46.5	0.5	66	8.5	121	89.
47	4.3	67	1.1	122	100.
50	5.4	68	0.4	123	7.9
51	3.4	74	1.9	124	0.7

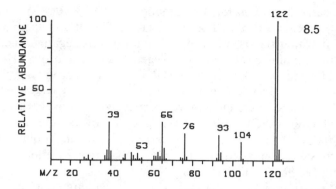

8.7 Charge-site reactions (one bond cleaved without rearrangement)

$$\text{OE}^{\ddagger} \quad \begin{array}{c} \text{AB}\overset{\frown}{}\overset{+\cdot}{\text{CD}} \\[4pt] \text{AB}\overset{\frown}{}\text{C}\!\!=\!\!\overset{+\cdot}{\text{D}} \end{array} \quad \xrightarrow{\;\;i\;\;} \quad \text{AB}^+ + \cdot\text{CD} \qquad (8.20)$$

$$\text{EE}^+ \quad \left.\begin{array}{c} \text{AB}\overset{\frown}{}\text{C}\overset{\smile}{}\overset{+}{\text{D}} \\[4pt] \text{AB}\overset{\frown}{}\overset{+}{\text{C}}\!\!=\!\!\text{D} \end{array}\right\} \quad \xrightarrow{\;\;i\;\;} \quad \text{AB}^+ + \text{C}\!\!=\!\!\text{D} \qquad (8.21)$$

Inductive (i) cleavage of a single bond requires charge migration (Table 8.1), and thus a different structural environment for the charge in the precursor and product ions. Because of the importance of charge stabilization in determining the favored products of unimolecular cation decompositions, single-bond i cleavages face substantial competitive restrictions. In only a few types of $\text{OE}^{\ddagger}$ precursor ions are such i cleavages competitive with radical-site reactions, and in EE^+ decompositions two-bond cleavage reactions are generally favored because the charge site does not migrate.

$OE^{\ddagger}$ single-bond inductive cleavages. Sigma ionization of an alkane requires ionization energies (I) of 10 to 11 eV. An electronegative substituent such as —F, —Cl, or —CN which does not lower this appreciably does not provide a highly favorable site for charge localization in the resulting molecule, $[\text{R}\!-\!\overset{+\cdot}{\text{Y}}] < [\overset{+\cdot}{\text{R}}\!-\!\text{Y}]$. This facilitates charge migration from such a site, $\text{R}\!-\!\overset{+\cdot}{\text{Y}} \to \text{R}^+ + \text{Y}\cdot$. Such reactions (4.23 and 4.24) can also be viewed as a sigma-bond cleavage aided by inductive electron withdrawal by the electronegative substituent, $\overset{+\cdot}{\text{R}}\!-\!\overset{\leftrightarrow}{\text{Y}} \to \text{R}^+ + \text{Y}\cdot$.

True charge migration for the i reaction $R—\overset{+\cdot}{Y} \rightarrow R^+ + Y\cdot$ requires that $I(RY) > I(RH)$, but that $I(Y\cdot) > I(R\cdot)$. The most common functional groups for which this is true are still quite electronegative, including $Br\cdot$, $I\cdot$, and functionalities that contain oxygen, such as hydroxy ($HO\cdot$), alkoxy ($RO\cdot$), aldehydo ($HCO\cdot$), carboxy ($HOOC\cdot$), and carboxyalkyl ($ROOC\cdot$). For these $I(Y\cdot) < 10$ eV, but $I(Y\cdot) > I(R\cdot)$ for a larger alkyl radical. Branched $R\cdot$ radicals are even more competitive for the charge; in the mass spectra of $C_4H_9—SCH_3$ (see Equation 8.17), by far the largest value of $[C_4H_9^+]/[CH_3S^+]$ is found for the *tert*-butyl isomer, in qualitative agreement with Stevenson's Rule. The charge migration reactions of Equations 8.16 and 8.19 show similar effects.

These restrictive energy requirements for charge-migration reactions cause them to be poorly competitive if a reaction leading to a more stable ion is possible. For less electronegative functional groups, this can be true because a homolog of the group shows a much lower radical ionization energy. For $R—CH_2—NH_2^{+\cdot}$ the formation of $RCH_2^+ + \cdot NH_2$ is negligible, despite the fact that $I(\cdot NH_2) \gg 8$ eV, because $I(\cdot CH_2NH_2) = 6.2$ eV; $CH_2{=}NH_2^+$ is the base peak of the spectra of $n\text{-}C_nH_{2n+1}NH_2$ compounds for $n \leq 16$. Consecutive as well as competitive reactions can cause such simplistic interpretations to be misleading, in some cases making the apparent product ion of a charge migration reaction unexpectedly abundant. In the spectrum of $n\text{-}C_6H_{13}—O—n\text{-}C_6H_{13}$ (Figure 3.9), $[C_6H_{13}^+] \gg [C_6H_{13}O^+{=}CH_2]$, despite the fact that $I(\cdot OC_6H_{13}) > 8$ eV and $I(C_6H_{13}OCH_2\cdot) < 7$ eV, which should influence the competition between reactions 8.22 and 8.23. A substantial part of this discrepancy must be due to the secondary decomposition (8.24), which can be preferentially eliminated at low electron-ionization energies.

$$C_6H_{13}—OCH_2—C_5H_{11}^{+\cdot} \overset{\alpha}{\longrightarrow} C_6H_{13}O^+{=}CH_2 + \cdot C_5H_{11} \quad (8.22)$$

$$\overset{i}{\longrightarrow} C_6H_{13}^+ + \cdot OC_6H_{13} \quad (8.23)$$

$$C_6H_{13}O^+{=}CH_2 \overset{i}{\longrightarrow} C_6H_{13}^+ + O{=}CH_2 \quad (8.24)$$

EE$^+$ single-bond inductive cleavages. In general EE^+ ion decompositions have higher critical energies than $OE^{+\cdot}$ decompositions: two-step pathways such as 8.22 plus 8.24 are 0.2 to 0.8 eV less favorable than the one-step counterpart 8.23. Despite this, reaction 8.24 is important because a substantial fraction of the $C_6H_{13}O^+{=}CH_2$ ions are formed with sufficiently high internal energies using 70 eV electrons. Losses of small stable molecules from EE^+

ions are favored; the common reaction of acyl ions, $RCO^+ \rightarrow R^+ + CO$, is not surprising in view of $I(CO) = 14$ eV. For reaction 8.24, $I(CH_2O) = 10.9$ eV, much larger than $I = \sim 8$ eV for the complementary primary alkyl radicals (C_3 or larger). Formation of abundant CH_3^+ (and H^+) ions from any reactions, including this, is unusual because $I(CH_3\cdot) = 9.8$ eV (see Unknown 5.2), which raises the reaction critical energy very significantly. However, such larger EE^+ ions from α-cleavage reactions, especially those initiated by less electronegative functionalities, often have more facile competitive decompositions in which the charge does not migrate. In contrast to the behavior of n-alkyl ethers (Equations 8.22 to 8.24), the mass spectra of n-C_4H_9—O—$CH(CH_3)_2$ obtained at both high and low electron ionizing energies (Djerassi and Fenselau 1965) show that the further decomposition of the α-cleavage product n-C_4H_9—O^+=$CHCH_3$ to form $C_4H_9^+$ + O=$CHCH_3$ is substantially less favorable than that to form C_4H_8 + HO^+=$CHCH_3$ (Section 8.10). From reaction 8.7, this is consistent with the proton affinity (PA) values of the competing molecules C_4H_8 or C_6H_{12} (PA = ~ 8.3 eV) versus CH_2O (PA = 7.9 eV) or CH_3CHO (PA = 8.3 eV).

Most stable even-electron ions have some multiple-bond character to the charge-site atom: R_2C=Y^+. Common exceptions are ions in which this atom is carbon, R_3C^+. (CH_3S^+ is moderately stable, but RCH_2S^+ ions readily rearrange to RCH=S^+H; Dill *et al.* 1979). Further dissociation of such EE^+ alkyl ions can occur by reaction 8.25. Figure 8D shows the spectra of n-C_7H_{15}—$CH(CH_3)_2$ and n-$C_5H_{11}CH(CH_3)$—$CH(CH_3)_2$. The $C_4H_9^+$ base peak in the latter spectrum is due in part to dissociation of the α-cleavage EE^+ product by reaction 8.26. For the mass spectrum of 2-methylnonane (Figure 8D) reaction 8.25 would involve $(M - C_3H_7)^+ \rightarrow C_5H_{11}^+ + C_2H_4$. This is less important because primary alkyl ions, such as the n-heptyl ion formed here, undergo isomerizations that are generally much more rapid than dissociation reactions such as 8.25 (Lavanchy *et al.* 1979).

$$R\overset{\frown}{-}CH_2\overset{\downarrow}{-}CR'_2{}^+ \xrightarrow{\ i\ } R^+ + CH_2=CR'_2 \qquad (8.25)$$

$$C_4H_9\overset{\frown}{-}CH_2\overset{\downarrow}{-}CH(CH_3)^+ \xrightarrow{\ i\ } C_4H_9^+ + CH_2=CHCH_3 \qquad (8.26)$$

EE^+ single-bond cleavage without charge migration (violations of the "even-electron rule"). Of the complementary ion products possible from bond cleavage in an EE^+ ion, the EE^+ ion from charge migration $RY^+ \rightarrow R^+ + Y$ is usually favored over charge retention $R\cdot + {}^+\dot{Y}$ ($OE^{\dot{+}}$). (The collisional-activation mass spectra, Section 6.4, of $C_2H_5O^+$=CH_2 and CH_3CH=O^+CH_3 show $OE^{\dot{+}}$ ions as representing 5 and 10 per cent, respectively, of the total ion abundance.) However, not all ionization energies of radicals are below

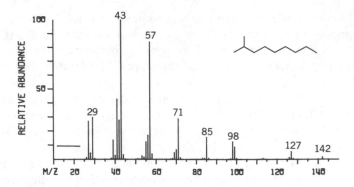

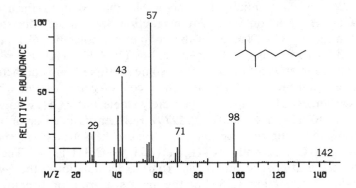

Figure 8D. Mass spectra of 2-methylnonane (top) and 2,3-dimethyloctane (bottom).

those of the complementary even-electron neutral species. Thus in the mass spectra of 1,3- and 1,4-dinitrobenzene the $C_6H_4^{+\cdot}$ (loss of N_2O_4) and $C_6H_3^+$ (loss of HN_2O_4) products are of nearly equal high abundances (only NO^+ is a more abundant product ion). However, this is due not only to the fact that $I(\text{benzyne}) < I(NO_2)$, but also to the fact that other reactions which involve EE^+ formation as well as charge retention, such as the loss of HNO_2 from $O_2N—C_6H_4^+$ to form $C_6H_3^+$, are not more competitive. In a very recent survey Karni and Mandelbaum (1980) cite a large number of cases of EE^+ ion decompositions producing $OE^{+\cdot}$ and OE° products.

Sample pressure studies show that the m/z 99 abundance given in Unknown 8.6 is not due to ion-molecule reactions. Exact mass measurements require the C_3H_4N and C_3H_7O assignments shown. $OE^{+\cdot}$ and EE^+ ions may be a problem in Unknown 8.7.

Unknown 8.6

m/z	Rel. abund.	m/z	Rel. abund.
15	4.8	54	76. C_3H_4N
27	23.	55	14.
28	18.	56	1.3
29	39.	57	13.
30	1.6	58	1.0
31	100.	59	57. C_3H_7O
32	1.3	60	2.1
39	1.2	68	1.7
40	2.5	69	0.4
41	13.	70	3.0
42	5.9	71	1.0
43	6.1	72	2.0
44	1.5	84	9.5
45	13.	85	0.5
46	0.3	98	3.5
52	3.5	99	0.5
53	0.7		

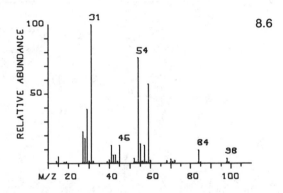

8.6

Unknown 8.7

m/z	Rel. abund.	m/z	Rel. abund.	m/z	Rel. abund.
15	2.1	65	1.6	120	6.5
28	2.7	74	4.0	121	0.6
30	1.4	75	6.0	122	1.0
38	1.2	76	12.	134	0.6
39	1.3	77	3.8	135	1.2
43	26.	78	0.5	150	100.
44	0.6	91	3.2	151	8.3
50	9.0	92	12.	152	0.9
51	2.9	93	1.0	165	18.
52	0.4	104	32.	166	1.6
63	2.3	105	2.4	167	0.2
64	2.1	119	1.1		

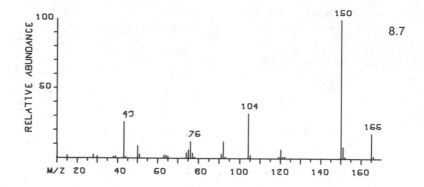

8.7

8.8 Decompositions of cyclic structures (two and three bonds cleaved)

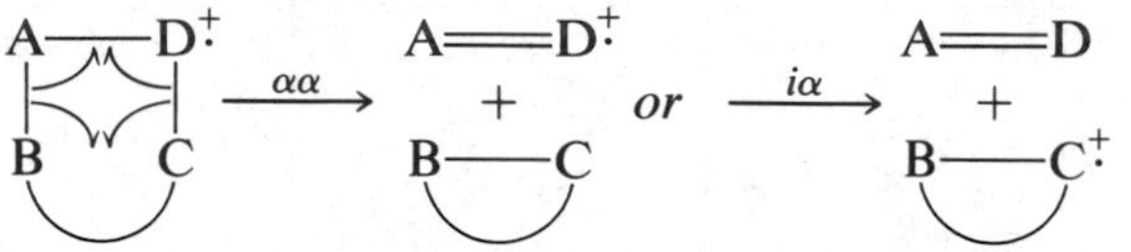

$$
\text{(8.27)}
$$

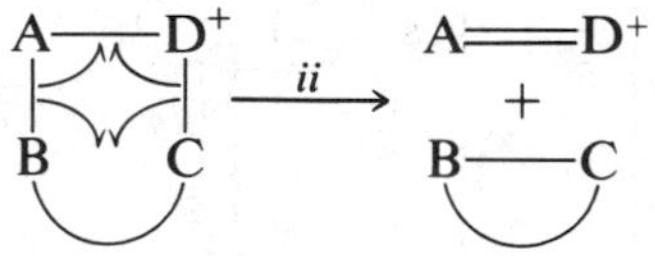

$$
\text{(8.28)}
$$

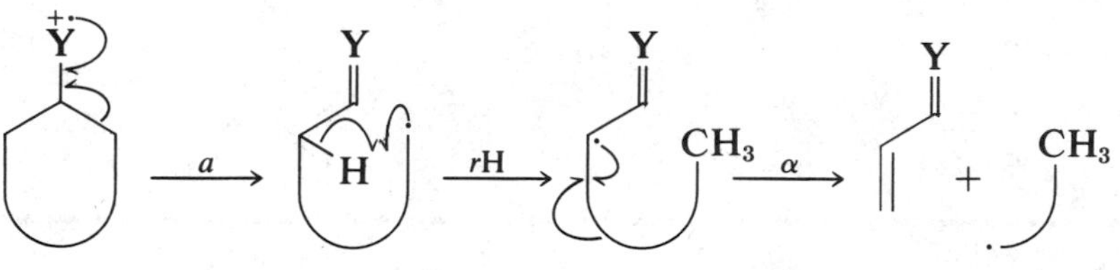

$$
\text{(8.29)}
$$

Cleavage of at least two ring bonds in a cyclic ion is necessary to form product ions by ring fragmentation. For an OE^{+} precursor this leads to an OE^{+} product ion; cleavage of a third bond is necessary to produce an EE^{+} product. The stability of the latter, however, often makes formation of such products favorable.

Two bonds cleaved. Examples were given in Section 4.8 in which the cleavage of two bonds in an OE^{+} cyclic ion led to abundant products. These involved ionization in the ring, such as of the σ-electrons in cyclohexane (reaction 4.30), n-electrons in dioxane (Unknown 4.14), and π-electrons in cyclo-hexene (retro-Diels-Alder reaction), with charge retention and charge migration (Equations 4.31 and 4.32). However, cases in which an abundant product is consistent with the latter are often better explained by an alternative ionization site initiating a charge-retention reaction. Reaction 8.30 is such an alternative for 4.32, $R = C_6H_5$.

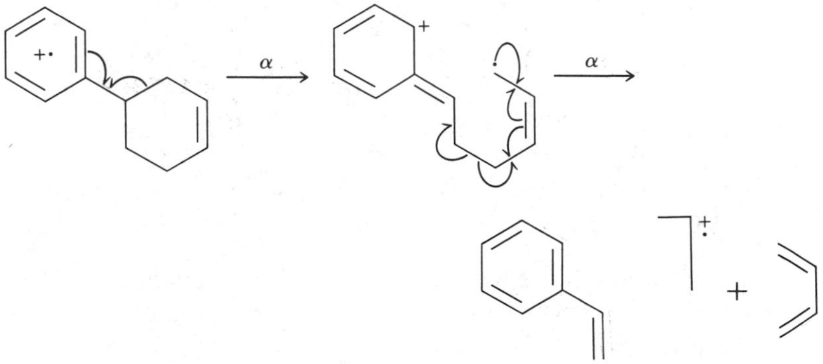

$$(8.30)$$

As noted for rearrangements (Section 4.9), an important driving force for reactions forming OE^{+} ions is stabilization of the radical site; a major reason that hydrogen is rearranged from the γ-position in reaction 4.33 is that this makes possible formation of a resonance-stabilized radical product, $\cdot CH_2—CR{=}O^{+}H \leftrightarrow CH_2{=}CR—O^{+}_{\cdot}H$. The diene product OE^{+} ion of the retro-Diels-Alder reaction (4.31), or the styrene $OE^{+}_{\cdot}$ ion of Equation 8.30, contains a conjugated π-system providing such radical (and charge) stabilization, but the $C_4H_8^{+}_{\cdot}$ product from the analogous cyclohexane decomposition (4.30) is not resonance-stabilized (it has the same structure as the ion from cyclobutane ionization; Nishishita et al. 1977). Such radical stabilization can be achieved for alicyclic compounds containing a heteroatom in a five-membered ring (8.31) or adjacent to a four-membered ring (8.32). In both cases the product ion is isoelectronic with an allylic radical. For larger

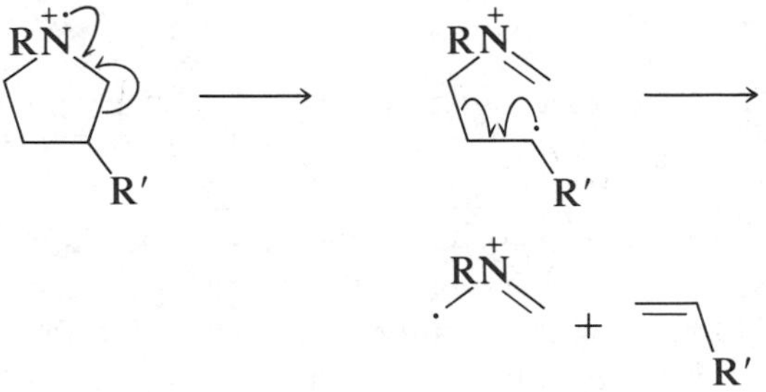

$$(8.31)$$

$$R = H, R' = CH_3: m/z\ 43,\ 100\%$$
$$R = CH_3, R' = H: m/z\ 57,\ 80\%$$

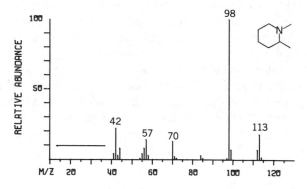

$$R = H: \ m/z \ 42, \ 100\%$$
$$R = CH_3: \ m/z \ 70, \ 100\%$$

rings such olefin loss would place the radical site of the homologous product ion too many atoms away from the double bond for such stabilization; the reaction analogous to 8.31 for the six-membered-ring isomer 1,2-dimethyl-piperidine gives (Figure 8E) $(M - C_2H_4)^{\ddagger}$ in 1 per cent abundance. The reactions analogous to 8.32 for larger ring cycloalkanones are much less important; $(M - C_3H_6)^{\ddagger}$ has an abundance of 10 per cent in the spectrum of 2,5-dimethylcyclopentanone and 4 per cent in that of 3,3,5-trimethyl-cyclohexanone (Figure 8F).

Figure 8E. Mass spectrum of 1,2-dimethylpiperidine.

The cyclic ethers have a lower tendency for such olefin elimination, at least partly because of their favorability for charge-site initiation (see also the answer to Unknown 4.14, Chapter 11). 3-Methyltetrahydrofuran behaves quite differently from its nitrogen analog in Equation 8.31, yielding only 3 per cent $CH_2—O^+{=}CH_2$, m/z 44, but 100 per cent $C_4H_8^{\ddagger}$ formed by CH_2O loss. The presumed mechanism is shown in reaction 8.33 for the six-

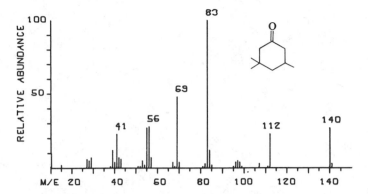

Figure 8F. Mass spectrum of 3,3,5-trimethylcyclohexanone.

membered-ring analog, tetrahydropyran; the mechanism could also be written as an α- followed by an i-cleavage. This spectrum also shows an abundant $(M - CH_2O - C_2H_4)^{+}$ ion (8.33), but little of the double α-cleavage (8.34) analogous to 8.31. In the five- and six-membered-ring nitrogen analogs 3-methylpyrrolidine (8.31) and piperidine, the corresponding $(M - H_2C=NH)^{+}$ ion abundances are, respectively, 2 and 40 per cent (although part could be loss of C_2H_5). A reaction of cycloalkanones (Figure 8F) similar to 8.33 eliminates carbon monoxide (8.35).

m/z 56, 55%

$$CH_2^{+} \quad (8.33)$$

m/z 28, 65%

m/z 58, 1% m/z 30, 5%   (8.34)

$$\xrightarrow{i,\alpha} \quad CO \; + \quad \boxed{} \qquad (8.35)$$

m/z 112, 25%

$$(8.36)$$

m/z 98, 100% m/z 70, 12%

Cyclic EE$^+$ ions also undergo ring cleavage in a similar manner; EE$^+$ product ions are produced without charge migration. The base peak in the mass spectrum of 1.2-dimethylpiperidine (Figure 8E) arises from the α-cleavage loss of methyl (the "largest alkyl loss" rule requires that the alkyl *be lost*), 8.36. The high stability of this cyclic ion is reflected in the relatively low abundance of the m/z 70 produced by EE$^+$ retro-Diels-Alder cleavage (8.36; see also Equation 8.13). Although $(M - CH_3)$ is also the base peak in the oxygen analog 2-methyltetrahydropyran, its retro-Diels-Alder decomposition yields a much less abundant (4 per cent) ion; the i,α-reaction 8.33 eliminating CH_3CHO gives $C_4H_8^{+}$ in 30 per cent abundance.

Three bonds cleaved. The mass spectra of functionalized cyclic ($\geq$ five-membered ring) compounds commonly show a ring-cleavage decomposition pathway leading to an EE$^+$ ion which is quite useful for structural characterization (Figures 8G and 8F, reactions 8.37 and 8.38). Radical-site-initiated ring-opening separates the radical and charge sites, similar to the first steps in reactions 8.31 and 8.32. The second step of those reactions moved the unpaired electron so that it enjoyed allylic stabilization; this can also be done by hydrogen rearrangement, step two in reactions 8.37 and 8.38. For compounds favorable for the olefin elimination defined in 8.31 and 8.32, this rearrangement reaction (loss of an alkyl radical containing only one ring carbon) is not competitive; this yields <3 percent product ions for the compounds illustrated in 8.31 and 8.32 except 2,2,4,4-tetramethylcyclo-butanone, for which $[(M - C_3H_7)^+] = 8$ per cent. Otherwise, ring size provides little restriction for the H-rearrangement reaction (8.37 and 8.38); cyclohexylamines behave like 8.37, and for unsubstituted cycloalkanones with five- through ten-membered rings, the resulting $C_3H_3O^+$ is formed in >90 per cent abundance. This pathway appears to be much less effective for cyclic compounds with heteroatoms in the ring, such as 1,2-dimethyl-piperidine (Figure 8E) or tetrahydropyran.

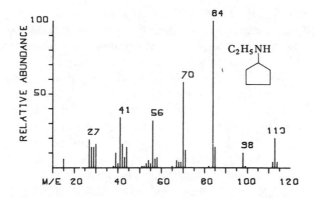

Figure 8G. Mass spectrum of *N*-ethylcyclopentylamine.

$$(8.37)$$

m/z 84, 100%

$$(8.38)$$

m/z 83, 100%

This α,rH reaction is of high utility for structural characterization of particular compound classes such as steroids (Section 9.6). However, it must be used with caution in interpreting an unknown mass spectrum because many decomposition pathways are possible for such alicyclic compounds (Kingston et al. 1975; Schwarz et al. 1979), such as the loss of RC_2H_3O from 2-R-cyclohexanones (Seibl and Gaumann 1963), the loss of H_2O from cyclo-alkanones, and reaction 8.39. Of particular mechanistic interest is the recyclization of ring-opened alicyclic ions to increase or decrease the original ring size, the "Fetizon-Seibl rearrangement" (Seibl and Gaumann 1963; Audier et al. 1975). In this reaction (8.40) the second hydrogen rearrangement leads to reformation of the ring, and competes with α-cleavage loss of alkyl (8.38). Structural features increasing the lability of the hydrogen taking part in this second rearrangement facilitate the reaction.

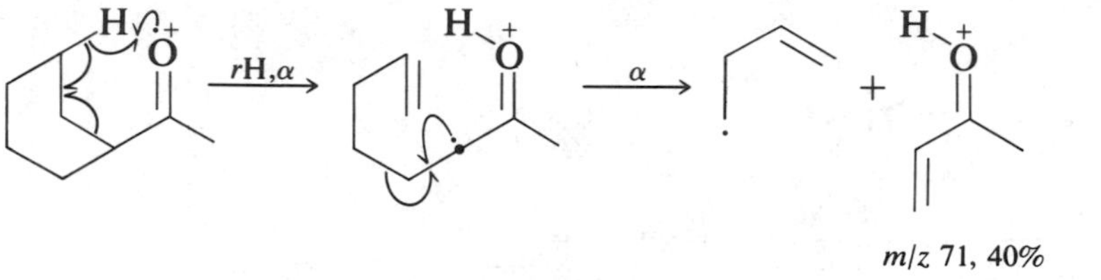

$$m/z\ 71,\ 40\% \tag{8.39}$$

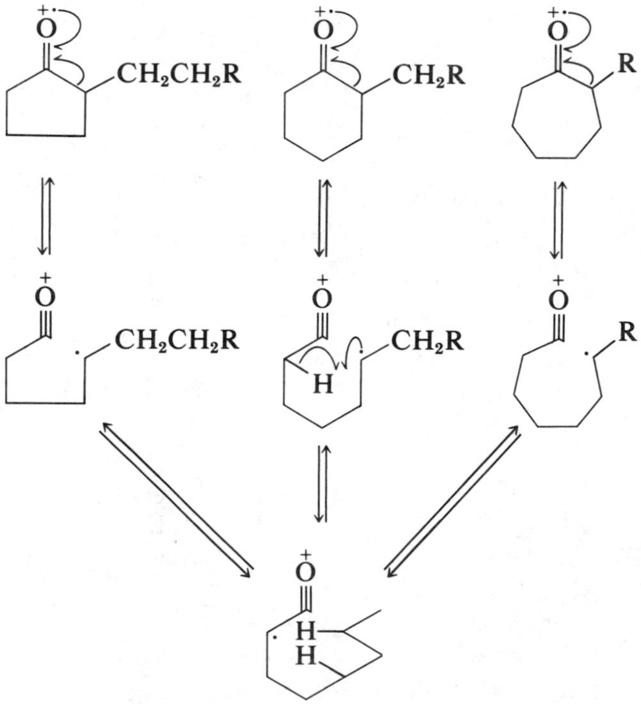

$$\tag{8.40}$$

When the pesticide DDE, $Cl_2C=C(p\text{-}C_6H_4Cl)_2$, is irradiated in air, a compound is formed whose mass spectrum is shown as Unknown 8.8.

Unknown 8.8

m/z	Rel. abund.
149	11.[a]
150	37.
151	5.1
152	0.3
184	6.5
185	20.
186	4.4
187	6.9
188	1.0
213	4.9
214	0.7
215	1.8
220	22.
221	2.8
222	15.
223	2.3
224	2.3
225	0.3
248	100.
249	14.
250	64.
251	9.2
252	9.9
253	1.5

[a]Data below m/z 60 were not recorded; the remaining data are shown in the bar graph.

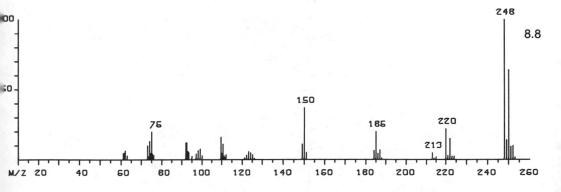

Unknown 8.9 is the spectrum from a trace of white powder found in the pocket of a suspected smuggler. Is the mass spectrum consistent with that which you would expect from 2β-methoxycarbonyl-3β-benzoyloxytropane, or cocaine (Figure 8H)?

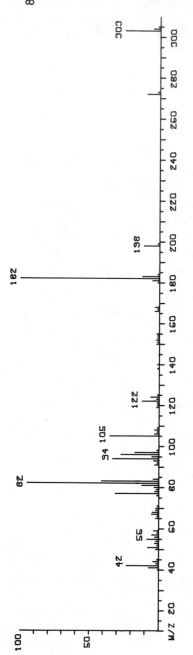

8.9

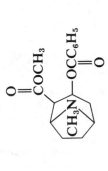

Figure 8H.

8.9 Hydrogen rearrangements

$$\begin{array}{ccc} \text{H} \quad \overset{\cdot +}{\text{D}} & \text{H}\!\!-\!\!\text{D}^+ & \text{H}\!\!-\!\!\text{D} \\ \overset{\alpha\alpha}{\longrightarrow} & + \quad \text{or} \overset{\alpha i}{\longrightarrow} & + \\ \text{B} \quad \text{C} & \text{B}\!\!-\!\!\text{C} & \text{B}\!\!-\!\!\text{C}^{\cdot +} \end{array} \qquad (8.41)$$

$$\begin{array}{cc} \text{H} \quad \overset{+}{\text{D}} & \text{H}\!\!-\!\!\text{D}^+ \\ \overset{ii}{\longrightarrow} & + \\ \text{B} \quad \text{C} & \text{B}\!\!-\!\!\text{C} \end{array} \qquad (8.42)$$

$$\text{H} \quad \overset{+\cdot}{\text{C}}\!\!-\!\!\text{D} \xrightarrow{\alpha, rH} \text{B}\!\!=\!\!\overset{+}{\text{C}}\text{H} + \cdot\text{D} \qquad (8.43)$$
$$\text{B}$$

Rearrangements are entropically unfavorable; the ion is in the conformation necessary for reaction only a very small fraction of the time. Thus this "tight activated complex" characteristic of rearrangements requires an offsetting energetic favorability for significant product-ion formation. For such reactions of molecular ions, a sufficient proportion must be formed with the required internal energy, i.e., there should be no corresponding "hole" in the $P(E)$ function (Section 7.4). The first step of an $OE^{+\cdot}$ rearrangement separates the radical and charge site so that they can act independently, like the first step in ring cleavages (Section 8.8). Cleaving a second bond produces an $OE^{+\cdot}$ product, for which radical stabilization as well as charge stabilization is important for reaction favorability. A second H-rearrangement can also occur to yield a stable EE^+ product. Although the reactive conformation must be sterically possible, steric differences in competing rearrangements are generally less influential than energetic factors. However, steric accessibility probably plays an important role in the generally observed rearrangement favorability of hydrogen versus that of larger functionalities. Review Sections 4.9 and 4.10 about the differences and similarities in radical- and charge-site rearrangements; both are included in this section.

Hydrogen rearrangements are the most common, best understood, and generally most useful for deducing structure. One of the earliest isotopic labeling studies of a mass-spectral reaction was of the abundant $C_2H_4O_2^{+\cdot}$ ion formed by the loss of C_2H_4 from acetic acid (Happ and Stewart 1952; reaction 4.33). Mechanistic details of such rearrangements have been of particular interest because of their close photochemical analogies. Nicholson (1954) first pointed out the strong parallels between the Norrish Type 2 rearrangement of ketones and the characteristic olefin loss in their mass spectra. The literature through the early 1970s has been comprehensively reviewed (Bursey *et al.* 1973; Kingston *et al.* 1974, 1975). Because rearrangements involve two or more reaction steps, prediction of relative pathway probabilities becomes more difficult as the number of functionalities is increased; these affect the lability of possible hydrogens transferred, the hydrogen affinity of possible receptor sites, and the favorability of dissociation reactions of the rearranged intermediate. When predicting the mass spectra of complex molecules, always check reference spectra of closely related structures.

Hydrogen lability. The first step of a radical-site rearrangement moves the site of the unpaired electron. Thus its stabilization at the new site influences the over-all competitiveness of the rearrangement. Hydrogen rearrangement is favored from more highly branched carbon atoms, from those with adjacent unsaturation (e.g. benzylic, Equation 8.44), or from electronegative atoms such as oxygen. For example, rearrangement formation of $C_7H_8^{+\cdot}$ by loss of CH_2O from $HOCH_2CH_2C_6H_5^{+\cdot}$ and $H_3COCH_2C_6H_5^{+\cdot}$ occurs to give abundances of 60 and 10 per cent, respectively (<10 per cent $C_7H_8^{+\cdot}$ is formed from $H_3CCH_2CH_2C_6H_5\cdot$).

Activation of the hydrogen can lead to rearrangements through transition states of unusual ring size. In methyl ester spectra an activated hydrogen results in characteristic $(M - CH_3OH)^{+\cdot}$ peaks. For ε-H rearrangement this is accompanied by an $(M - 76)^{+\cdot}$ peak (reaction 8.44; Meyerson and Leitch 1966); for rearrangement from a distant functionality (8.45), a variety of secondary reactions are possible. For this reaction hydrogen activation is caused by groups such as keto, amino, ether, hydroxy, and trimethylsilyloxy, as well as by chain branching and unsaturation.

Hydrogen rearrangements in EE^+ ions can involve proton transfer; the tendency for this should reflect the relative proton affinities of the old and new sites. This can be very common in chemical ionization; Longevialle *et al.* (1979) have shown that the cross section for CI protonation of hydroxyl is equivalent to that of an amino group present in the same molecule. Although the subsequent transfer of H^+ from $-OH_2^+$ to $-NH_2$ is fast, molecular-geometry restrictions can make the loss of H_2O competitive.

Site receptivity. The favorability of the first H-rearrangement step also depends on the relative change in the charge-site stabilization which it effects. For example, Benoit *et al.* (1977) find the first step in alkanoate rearrangements $H-CHR'CH_2-OCOR^{+\cdot} \rightarrow \cdot CHR'CH_2-OC(OH)R^+$ (reaction 4.46) to be 0.6 to 1.1 eV exothermic (and thus *not* rate-determining), indicative of substantially increased charge stabilization. If two sites in the same $OE^{+\cdot}$ ion are competing for the rearranging hydrogen (8.46), under equivalent steric conditions their proton affinities should determine the relative transfer probabilities. Table A.3 shows, for the common functionalities, that those of nitrogen should in general be the most receptive sites, followed by sulfur-

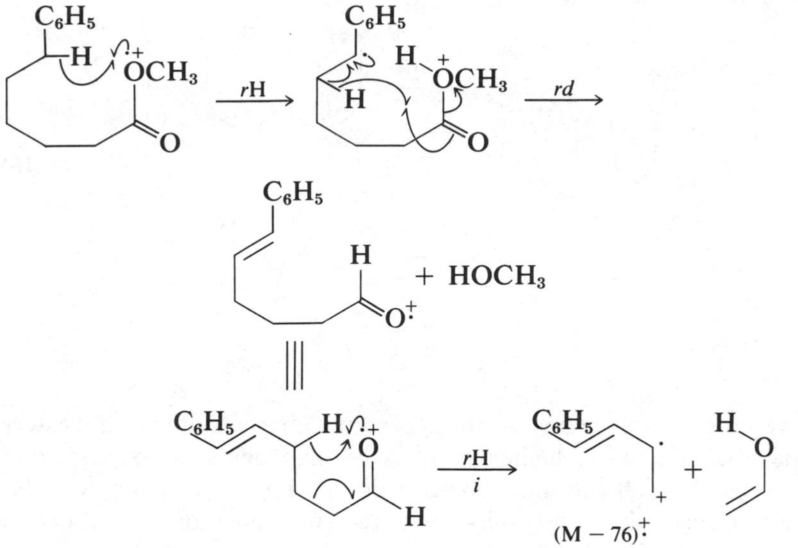

$$+ \text{ HOCH}_3 \qquad\qquad (8.44)$$

$$(M - 76)\overset{+}{\cdot}$$

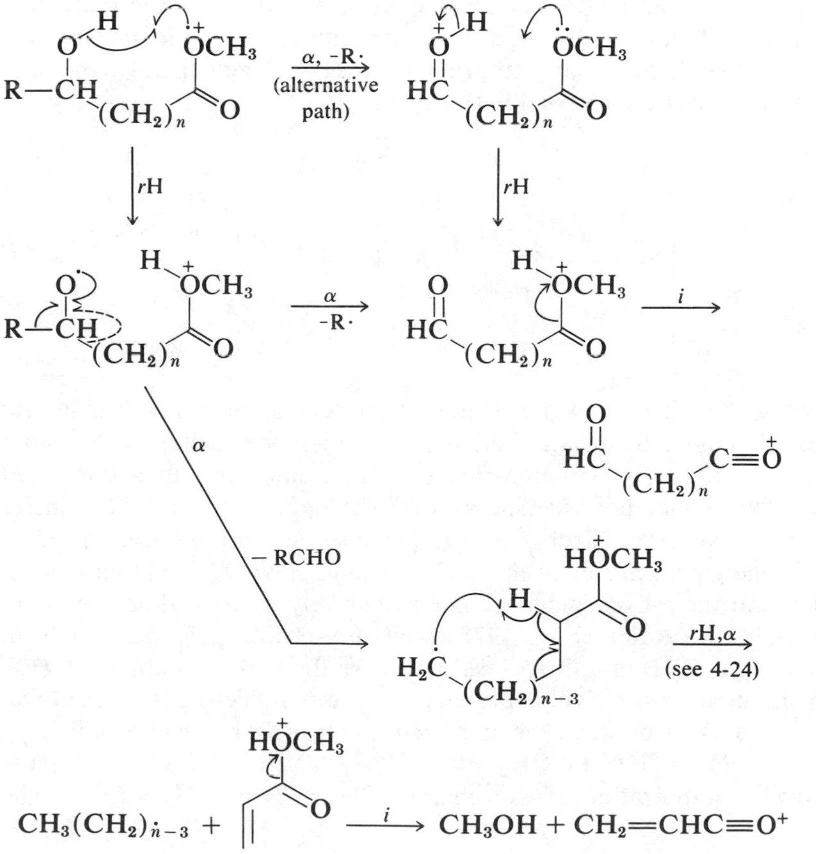

$$(8.45)$$

$$\text{CH}_3(\text{CH}_2)\overset{\cdot}{}_{n-3} + \quad \longrightarrow \quad \text{CH}_3\text{OH} + \text{CH}_2{=}\text{CHC}{\equiv}\text{O}^+$$

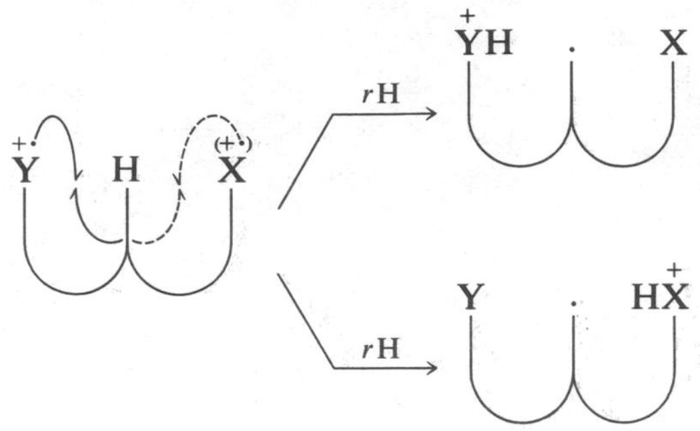

$$\text{(8.46)}$$

and by oxygen-containing ones. Among the latter, ethers, ketones, and esters should have a somewhat higher receptivity than acids, aldehydes, and alcohols, as well as alkene and aromatic double bonds. In phenetole, H-transfer is to oxygen through a four-membered-ring transition state (Section 7.4), not to the *ortho*-position through the usually more favorable six-membered ring (8.47). Steric factors affecting hydrogen approach to the site also appear important; γ-H rearrangements to an aromatic ring (Equation 8.8) are hindered seriously by larger *ortho*-substituents. It should be stressed, however, that such factors are important only if H-rearrangement is the rate-determining step of the over-all pathway.

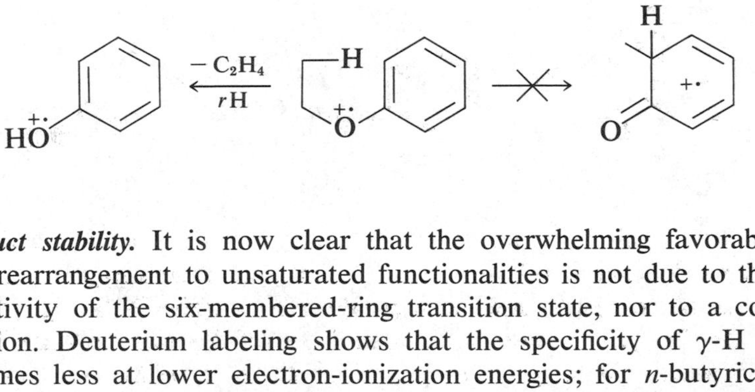

$$\text{(8.47)}$$

Product stability. It is now clear that the overwhelming favorability for γ-H rearrangement to unsaturated functionalities is not due to the steric selectivity of the six-membered-ring transition state, nor to a concerted reaction. Deuterium labeling shows that the specificity of γ-H transfer becomes less at lower electron-ionization energies; for *n*-butyric acid at threshold energies, transfer of the β-H is actually favored over that of γ-H, so that all but the α-hydrogens become extensively scrambled before dissociation (8.48; McAdoo *et al.* 1978). Initial γ-H and β-H transfer lead, respectively, to C_2H_4 and $CH_3\cdot$ loss; these are the two most abundant (100 and 25 per cent, respectively) primary products, reflecting their high stability. In the OE^{+} product the unpaired electron is resonance-stabilized ($\cdot CH_2{-}C(OH){=}O^{+}H \leftrightarrow CH_2{=}C(OH)_2^{+}$), and the EE^{+} product provides similar stabilization of the charge: $CH_2{=}CHC(OH){=}O^{+}H$. The

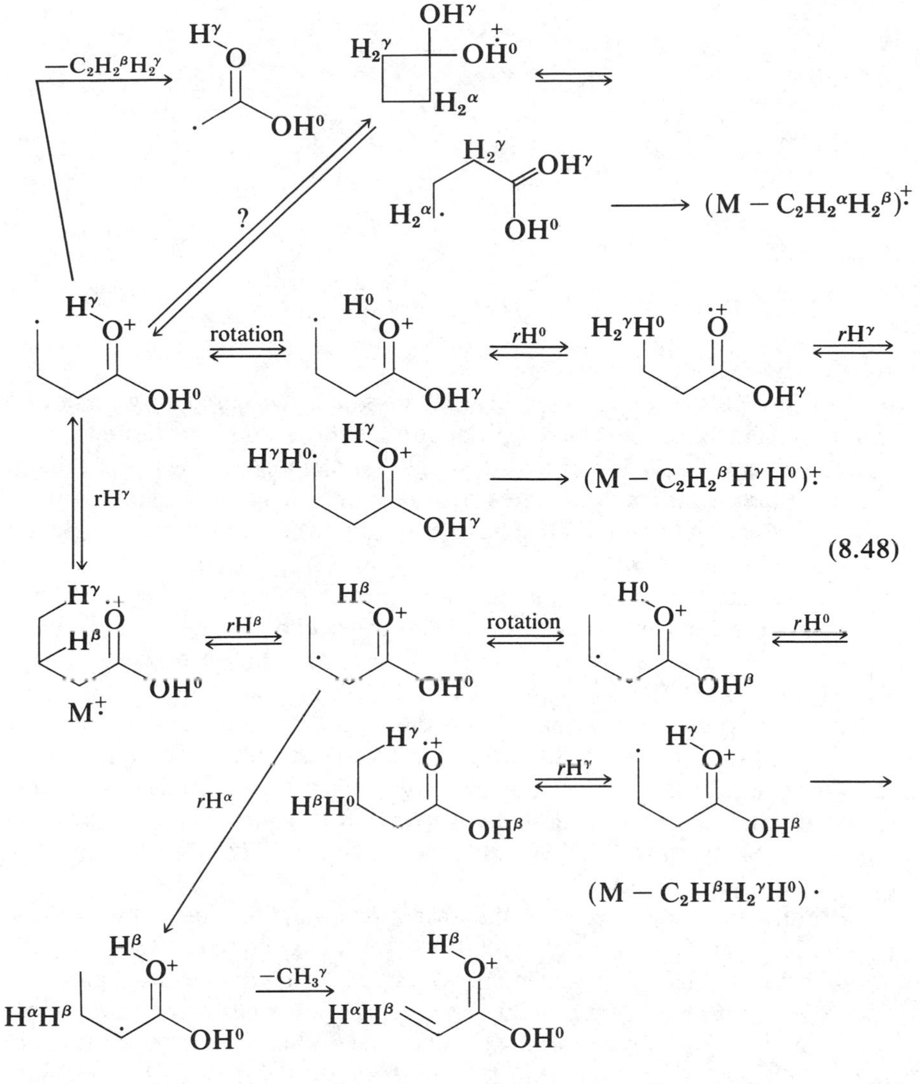

$$(8.48)$$

latter-type product $CR_2{=}CR'CR''{=}Y^+H$ can be formed also by initial H-transfer from a variety of sites, such as those shown in reaction 8.49. This reaction is actually useful for deducing structure, since the product includes the substituents on the carbon *beta* to the functionality (the γ-H rearrangement indicates those in the α-position). Thus in α,β-unsubstituted methyl alkanoates, the large m/z 87 peak (Figure 3M) arises from δ-H or ε-H transfer in reaction 8.49; the stability of the EE^+ product compensates for the extra rearrangement step required (Dinh-Nguyen *et al.* 1961; Kraft and Spiteller 1969).

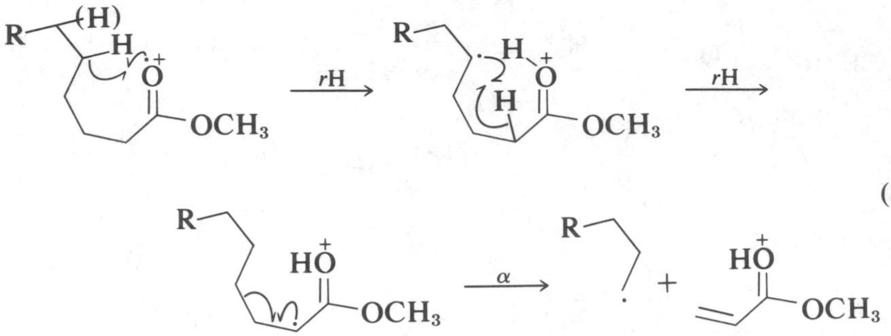

(8.49)

A variety of cases were given earlier in which the competitiveness of a rearrangement is largely due to product stability. The unsubstituted cyclo-alkanones (reaction 8.38) of five- through ten-membered rings, and thus of widely different transition states for rearrangement, all give abundant $CH_2=CH-C\equiv O^+$ ions. Despite the extra rearrangement step, the ester pathway $H_2-R-OCR'=O^{\ddagger} \rightarrow R\cdot + R'C(OH)_2^+$ (reaction 4.46) produces characteristic product ions. A similar double-hydrogen rearrangement gives significant peaks for higher-molecular-weight ketones (e.g., $H_2-R-CH_2COCH_3^{\ddagger} \rightarrow R\cdot + (CH_3)_2C=O^+H$, m/z 59), and esters (e.g., $H_2-R-CH_2COOCH_3^{\ddagger} \rightarrow R\cdot + C_3H_7O_2^+$, m/z 75); apparently the insignificant occurrence for low mass $M^{\ddagger}$ is due to the low probability of forming the required low energy $M^{\ddagger}$ ions (Section 7.4). In the latter double-hydrogen rearrangements, the product-ion mass is higher than that of the simple cleavage product because both hydrogens are transferred to the product ion; this is not true in 8.37, 8.38, and 8.48, since one H-transfer is *away* from the incipient product ion.

Note that the general rule for the formation of an $OE^{\ddagger}$ ion by cleavage of two bonds in an $OE^{\ddagger}$ precursor requires that the two bonds be *between* the potential products. A rearrangement which is internal to one of the products does not count as a bond cleaved, such as *rH* followed by α-cleavage loss of R· (second line of reaction 8.45), or the threshold formation of the more stable $RCH=SH^+$ isomer from RCH_2SR' (8.50) through a "displacement reaction" (Section 8.10). Such a rearrangement thus effects an isomerization of one of the products, which can increase product stability; extensive isomerizations of this type are discussed under "Random Rearrangements" at the end of this Section.

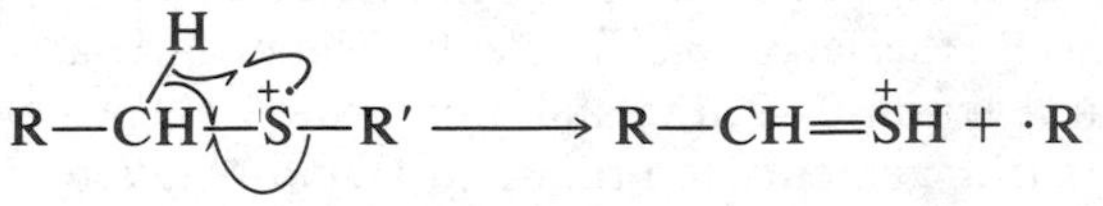

$$R-CH-\overset{+}{S}-R' \longrightarrow R-CH=\overset{+}{S}H + \cdot R \qquad (8.50)$$

Consecutive rearrangements. An $OE^{+\cdot}$ product ion produced by rearrangement still has an unpaired electron, which can produce a secondary product ion by a further radical-site reaction. For example, the γ-H rearrangement of alkanones produces an $OE^{+\cdot}$ product with an α-position radical site, which can then lead to secondary reactions of α-cleavage (8.51) or a "consecutive" rearrangement (8.52).

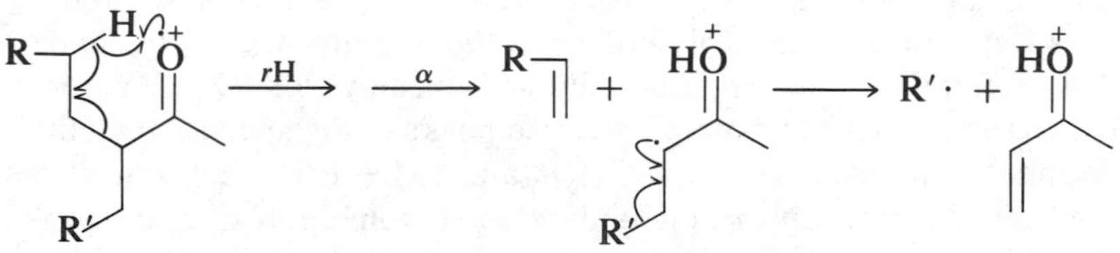

$$(8.51)$$

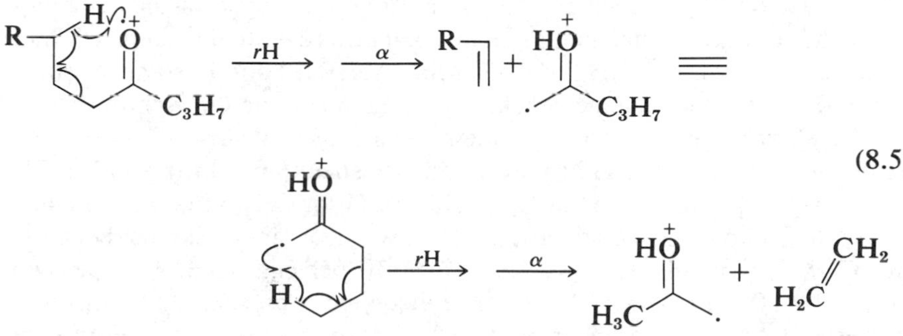

$$(8.52)$$

Stereochemistry. The tight activated complex typical of rearrangement reactions is the most severe general limitation to their competitiveness (reread Section 8.2). However, important decomposition pathways can even involve multiple rearrangement steps (8.44, 8.45, 8.48, 8.49). An unusual example is shown in 8.53 (Abbott *et al.* 1979).

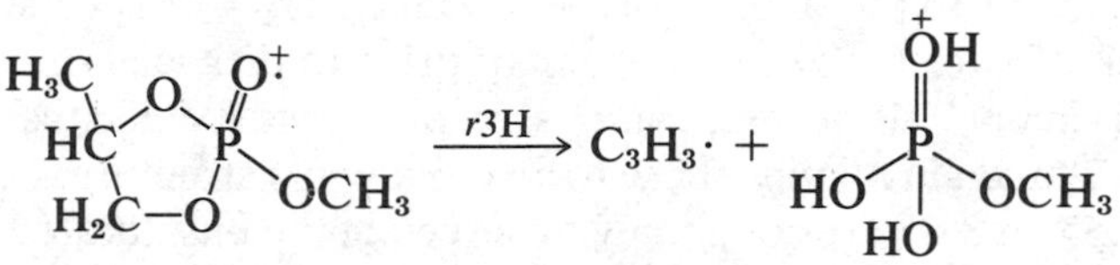

$$(8.53)$$

Random rearrangements. A lack of reactive centers in an ion results in high critical energies for the dissociations that do take place, and more competitive rates for the possible prior and concomitant rearrangements. The higher the tendency for such accompanying rearrangements, the less the product ions will be characteristic of the *specific* precursor structure; however, the resulting "low mass ion series" (Section 5.2) can still indicate *general* structural features. Such "random rearrangements" are typical of molecular ions from high-ionization-energy molecules such as alkanes, including those substituted with halogen, cyano, and nitro groups. The spectrum of $(CH_3)_3CH$ shows a $C_2H_5^+$ ion, and that of CCl_2FCCl_2F shows a CCl_3^+ ion; these peaks are more abundant, however, in n-C_4H_{10} and CCl_3CClF_2, respectively. Chain branching and unsaturation decrease such scrambling tendencies but do not necessarily eliminate them; neopentane (Figure 3A) yields $C_2H_5^+$ and $C_3H_7^+$ peaks and perfluorobenzene yields CF_3. Hydrogen scrambling is even more common. The base peaks in the spectra of methylcyclohexane and toluene are $(M - CH_3)^+$ and $(M - H)^+$, respectively, as expected from the mechanisms cited above, yet from $C_6H_{11}CD_3$ the peaks $(M - CHD_2)^+$ and $(M - CH_2D)^+$ are larger than $(M - CD_3)^+$, and from $C_6H_5CD_3$ $(M - H)^+$ is more abundant than $(M - D)^+$.

The most readily formed EE^+ ions are generally those of high stability.; thus a variety of rearrangements can be competitive with or accompany their further decompositions (8.54 to 8.57; McLafferty 1980b). In some favorable cases EE^+ ion rearrangements closely resemble their $OE^{\overset{+}{\cdot}}$ counterparts; specific H-transfer through a six-membered-ring transition state (8.54) is found for protonated ethyl acetate (8.58a, Pesheck and Buttrill 1974). The EE^+ ions of 8.58b–d (R = H or C_2H_5; R' = CH_3 or C_6H_5; Benoit and Harrison 1976; Leung and Harrison 1977) show surprisingly similar behavior, with C_3H_6 loss involving the transfer of ~ 30 per cent α-H, ~ 20 per cent β-H, and ~ 50 per cent γ-H. Persuasive evidence is presented that this does not result from H scrambling in the C_3H_7 group prior to the H-transfer accompanying C_3H_6 loss. For such rearrangements in odd-electron ions which involve a saturated-ring transition state, the analogous C_3H_6 loss from $OE^{\overset{+}{\cdot}}$ propyl phenyl ether molecular ions shows a similar H-site distribution, as does C_4H_8 loss from n-$C_4H_9OC_6H_5$ compared to that from $CH_3CH_2CH_2CH_2\overset{+}{O}{=}CH_2$ (8.56; Budzikiewicz *et al.* 1967). The high specificity shown by the EE^+ displacement rearrangement (8.59) of threshold energy ions is due to the stability of the product $CH_2{=}CH_2$ versus $CH_3CH{:}$ and the lack of sites for H rearrangement prior to dissociation.

However, many EE^+ decompositions are even less specific than these, and care must be taken in relating these to an unknown's structure. For example, reactions 8.54 to 8.57 are commonly observed at the unsaturated site of EE^+ ions, such as those formed by α-cleavage; an example of 8.56 (reaction 4.44) was discussed in Section 4.10. Like the radical-site rearrangements (Section 4.9), if the unsaturation is in the ring of the rearrangement transition state, this needs to be six-membered so that the double-bond can move away from the H-receptor site (8.54 and 8.55). With only a single bond at this site (8.56 and 8.57), cleavage of this bond is required (possibly in concert with H-rearrangement), and a variety of transition-state ring sizes is observed.

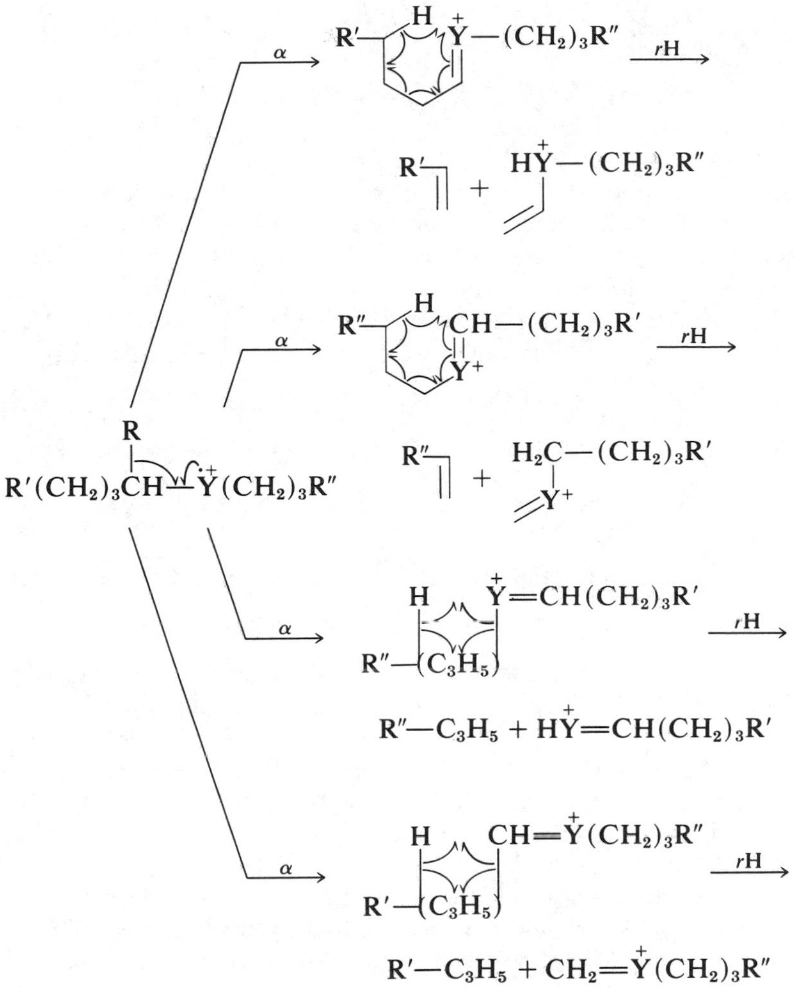

$$\xrightarrow{\alpha} \quad R'\overset{H}{\diagdown}\overset{+}{Y}-(CH_2)_3R'' \xrightarrow{\ rH\ }$$

$$(8.54)$$

$$R'\overset{|}{\|} \ + \ \overset{+}{HY}-(CH_2)_3R''$$

$$\xrightarrow{\alpha} \quad R''\overset{H}{\diagdown}CH-(CH_2)_3R' \xrightarrow{\ rH\ } \\ \overset{|}{\underset{Y^+}{\|}}$$

$$(8.55)$$

$$R''\overset{|}{\|} \ + \ \overset{H_2C-(CH_2)_3R'}{\underset{Y^+}{\diagup}}$$

$$R'(CH_2)_3\overset{R}{\underset{|}{CH}}\overset{.+}{Y}(CH_2)_3R''$$

$$\xrightarrow{\alpha} \quad R''-\left(C_3H_5\right)\overset{H}{\underset{}{\diagup}}\overset{+}{Y}=CH(CH_2)_3R' \xrightarrow{\ rH\ }$$

$$(8.56)$$

$$R''-C_3H_5 + H\overset{+}{Y}=CH(CH_2)_3R'$$

$$\xrightarrow{\alpha} \quad R'-\left(C_3H_5\right)\overset{H}{\underset{}{\diagup}}CH=\overset{+}{Y}(CH_2)_3R'' \xrightarrow{\ rH\ }$$

$$(8.57)$$

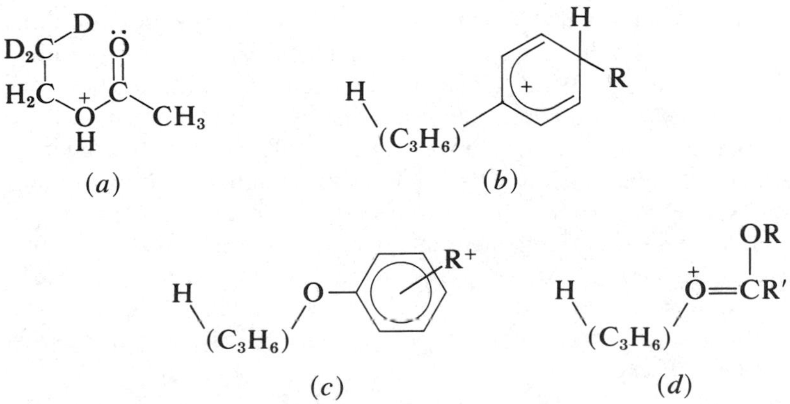

$$R'-C_3H_5 + CH_2=\overset{+}{Y}(CH_2)_3R''$$

(a)

(b)

$$(8.58)$$

(c)

(d)

$$H_2C \xrightarrow{\overset{\overset{\displaystyle H}{\overset{\displaystyle |}{H}} \overset{+}{O}-C_2H_5}{\underset{|}{\,}}} CH_2 \xrightarrow{\ rH\ } H_2C{=}CH_2 + H_2O{-}C_2H_5 \qquad (8.59)$$

Although abundant products are formed by reactions like 4.44 which formally correspond to a four-membered-ring intermediate, those are unfavorable from orbital symmetry considerations (Pescheck and Buttrill 1974; Williams 1977). The general preference for reaction 8.56 over 8.57 for amines is useful for deducing structure (Figure 4D versus Unknown 4.7; Unknowns 4.20 and 6.2). In keeping with this preference, for metastable ion decomposi-

tions Uccella et al. (1971) find that $CH_3CD_2\overset{+}{N}H{=}CH_2$ specifically loses

$C_2H_2D_2$ (reaction 8.56), but that $CH_3CD_2CH{=}\overset{+}{N}H_2$ (reaction 8.57) and

$CD_3C(CH_3){=}\overset{+}{N}H_2$ undergo complete H/D scrambling in the loss of ethylene. For oxygenated ions, four isomeric $C_3H_7O^+$ ions show favored

loss of C_2H_4 to yield $COH_3{}^+$. For $CH_3CH_2\overset{+}{O}{=}CH_2$ and $CH_3CH_2CH{=}\overset{+}{O}H$

this should be due to reactions 8.56 and 8.57, respectively; for $(CH_3)_2C{=}\overset{+}{O}H$

and $CH_3CH{=}\overset{+}{O}CH_3$ the rearrangements involved are more complex (Tsang and Harrison 1971; McLafferty and Sakai 1973). Sulfur-containing EE^+ ions of appropriate structure undergo reactions 8.54 through 8.57 plus a number of others. For example, the σ-bond cleavage product $R{-}CHR'{-}S^+$ is unstable (except for H_3CS^+; Dill et al. 1979), rearranging to $R{-}CR'{=}S^+H$ (van de Graaf and McLafferty 1977). Note the similar concerted $OE^{\ddot+}$ reaction (8.50).

Ions which have no orbitals available for H-transfer, such as the saturated EE^+ ion of reaction 8.59, should not undergo hydrogen scrambling. This is also generally true for alkane molecular ions as well as protonated alkanes (Houriet et al. 1977). This is not true, however, for EE^+ alkyl ions; the $n\text{-}C_6H_{13}{}^+$ ions formed by chemical ionization of $n\text{-}C_6H_{14}$ undergo total scrambling prior to dissociation. This probably involves reactions such as hydride transfer to the saturated cation site, analogous to solution rearrangements (Lavanchy et al. 1979). Isomeric n-octene $OE^{\ddot+}$ molecular ions (Borchers et al. 1977) of sufficient internal energy to decompose undergo isomerization to a mixture of interconverting structures within 10^{-9} seconds, but isomerization of the nondecomposing ions is incomplete in 10^{-6} seconds. Thus radical-site H-migrations appear to be facile in such unsaturated $OE^{\ddot+}$ ions. Finally, hydrogen rearrangement can occur in an EE^+ ion to a site that has neither a charge nor an unsaturated bond (reactions 4.45, 8.60, and 8.61). HY elimination in 4.45 and 8.60 produces a double bond in conjugation with one in the precursor ion; product stability appears to be the main driving force for 8.61.

$$\underset{H_2\overset{+}{N}\curvearrowleft CHCH_2CH_2SCH_3}{\overset{\displaystyle COOR}{|}} \xrightarrow{-ROC\cdot} H_2\overset{+}{N}=CH-\underset{}{\overset{H \; \overset{..}{S}CH_3}{CH}}CH_2 \xrightarrow{rH} \quad (8.60)$$

$$H_2\overset{+}{N}=CHCH=CH_2 + HSCH_3$$

$$\underset{R\curvearrowleft CH\curvearrowright \overset{..}{O}H}{\overset{\displaystyle H_3C}{|}} \xrightarrow[\alpha]{-R\cdot} \overset{H_3C \; H}{HC}=O^+ \xrightarrow{rH} CH_4 + HC\equiv O^+ \quad (8.61)$$

In Unknown 8.3 the EE^+ rearrangement $C_4H_{10}NO^+ \rightarrow C_4H_8N^+$ would have been an important indicator of hydroxyl in a total unknown. A similar reaction can aid in elucidating Unknown 8.10, which is an α-amino acid isolated from a natural source as the ethyl ester.

Unknown 8.10

m/z	Rel. abund.	m/z	Rel. abund.	m/z	Rel. abund.
29	22.	58	1.2	104	48.
30	11.	59	1.8	105	2.6
31	0.8	60	0.8	106	2.3
39	1.8	61	100.	114	1.1
40	0.4	62	3.0	116	3.9
41	2.3	63	4.3	117	0.2
42	6.7	74	12.	129	6.1
43	7.9	75	8.0	130	0.4
44	2.1	76	0.6	131	6.3
45	6.2	83	2.1	132	0.4
46	4.9	85	0.7	133	0.3
47	5.3	86	0.5	148	3.2
48	1.6	87	2.2	160	0.6
49	1.3	88	3.5	177	12.
54	2.4	100	5.6	178	1.2
55	2.7	101	0.6	179	0.7
56	48.	102	5.4		
57	5.1	103	1.8		

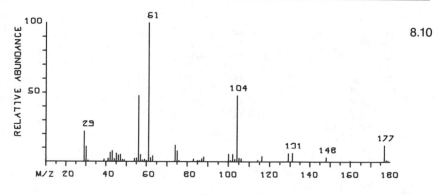

8.10

Unknown 8.11

m/z	Rel. abund.
17	9.8
18	32.
19	9.2
27	11.
28	9.0
29	38.
30	5.3
30.4	0.3
31	56.
32	2.9
33	2.4
42	11.
43	89.
44	53. $C_2H_4O^{+\cdot}$
45	9.9
46	0.5
60	10.
61	100.
62	5.2
63	0.6
73	0.6
74	1.3 $C_3H_6O_2^{+\cdot}$
75	0.9

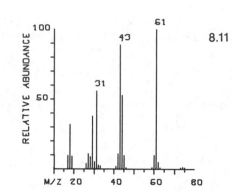

Unknown 8.12

m/z	Rel. abund.	m/z	Rel. abund.
27	15.	71	0.5
28	21.	72	60.
29	40.	73	2.3
30	1.3	74	0.4
31	1.3	77	0.4
32	1.0	78	20.
33	2.6	79	0.5
41	2.4	96	1.4
42	3.9	97	17.
43	3.4	98	0.4
44	36.	113	0.6
45	3.3	114	4.5
46	2.6	126	100.
47	4.0	127	3.6
50	2.9	128	0.3
51	2.7	140	7.0
56	2.7	141	70.
69	60.	142	3.4
70	4.7	143	0.2

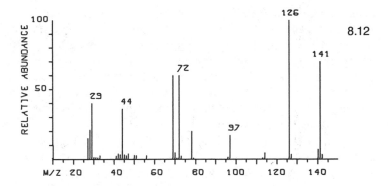

8.10 Other rearrangements

$$\overset{+\cdot}{\underset{B}{\overset{A}{\big|}}} C\text{—}D \xrightarrow{\;rd\;} \overset{+}{\underset{B}{\overset{A}{\big|}}}\!\!\!>\!C + D \tag{8.62}$$

$$\overset{+}{\underset{B}{\overset{A}{\big|}}} C\text{—}D \xrightarrow{\;rd\;} \underset{B}{\overset{A}{\big|}}\!\!\!>\!C + D^+ \tag{8.63}$$

$$\underset{B\ \ C}{\overset{A\ \ \overset{+(\cdot)}{D}}{\big|\ \ \big|}} \xrightarrow{\;re\;} \underset{B\text{—}C}{\overset{A\text{—}\overset{+(\cdot)}{D}}{}} + \quad or \quad \underset{B\text{—}\overset{+(\cdot)}{C}}{\overset{A\text{—}D}{}} \tag{8.64}$$

Because such a wide range of internal energies ($>$15 eV) is deposited in molecular ions formed by 70 eV electron ionization, a great diversity of rearrangement reactions are observed. As a general rule, the non-hydrogen rearrangements are useful for deducing structure of unknown spectra only if the spectra of closely related compounds are available; the interpreter must be on guard against drawing misleading conclusions from ion products of such reactions. Many of these are steps in "random rearrangements" (Section 8.9), and so are useful structurally only for "ion series" type information. For example, alkyl ions undergo facile skeletal as well as hydrogen rearrangement (Lavanchy *et al.* 1979). Those rearrangements which can be useful for deducing structure have been classified here as "displacement" and "elimination" rearrangements. These involve a bond-forming step between two parts of the ion in combination with the cleavage of a bond to expel an exterior part of the ion ("displacement, *rd*"), or the cleavage of two bonds to eject an interior fragment ("elimination, *re*"). The review by Bosshardt and Hesse (1974) is recommended.

Displacement reactions. These reactions are favorable energetically, since a new bond is formed in compensation for the one cleaved. Offsetting this are the steric requirements for forming the new bond which depend critically on the degree of substitution of the two atoms involved as well as the other factors discussed earlier. Thus there is a substantial steric advantage if one of the atoms is monovalent, a major reason for the ubiquity for other hydrogen rearrangements. Displacement reactions involving hydrogen migration can be important (reaction 8.50), but among the most striking examples of *rd* reactions are those involving the monovalent elements chlorine and bromine (reactions 4.42 and 8.65, $Y = Cl$, Br). Despite the low general tendency of halogens to initiate radical-site reactions, for $Y = Cl$ the peak corresponding to $n = 4$ (m/z 91) is the base peak for $R = C_2H_5$ through C_8H_{17} (Figure 3W), and the homologous $n = 5$ peak is usually 5 to 20 per cent as abundant. For $Y = Br$, the corresponding m/z 135/137 ions give major peaks (Figure 3X and Unknown 8.1). Chain branching or other substitution drastically reduces this displacement reaction; apparently its energy requirements are only lower than those of other secondary C—C bond cleavages. The reaction can be aided by substituents which stabilize the R· product; the mass spectra of methyl ω-aminoalkanoates show significant $(M - \cdot CH_2COOCH_3)^+$ peaks.

$$R\overbrace{}^{} \quad \begin{matrix} +\cdot \\ Y \\ \diagdown \diagup \\ (CH_2)_n \end{matrix} \quad \xrightarrow{rd} \quad R\cdot \; + \quad \begin{matrix} + \\ Y \\ \diagup \diagdown \\ (CH_2)_n \end{matrix} \qquad (8.65)$$

The spectrum of *n*-tridecanenitrile, Figure 3Y, shows an important ion series m/z 82, 96, 110, 124. . . . Reaction 8.65 can account for the series maximum at m/z 110 ($n = 6$), with the large ring size reflecting the wide bond angle for C—C≡N. The tendency for Equation 8.65 to proceed through a five-membered ($Y = Cl$, Br, SH) or six-membered ($Y = NH_2$) ring is discussed in Section 8.2; a three-membered ring intermediate can also be important, such as in the formation of cyclic $C_2H_4SH^+$ (protonated ethylene sulfide) from $R—CH_2CH_2SH$ (van de Graaf and McLafferty 1977).

Green *et al.* (1978) have pointed out that this reaction closely parallels homolytic substitution at saturated carbon in free-radical chemistry, and involves inversion of configuration at the carbon attacked. Another example with a close parallel in solution chemistry is the displacement loss of Br· from 2-phenylethylbromide to form the ethylenebenzenium ("phenonium") ion (8.66; Koppel and McLafferty 1976). Reaction 8.67 (Levsen *et al.* 1977) was shown to involve cyclization to nitrogen, not oxygen; note that a vinylic bond is cleaved. Cyclization to oxygen apparently takes place in 8.68.

Displacement can occur from an atom other than the one attacked by the migrating group. For example, Resink *et al.* (1974) have shown that the loss of $C_2H_5\cdot$ and $CH_2\text{=}CO$ from ethyl 3-phenylpropionate involves a displacement at oxygen that moves double-bond character to an adjacent C—O bond to displace the alkyl group on that oxygen (8.69). Note that the cleavage R'—OCOR· → R'· + $^+$OCOR is usually unfavorable ($[C_2H_5^+]=$ 2 per cent from $C_2H_5OCOC_6H_5$).

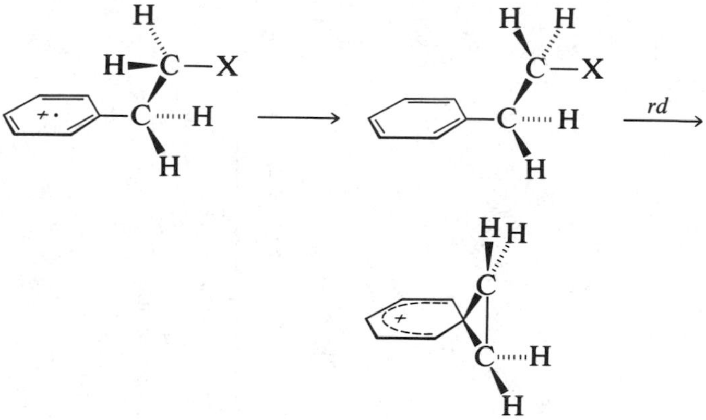

(8.66)

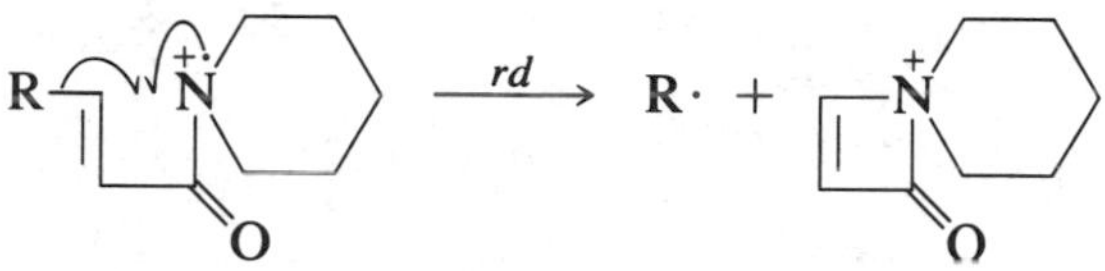

(8.67)

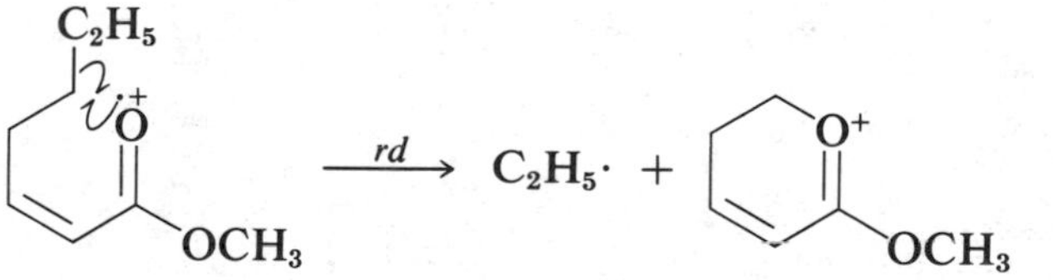

(8.68)

m/z 113, 55%   (m/z 99, 127: < 2%)

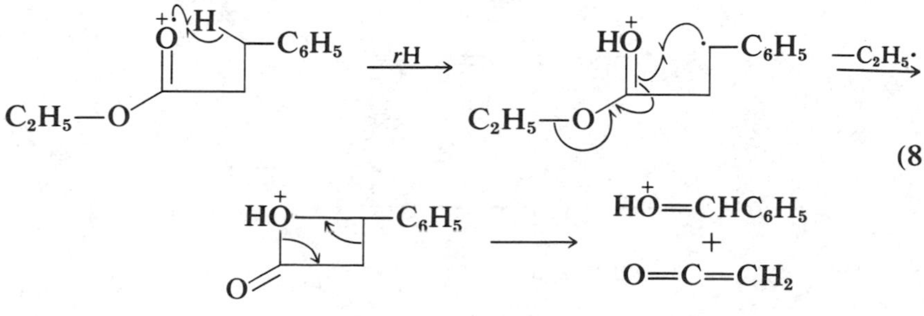

(8.69)

m/z 149, 3%   m/z 107, 44%

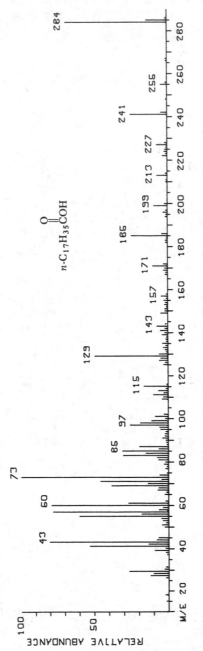

Figure 8I. Mass spectrum of *n*-octadecanoic acid.

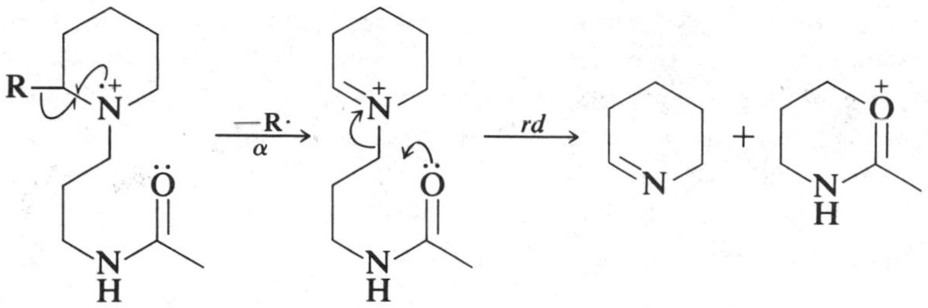

$$(8.70)$$

Charge site displacement reactions also appear possible (8.70; Seiler and Hesse 1968); note the similarity of this and displacement 8.68. In the mass spectra of fatty-acid derivatives (n-alkyl-Y), similar displacement reactions may be the cause of a characteristic periodicity of $(CH_2)_4$ in the ion series beginning with the product of reaction 8.49. Thus the mass spectrum of n-octadecanoic acid (Figure 8I; see also Figure 3L) shows maxima in this series at m/z 73, 129, 185, and 241. The methyl alkanoates (Figure 3M) show maxima at 87, 143, 199, 255, . . .

Unknown 8.13. Formulate a displacement rearrangement which will rationalize an abundant ion in the mass spectrum of β-ionone, Unknown 4.13. (For the answer see Equation 9.29.)

Elimination reactions. For these reactions the general mechanistic formulation ABCD$^+$ (OE‡ or EE$^+$) $\rightarrow$ AD$^+$ + BC would include hydrogen rearrangements for compounds with A − H. For the non-H elimination reactions discussed here, some types of A groups exhibit a high tendency for migration, small stable neutrals are commonly eliminated as BC, and the new AD$^+$ ion can show a substantial increase in stability over the precursor.

Rearrangement of oxygen functionalities is relatively common, with methoxy migration probably the best documented non-H rearrangement. This has been found in a wide variety of polymethoxy compounds such as derivatized sugars and other alicyclic compounds (Budzikiewicz *et al.* 1967; Winkler and Grutzmacher 1970). In cycloalkanes with three or more methoxy groups, reaction 8.71 gives rise to a major peak in the spectrum. Trimethyl-

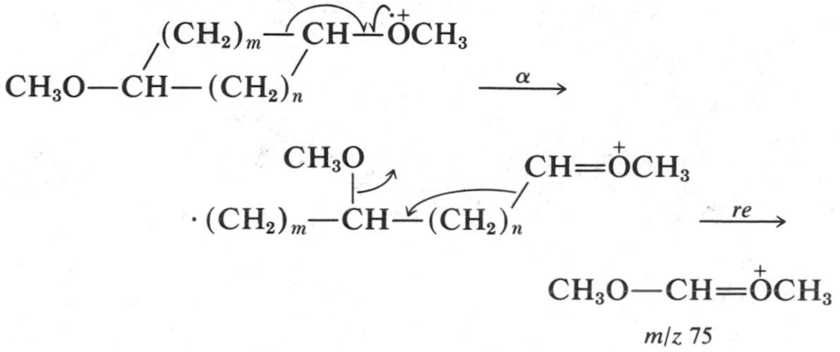

$$(8.71)$$

m/z 75

silyl group migrations are common; although these usually occur through five- to eight-member-ring transition states (Brooks 1979), the long-range rearrangement of 8.72 gives a base peak in the spectrum. Examples of eliminations from EE^+ ions are shown in reactions 8.73 through 8.75.

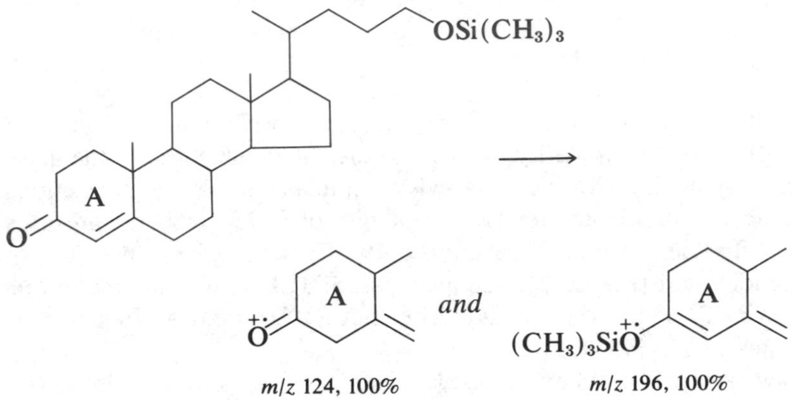

OSi(CH3)3
A
O
and
(CH3)3SiO
m/z 124, 100%
m/z 196, 100%

(8.72)

(CH3)3Si
O=Si(CH3)2
O
CH2
(CH2)n
re
O
CH2 + (CH3)3SiO=Si(CH3)2

(8.73)

CH3O CH=OCH3
H2C CHOCH3
re
CH2=CHOCH3 + CH3O—CH=OCH3

(8.74)

CH2OAc
O
OAc
AcO
OAc
re
(CH3C)3O+

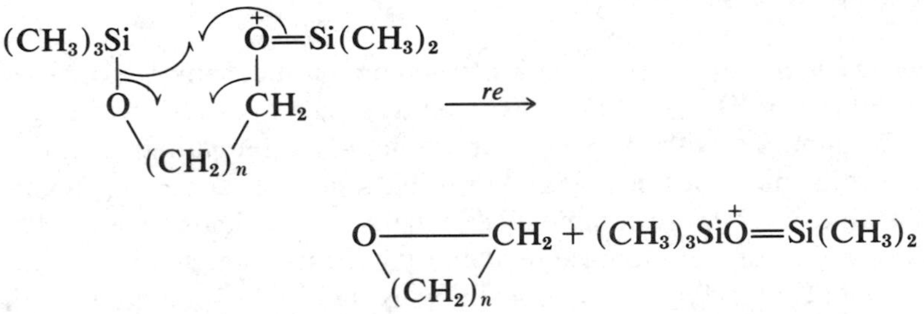

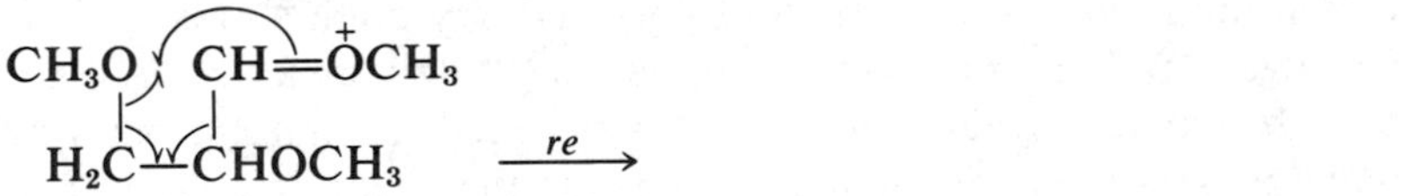

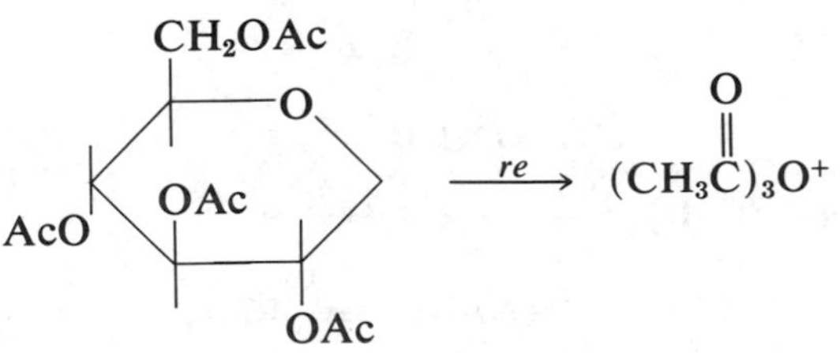

(8.75)

Elimination also often follows initial bonding between two large groups, particularly if these are aromatic. A variety of 2'-azabenzanilides were found (Broxton *et al.* 1977) to eliminate CO and then H after initial cyclization between the aromatic rings (8.76); the ring nitrogen provides a radical-like site for the initial attack, and the larger ring system formed provides stabilization for the unpaired electron as well as the charge. The elimination of H_3PO_4 from triphenyl phosphate must involve a multiplicity of such steps (8.77).

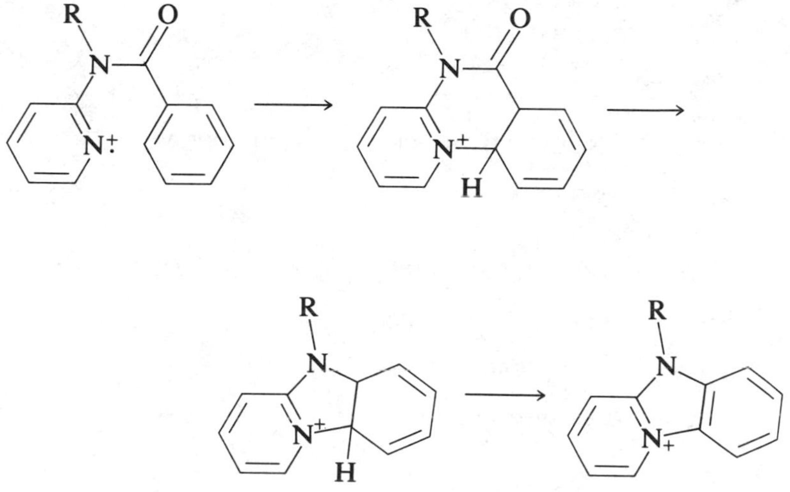

(8.76)

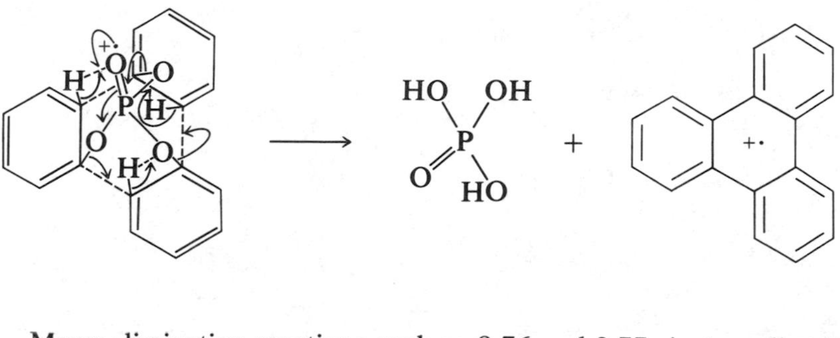

(8.77)

Many elimination reactions such as 8.76 and 8.77 eject small stable radicals or molecules such as H, CO, H_3PO_4, H_2O, H_2S, and HCl. Table 8.2 gives examples of such reactions.

Table 8.2 *Examples of Other Rearrangement Reactions*

Neutral m/z	Eliminated Formula	Molecule (R = alkyl, Ar = aryl)
15	$CH_3\cdot$	$ArCH{=}CHAr$
18	H_2O	$C_{10}H_{21}OC_{10}H_{21}$
28	CO	$ROCOR$, $RCCH_2COR$, $ArOH$, $ArCAr$, $ArOAr$, $ArCCl$
	N_2	$ArN{=}NAr$
	C_2H_4	$C_6H_5CH_2CH_2CH_2Br$
29	$CHO\cdot$	C_6H_5COH, $C_6H_5CH{-}CHC_6H_5$, $C_6H_5CH{=}CHCOC_6H_5$
30	$NO\cdot$	$ArNO_2$
	CH_2O	$ROCH_2OR$, $C_3H_7{-}CH_2O{-}CC_2H_5$, $CH_3CCH_2OSO_2CH_3$
31	$CF\cdot$	$CHF{=}CFBr$, $CF_2{=}CFCl$
32–4	S, $\cdot SH$, SH_2	RSR, $RSAr$, $ArSAr$, $RSSR$
41	CH_3CN	$C_6H_5C(CH_3){=}N{-}OH$,
43	HNCO	$ROCNHR$
	$C_3^*H_7\cdot$	$C_{14}H_{29}(C^*H_2)_3COCH_3$, $R(C^*H_2)_3R$
44	CS	$ArSAr$
	CO_2	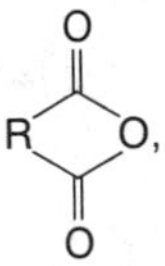 $R{\overset{O}{\diagup}}{\overset{O}{\diagdown}}O$, $ROCOR'$, $ROCSR$, $ArCOC(CH_3)_3$, $ArOCC(CH_3)_3$,
		phthalimide NR structure
	C_2H_4O	$C_2H_5{-}CH(CH_3)O{-}CR$

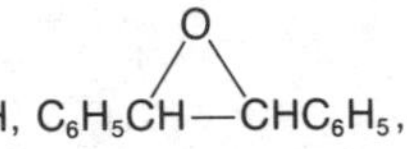

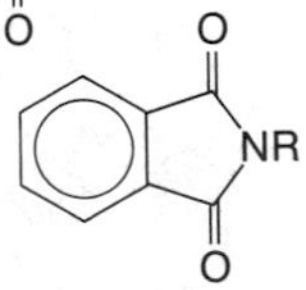

Table 8.2 *Examples of Other Rearrangement Reactions*

Neutral *m/z*	Eliminated Formula	Molecule (R = alkyl, Ar = aryl)
45	$HCO_2\cdot$	$\overset{\displaystyle O}{\overset{\|}{ROC}}CH{=}CH\overset{\displaystyle O}{\overset{\|}{C}}OR$, $C_6H_5\overset{\displaystyle O}{\overset{\|}{OC}}CH_2C_6H_5$
50	CF_2	$C_6H_5CF_3$
54	C_4H_6	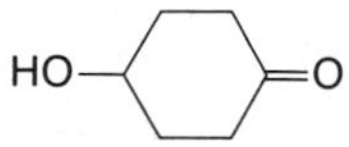 HO—⬡=O
60	COS	$R\overset{\displaystyle O}{\overset{\|}{SC}}OR$
	C_2H_4S	$C_2H_5{-}CH_2CH_2S{-}\overset{\displaystyle O}{\overset{\|}{C}}R$
64	SO_2	RSO_2R, $ArSO_2OR$

9

MASS SPECTRA OF COMMON
COMPOUND CLASSES

Now we shall attempt to apply the principles set forth above to the interpretation of common types of organic compounds. It is important for you to gain an appreciation of the reliability — and lack of reliability — of the various types of mass-spectral information in providing evidence concerning various structural features. Note that there is an exponential increase in the number of possible fragmentation pathways with an increase in the number of reaction-initiating sites; this can be seen by comparing the behavior of esters with that of either ketones or ethers.

This chapter is intended to illustrate how the mechanisms can be applied to particular compounds, *not* to give a comprehensive catalog of the mass-spectral behavior of organic structures. Detailed correlations have now been published of the mass spectra of a wide variety of molecular classes. In many of these studies, isotopic labeling and other special techniques have been employed to elucidate mechanistic pathways. An excellent summary of this literature to 1967 is available in the comprehensive volume of Budzi-kiewicz, Djerassi, and Williams (1967). This is recommended as a basic reference; the compound classifications in this chapter are patterned after it. A wealth of specific material is to be found in the specialized journals *Organic Mass Spectrometry* (Heyden, London) and *International Journal of Mass Spectrometry and Ion Physics* (Elsevier, Amsterdam), and thorough referencing of the current literature is given in *CA Selects: Mass Spectrometry* (Chemical Abstracts, Columbus, Ohio).

In dealing with compounds that contain more than one functional group, remember that there are great differences in the over-all ability of a particular functional group to influence the fragmentation of a molecular ion. For example, although in the spectrum of $C_2H_5CH(OCH_3)(CH_2)_6CH_3$ the α-cleavage peak, $C_2H_5CH{=}\overset{+}{O}CH_3$, is more than twice as high as any other in the spectrum, it is <10 per cent of the height of the $C_2H_5CH{=}\overset{+}{N}H_2$ peak in the spectrum of $C_2H_5CH(OCH_3)(CH_2)_4CH(NH_2)C_2H_5$ (Section 8.6). Also, synergistic effects are common; a molecule containing two functional groups can undergo reactions that are not found for those with either group alone. To reiterate the precautions given above, always check the actual mass-spectral behavior of compounds closely related to your proposed structure.

9.1 Hydrocarbons

Hydrocarbons, unfortunately, exhibit a substantial tendency to undergo *random rearrangements* (Section 8.9) in which the hydrogen-atom positions and, to a lesser extent, the carbon skeletal arrangement are scrambled. This is the most pronounced in alkanes, in which the σ-electron ionization (Section 8.4) and C—C bond cleavages are higher-energy processes, producing more highly excited ions which have a higher tendency for isomerization and rearrangement decompositions. This is reduced by the presence of chain branching and unsaturation; the addition of a polar functional group changes the spectrum dramatically (compare Figures 3B and 3P).

Saturated aliphatic hydrocarbons. As shown in the spectra of *n*-dodecane (Figure 3B) and *n*-hexatriacontane (Figure 5B), straight-chain alkanes show weak molecular ions and typical series of $C_nH_{2n+1}^+$, and to a lesser extent $C_nH_{2n-1}^+$, ions with abundance maxima around C_3 or C_4. Chain branching (Figures 3C, 3D, 4A, 8D) causes a decrease in $[M^{\ddot{+}}]$ and characteristic increases in the abundances of $C_nH_{2n+1}^+$ and $C_nH_{2n}^{\ddot{+}}$ ions through cleavage and charge retention at the branched carbon, with the loss of the largest alkyl group favored (reaction 9.1). (An apparent exception is the preferential

$$C_nH_{2n+1}R_2C—CH_2R' \xrightarrow{-e} C_nH_{2n+1}R_2C^{\ddot{+}}CH_2R' \qquad (9.1)$$

$$\xrightarrow{\sigma} C_nH_{2n+1}R_2C^+ + \cdot CH_2R'$$

$$\xrightarrow{rH} C_nH_{2n}CR_2^{\ddot{+}} + CH_3R'$$

loss of the isopropyl radical which can occur from simple 2-methylalkanes, although this does not produce particularly significant peaks in Figures 4A and 8D). Lavanchy *et al.* (1979) show that 80 per cent of $C_nH_{2n+1}^+$ ions in the spectra C_mH_{2m+2} alkanes are formed by direct σ-cleavage for $n \geq 0.5$ m, but that only ~ 15 per cent are formed in this way for $n \leq 0.5$ m. In Figure 9A only the $(CH_3)_3CC(CH_3)_2^+$ (*m/z* 99), $(CH_3)_3C^+$ (*m/z* 57), and CH_3^+ peaks occur by direct σ-cleavage. The rearrangement $C_nH_{2n+1}^+$ peaks that are larger than half the molecular size are obviously less abundant than those of smaller mass, such as $C_2H_5^+$ and $C_3H_7^+$. Peaks resulting from secondary cleavages are even less reliable indicators of structure for these reasons;

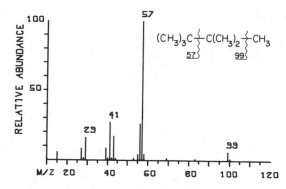

Figure 9A. Mass spectrum of 2,2,3,3-tetramethylbutane. $[M^{\ddot{+}}] = 0.03$ per cent.

nevertheless, the tertiary carbenium ion formed by initial σ-cleavage at a fully substituted carbon can undergo a secondary olefin loss of significant specificity, as in reaction 9.2 (compare Figure 8D).

$$R—CH_2CR_2'—R''^{\ddot{+}} \xrightarrow[-R''\cdot]{\sigma} R—CH_2CR_2'^{+} \xrightarrow{\alpha}$$

$$R^{+} + CH_2{=}CR_2' \tag{9.2}$$

Unknowns. Try the $C_{16}H_{34}$ isomeric unknowns 5.4 to 5.6 again.

Unsaturated aliphatic hydrocarbons. The addition of a double bond to an alkane increases the abundance of the $C_nH_{2n-1}^{+}$ and $C_nH_{2n}^{+}$ ion series, as shown in the spectrum of 1-dodecene (Figure 3E). Note the increasing importance of the $C_nH_{2n+1}^{+}$ alkyl ion series at lower masses. Adding a double bond increases $[M^{\ddot{+}}]$ only for compounds of lower molecular weight. *Cis* and *trans* isomers usually have very similar mass spectra.

Alkene ions exhibit allylic cleavage (reactions 4.15 and 9.3), but also show

$$R'—CH_2—CH{\overset{\cdot+}{—}}CHR \xrightarrow{\alpha}$$

$$R'\cdot + CH_2{=}\overset{+}{C}HR \longleftrightarrow \overset{+}{C}H_2—CH{=}CHR \tag{9.3}$$

a strong tendency to isomerize through migration of the double bond. The mass spectra of alkenes, and especially polyenes, tend to be independent of double-bond position unless the double bond is highly substituted or a number of double bonds act together, as in benzylic cleavage, $C_6H_5CH_2{\text{—}\!\!/\!\!\text{—}}R$. Thus allylic cleavage to produce the $C_5H_9^{+}$ ion in *allo*-ocimene is retarded because the cleaved bond is also vinylic, but is enhanced in myrcene in which the cleaved bond is doubly allylic (9.4).

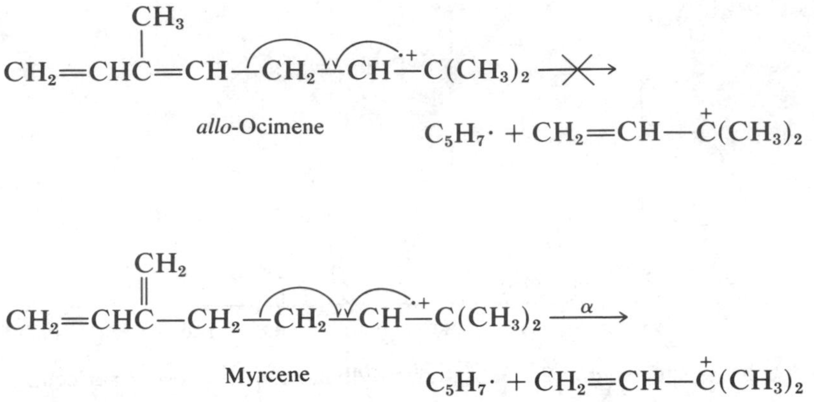

$$CH_2\!\!=\!\!CH\underset{\underset{\displaystyle CH_3}{|}}{C}\!\!=\!\!CH\!\!-\!\!CH_2\!\!-\!\!CH\!\!-\!\!C(CH_3)_2 \longrightarrow\!\!\!\!\times\!\!\!\!\longrightarrow$$

allo-Ocimene $\qquad C_5H_7\cdot + CH_2\!\!=\!\!CH\!\!-\!\!\overset{+}{C}(CH_3)_2$

$$(9.4)$$

$$CH_2\!\!=\!\!CH\underset{\underset{\displaystyle CH_2}{\|}}{C}\!\!-\!\!CH_2\!\!-\!\!CH_2\!\!-\!\!CH\!\!-\!\!C(CH_3)_2 \overset{\alpha}{\longrightarrow}$$

Myrcene $\qquad C_5H_7\cdot + CH_2\!\!=\!\!CH\!\!-\!\!\overset{+}{C}(CH_3)_2$

Henneberg and Schomburg (1968) report that the spectra of branched unsaturated alkenes, $RCH\!\!=\!\!C(CH_3)CH_2R'$ and $RCH_2C(CH_3)\!\!=\!\!CHR'$, show abundant $RCH_2{}^+$ ions which appear to arise by initial migration of the double bond away from the position of branching. In homosqualene, cleavage at the doubly allylic position produces an abundant $(M-83)^+$ ion, but the characteristic $(M-57)^+$ ion involves double-bond migration to the conjugated position, followed by allylic cleavage (9.5).

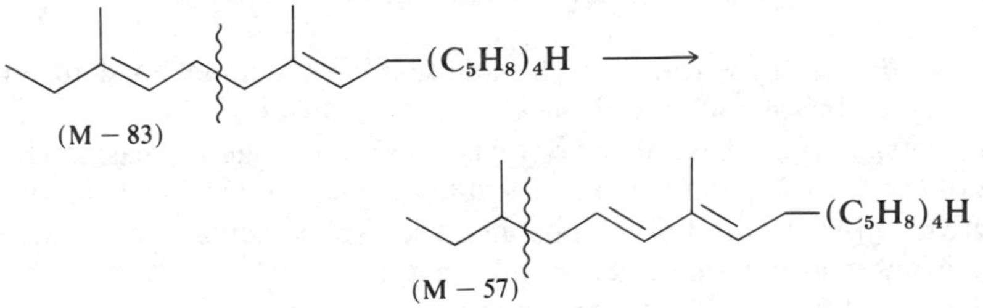

$(M-83)$ $\qquad (C_5H_8)_4H \longrightarrow$

$$(9.5)$$

$(M-57)$ $\qquad (C_5H_8)_4H$

Levsen *et al.* (1978) report that field-ionization mass spectra of alkenes show allylic cleavages which allow unequivocal location of the double bond.

Elimination of an olefin molecule through γ-hydrogen rearrangement with formation of an odd-electron ion (reaction 9.6; Sections 4.9 and 8.9) is found in the spectra of substituted alkenes. Stevenson's Rule is generally applicable, so that charge migration is common. 2-Methyl-1-pentadecene yields a base peak of $C_4H_8{}^{+}$: I (2-butene) = 9.2 eV; I $(C_{10}H_{21}CH\!\!=\!\!CH_2) = \sim 9.4$ eV. Note that the same bond is ruptured in this rearrangement as in simple allylic cleavage. However, ions can be produced in a variety of ways; important OE^+ ions are produced by other mechanisms. The *n*-alkenes yield a significant $C_nH_{2n}{}^{+}$ ion series, and OE^+ ions can be formed from specific skeletal rearrangements as in 9.5.

$$
\text{(9.6)}
$$

In general, the fragmentation behavior of acetylenic (Woodgate *et al.* 1972; Lifshitz 1977) and allenic (Wiersig *et al.* 1977) compounds resembles that of alkenes.

Unknown. You should now be able to distinguish more easily between Unknowns 5.8 and 5.9.

Saturated alicyclic hydrocarbons. Although the molecular ions of cycloalkanes are more abundant than those of their acyclic counterparts, their spectra are often more difficult to interpret. The base peak in the spectrum of 1-methyl-3-*n*-pentylcyclohexane corresponds to the loss of $C_5H_{11}\cdot$. *n*-Decylcyclohexane similarly gives $C_6H_{11}^+$ as the largest peak; $C_6H_{10}^{\ddot{\cdot}}$ is the second largest, similar to the behavior of larger alkanes (reaction 9.1). In the spectrum of *n*-undecylcyclopentane the $C_5H_9^+$ and $C_5H_8^{\ddot{\cdot}}$ are also the largest, but do not dominate the $C_nH_{2n}^{\ddot{\cdot}}$ series. Collisional activation studies on substituted cycloalkanes (Borchers *et al.* 1977) and ion photodissociation studies on unsubstituted cycloalkanes (Benz and Dunbar 1979) indicate that molecular ions of $\geq$ six-membered rings are stable to skeletal isomerization prior to dissociation, but that smaller rings undergo ring-opening to form initially 1-alkene ions. Skeletal decompositions must involve cleavage of at least two bonds, which may contribute to the increased randomization observed with these compounds. In the spectrum of methylcyclohexane (Figure 9B) isotopic labeling shows that the prominent $(M - CH_3)^+$ peak

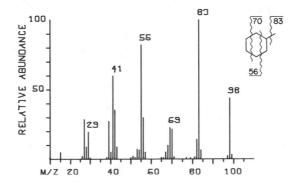

Figure 9B. Mass spectrum of methylcyclohexane.

is due to loss of the methyl substituent. However, only half of the corresponding peak in methylcyclopentane arises in this manner (indicating a greater degree of endocyclic cleavage in the five-membered ring), and cyclohexane itself shows a $(M - CH_3)^+$ ion which is 27 per cent of the base peak. The $(M - C_2H_4)^+$ peak of methylcyclohexane can be rationalized on the basis of initial ionization at the endocyclic branched sigma bond (9.7); labeling indicates that 60 per cent of the peak arises from the elimination of C_2H_4 originating in this position.

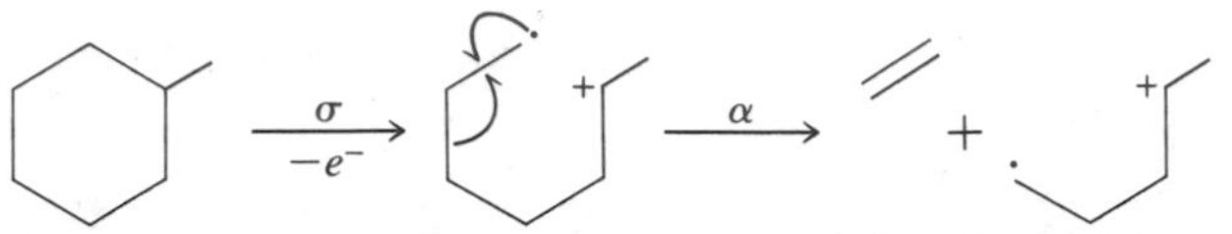

$$(9.7)$$

However, careful spectral correlation can give valuable structural information on such compounds; the detailed studies of Djerassi and his students (Tokes *et al.* 1968) on the steroid hydrocarbon skeleton provides a classic case. For example, in the mass spectrum of 5-α-pregnane (Figure 9C), the $(M - CH_3)^+$ arises from the loss of the angular methyl groups (mainly C19). Loss of C_2H_5 by the exocyclic fragmentation of the 17–20 bond is negligible because of favored 13–17 cleavage of the D ring producing the stable tertiary carbenium ion. Radical-site reactions of this species can proceed either through reciprocal hydrogen transfer (Equation 9.8) and 14–15 cleavage to yield the characteristic $OE^{\ddagger}$ m/z 218 ion, or through electron donation

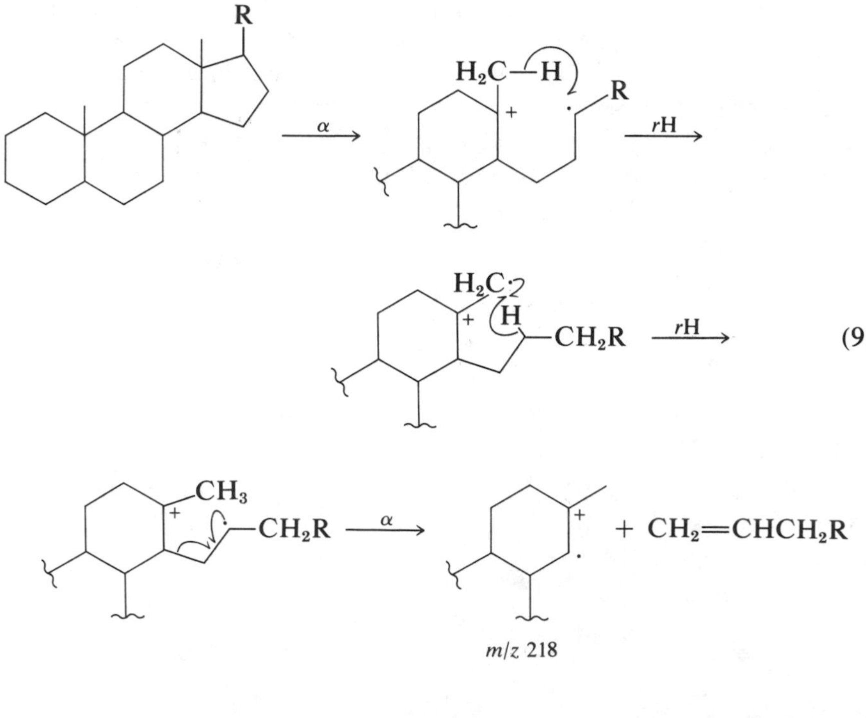

$$(9.8)$$

m/z 218

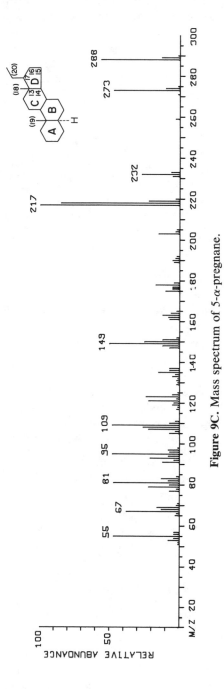

Figure 9C. Mass spectrum of 5-α-pregnane.

183

to the neighboring bond, resulting in the m/z 232 peak through 15–16 cleavage (9.9). A substantial part of the base peak at m/z 217 is due to C19 methyl loss from m/z 232; a possible mechanism is shown (9.9). Finally, the m/z 149 ion

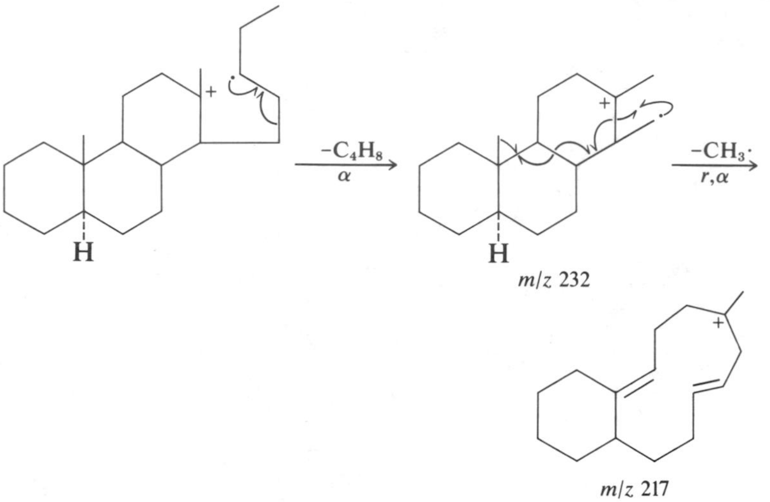

$$m/z\ 232 \tag{9.9}$$

$m/z\ 217$

is composed mainly of rings A and B, but is formed in a complex mechanism operating in conjunction with a triple hydrogen transfer. Despite this apparent degree of randomness, the stereochemistry of the A/B rings has a characteristic effect on this process; in the mass spectrum of the 5-β-(cis)-isomer, the $[m/z\ 149]/[m/z\ 151]$ is much lower than in that of the 5-α-isomer. This effect is relatively independent of the C17 substituent and has been used by Seifert *et al.* (1972) to establish the presence of stereoisomers of C_{22}—C_{24} steroid acids in highly complex petroleum mixtures; animal contribution to the genesis of this petroleum is proposed because the bile acids of mammals are genetically related to the particular steroid acids identified.

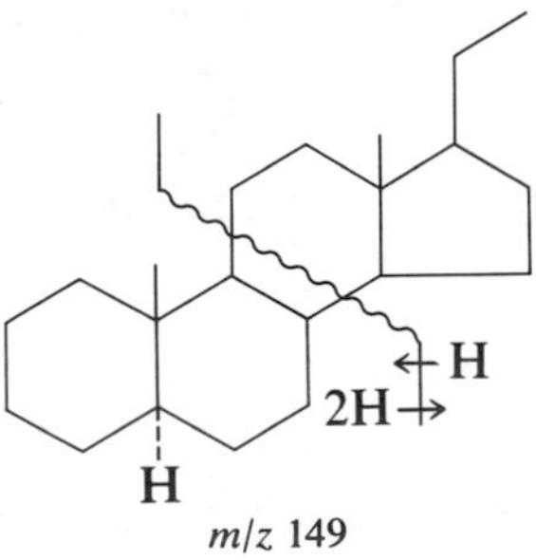

$m/z\ 149$

Unsaturated alicyclic hydrocarbons. For many types of cycloalkenes the position of the double bond has little effect on the mass spectrum. Ten isomers of C_5H_8, which can have no larger than a cyclopentene ring, show

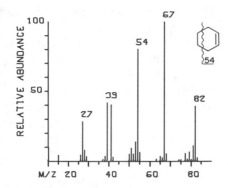

Figure 9D. Mass spectrum of cyclohexene.

similar spectra, as do a variety of C_7H_{10} (three rings + double bonds) isomers and other compounds with a high degree of unsaturation. Monoterpenoid hydrocarbons ($C_{10}H_{18}$, two rings + double bonds) containing a cyclohexane ring and an exterior double bond exhibit spectra which are consistent with initial isomerization to a cyclohexene structure, plus other rearrangements, such as methyl migration.

Cyclohexene derivatives, on the other hand, can undergo quite specific fragmentations, as illustrated by the spectrum of cyclohexene itself (Figure 9D). The base peak can be rationalized by a hydrogen abstraction mechanism transferring the unpaired electron to an allylic radical site (reaction 9.10) followed by methyl loss. The significant odd-electron ion of m/z 54 which arises through a retro-Diels-Alder reaction (4.31, 9.11) is a convenient characteristic of the cyclohexene structure. This reaction produces a diene

$$\text{(9.10)}$$

$$CH_3\cdot + \quad m/z\ 67$$

$$\text{(9.11)}$$

$$m/z\ 54$$

and monoene fragment; as in the case of many hydrogen rearrangement reactions, the products compete for the positive charge. Tetrahydro-cannabinol produces a significant m/z 246 peak corresponding to the ionized monoene through loss of C_5H_8; Reaction 9.12 is one possible rationalization of the electron shifts involved. Note that CH_3 loss from the OE^+ m/z 246 ion gives the base peak, consistent with the aromatic stabilization achieved in this EE^+ product ion. The retro-Diels-Alder reaction is also a common decomposition mode of even-electron ions, so that a substantial part of the m/z 231 peak may be formed by C_5H_8 loss from the $(M - CH_3)^+$ ion.

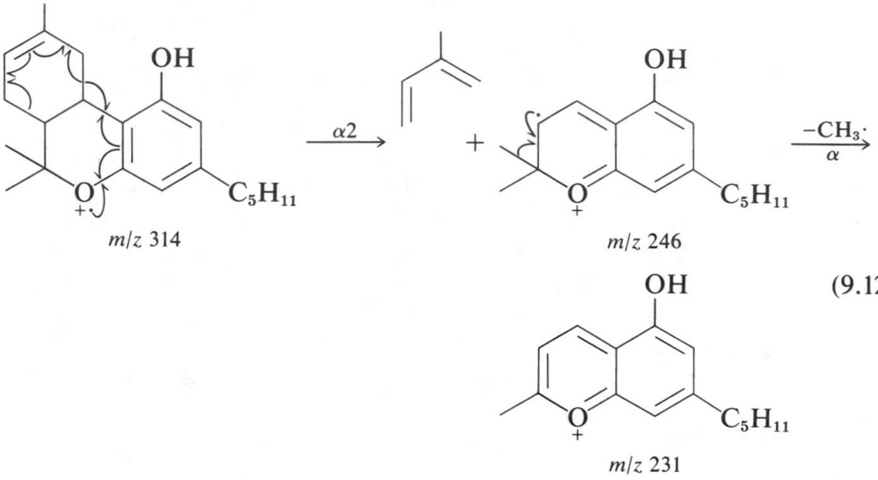

$$ (9.12) $$

Unknowns. The spectra of α- and β-ionone (Unknowns 4.12 and 4.13) illustrate such reactions.

Aromatic hydrocarbons. The spectra of benzene (Unknown 2.5), naphthalene (Unknown 5.1), isoquinoline (Unknown 5.10), *n*-butylbenzene (Unknown 4.18), *n*-hexylbenzene (Figure 3G), and *n*-octylbenzene (Unknown 5.7) are given above. The addition of an aromatic nucleus to an alkane substantially influences the spectrum. Molecular ion peaks are higher than those of the corresponding alkanes, and doubly charged ions are often abundant (1–5 per cent). Aromatic-ring cleavage processes usually require relatively high energies, and are accompanied by a high degree of hydrogen and even skeletal scrambling; this leads to the characteristic "aromatic ion series" (Table A.6). The ions $C_nH_{0.5n}$–$C_nH_{0.8n}$ ($n = 3$ to 6) often predominate if the ring contains electronegative substituents; the series $C_nH_{0.9n}$–$C_nH_{1.2n}$ is more typical of the presence of electron-donating substituents or heterocyclic compounds. Also the *ring* position of alkyl substitution usually has little influence on the spectrum (for example, *o*-, *m*-, and *p*-xylene have very similar mass spectra).

Another characteristic homologous ion series of phenylalkanes can be seen in Figure 3G at masses corresponding to $C_6H_5(CH_2)_n^+$ (m/z 77, 91, 105, 119, . . .); the abundances of the individual ions are much more dependent on the structure of the molecule than those of the low-mass ion series. However, ions stabilized by the aromatic nucleus can be formed through a variety of fragmentation pathways, so that the presence of a large peak, such as m/z 91, only signifies that the molecule contains one or more of several possible structural features.

The most characteristic ion of this series in Figure 3G corresponds to rupture of the benzylic bond of the largest alkyl group; thus 1-phenylalkanes exhibit an abundant $C_7H_7^+$ ion (Reaction 9.13). The resonance-stabilized

$$m/z\ 91 \tag{9.13}$$

benzyl ion is depicted as the $C_7H_7^+$ product, but in particular cases the similarly stabilized tropylium ion has been shown to be the chief product (for most interpretive mechanisms it is not important to make this distinction). For example, toluene molecular ions of energies sufficient to lose H· can undergo isomerization to the less-stable cycloheptatriene molecular ion, which by H loss forms the tropylium ion. Higher-energy toluene ions have a higher tendency to lose H· before such isomerization to form benzyl ions, but benzyl ions ($\Delta H_f = 9.2$ eV) of sufficiently high internal energy can isomerize (critical energy 2 eV) to the more stable tropylium ions ($\Delta H_f = 9.0$ eV). Thus from toluene or cycloheptatriene the $C_7H_7^+$ isomers formed are a critical function of the ionizing electron energy; at threshold energies pure tropylium ions are formed, changing to a minimum of 1:1 [tropylium]/[benzyl] at 15 eV, but 2:1 at 70 eV (McLafferty and Bockoff 1979; earlier work is reviewed by Grubb and Meyerson 1963). Such rearrangements in polyalkylaromatics can account for some competing cleavage at the ring-alkyl bond, such as for the prominent $(M - CH_3)^+$ peak in dimethylbenzenes. Olefin elimination from the primary benzyl ion product can also give $C_6H_5(CH_2)_n^+$ ions of significant abundance (Reaction 9.14). (It has been postulated that the m/z 119 ion of 9.14 decomposes through a "phenylated cyclopropane" intermediate; in the α-^{13}C-labeled analog, 65 per cent of the ^{13}C is lost in forming $C_7H_7^+$.)

$$m/z\ 119 \tag{9.14}$$

When the alkyl side chain is propyl or larger, hydrogen rearrangement through a six-membered-ring transition state can produce the characteristic $OE^{+\cdot}$ ion (reaction 4.36, 9.15). Substituents in the α-position tend to decrease this rearrangement in favor of benzylic cleavage (9.13).

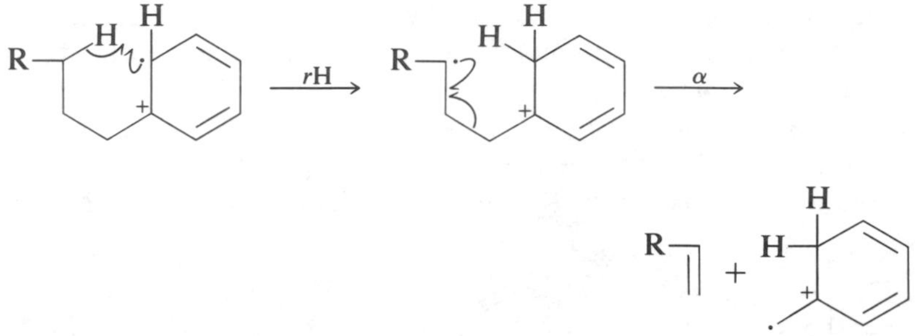

$$(9.15)$$

Note that the *same* bond is ruptured in Reaction 9.15 as in simple benzylic cleavage (Reaction 9.13), which is not the case in the analogous reactions of polar unsaturated functionalities (Equations 4.33 and 4.14). In some cases the ratio of these competing reactions can be rationalized in terms of the relative charge stabilization in their initiating canonical forms. The relative abundances of the corresponding rearrangement ion, $C_8H_{10}^{+\cdot}$, in the mass spectra of m-tolyl- and p-tolyl-1-propane are 5.5 and 0.9 per cent, respectively, compared to the abundances of the ion from benzylic cleavage, $C_8H_9^+$. The rearrangement is unfavorable when either the γ-position or the ortho-positions are completely substituted, the latter presumably because of steric interference (Equation 8.8).

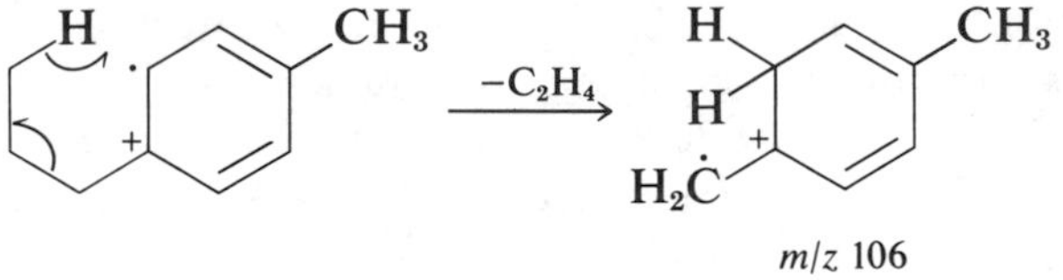

m/z 106

$$(9.16)$$

m/z 105

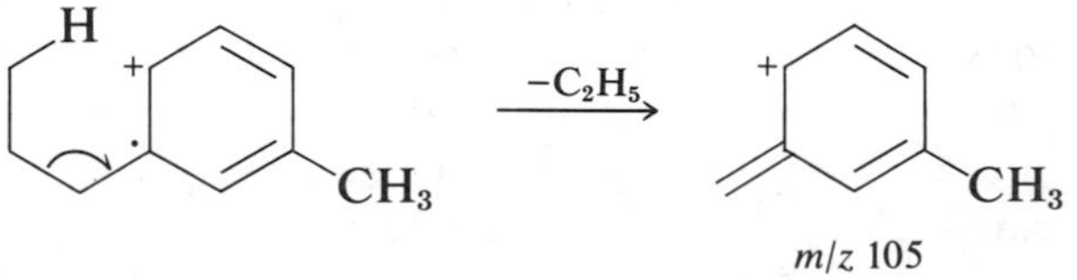

$$(9.17)$$

Polyaryl and polycyclic aromatic compounds. The behavior of higher aromatic hydrocarbons can be viewed as an extrapolation of that of alkylbenzenes described above. The molecular and doubly charged ions are of higher abundance. Similar characteristic alkyl side-chain cleavages occur; decomposition in the unsaturated system is subject to severe hydrogen and skeletal scrambling.

Unknowns. Try Unknowns 4.18 and 5.7 again.

9.2 Alcohols

Saturated aliphatic alcohols. Mass spectra have been given above for methanol (Unknown 1.3), isopropanol (Figure 6C), 3-methyl-3-hexanol (Figure 4B), 4-octanol (Unknown 4.19), 1-dodecanol (Figure 3H), 1-hexadecanol (Unknown 5.8), glycerol (Unknown 8.11), 2-aminoethanol (Unknown 4.2), and α-(aminomethyl-1-ethyl)-benzyl alcohol (Figure 4E).

Addition of a hydroxyl group to an alkane lowers the ionization energy, but the molecular ion abundance decreases (and often is not observable, as in the spectrum of Figure 9E) despite this stabilization because of the

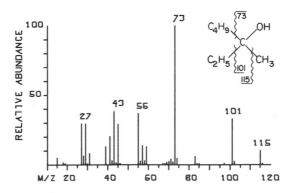

Figure 9E. Mass spectrum of 3-methyl-3-heptanol.

increased ease of decomposition through reactions initiated at the ionized hydroxyl group. Offsetting this lack of M^+, alcohols do undergo ion-molecule reactions to give a pressure-dependent $(M + H)^+$ peak which is useful for deducing molecular weight (Section 6.1). Thermal and catalytic dehydrogenation and dehydration of the sample, especially in metal reservoir systems, can give spurious peaks such as $(M - 2)^{+\cdot}$, $(M - 18)^{+\cdot}$, and $(M - 20)^{+\cdot}$, although these can also arise through electron-impact-induced fragmentation; the presence of the corresponding CA or metastable ion product (Section 6.4) provides a convenient proof of the latter. A substantial peak at m/z 18 has little meaning because of the high possibility of the presence of H_2O from desorption in the inlet system or as an impurity.

Both the radical site and the charge site of the hydroxyl group are intermediate in their capability for reaction initiation, so that mass spectra of alcohols exhibit many of the types of reactions outlined in Chapters 4 and 8. Tertiary alcohols contain the largest, and primary alcohols the smallest, total abundance of oxygen-containing ions in their spectra. In all but the spectra of 1-alkanols, α-cleavage (Reaction 4.17) is the most useful characteristic reaction. In the spectrum of 3-methyl-3-heptanol (Figure 9E) this produces the important peaks at m/z 73, 101, and 115 (9.18), whose abundance reflects the size of the alkyl groups lost. Formation of such $C_nH_{2n+1}O^+$ ions through the cyclization-displacement reaction (8.65) should be of little importance.

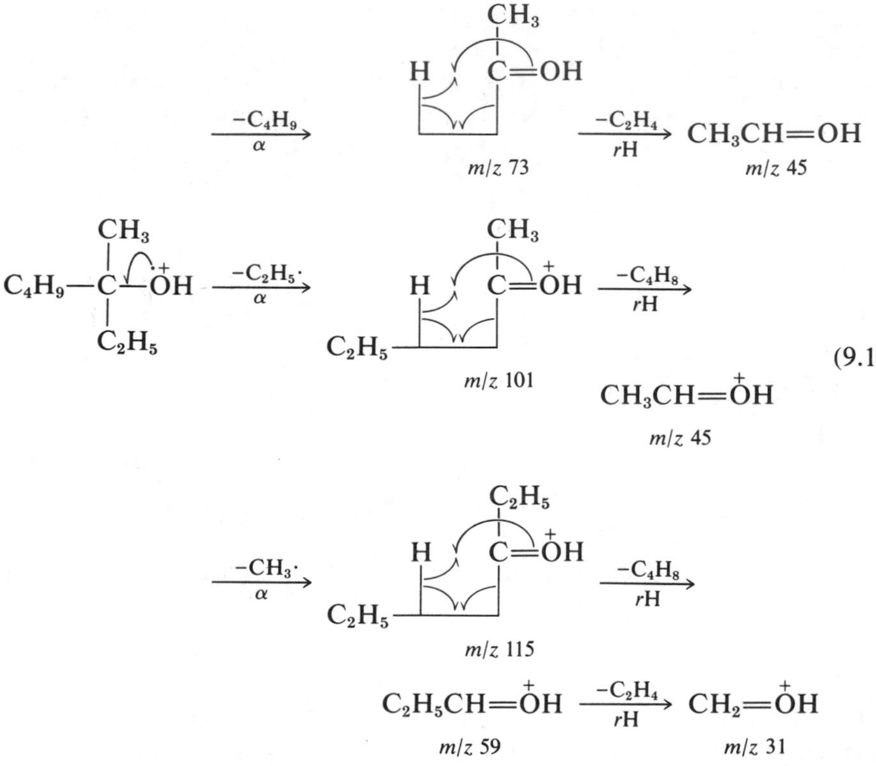

(9.18)

There are a number of possible paths for further decomposition of these EE^+ ions. (The only simple cleavage which could yield another EE^+ ion, Reaction 4.26, gives only H^+.) Rearrangement of hydrogen to the carbon atom of the oxonium double bond to eliminate C_nH_{2n} (Equation 8.57) can account for the m/z 31, 45, and 59 peaks; the most abundant of these is that predicted for the loss of the largest C_nH_{2n} molecule from the most abundant primary product ion. Further, rearrangement 8.54 could also yield m/z 59 and 73 ions.

The main competitive decomposition of the EE^+ $C_nH_{2n+1}O^+$ ions is dehydration, which contributes to the formation of the m/z 55, 83, and 97 peaks (Reaction 4.45). Part of the abundant m/z 29 and 43 peaks are due to CHO^+ and $C_2H_3O^+$, which also may arise from decomposition of the $C_nH_{2n+1}O^+$ ions; obviously, care should be exercised in using products of secondary EE^+ ion reactions, such as these and those of 9.18, as evidence of structure.

The other major decomposition pathway possible for alcohol molecular ions leads to the loss of H_2O and of $(H_2O + C_nH_{2n})$ $(n \geq 2)$, Reaction 4.38. As can be seen from the abundances of the predicted m/z 112 and 42 peaks, this reaction is less competitive in branched alcohols. In contrast, most of the spectra of straight-chain alcohols, such as 1-dodecanol (Figure 3H), show products which would result from such a primary dehydration of the molecular ion. The $C_nH_{2n+1}O^+$ ions which would arise from simple α-cleavage (m/z 31) and displacement (Reaction 8.65) are minor. The rest of the spectrum closely resembles that of the corresponding alkene, as can be seen by comparing the spectra of 1-hexadecanol and 1-hexadecene (Unknowns 5.8 and 5.9). The prominent $C_nH_{2n-1}^+$, $C_nH_{2n}^{+\cdot}$, and $C_nH_{2n+1}^+$ ion series of 1-alkanols (which could arise in part directly from $M^{+\cdot}$) can give evidence of branching or other structural features of the carbon skeleton; this fragmentation behavior generally follows that outlined for the corresponding hydrocarbons.

Unknowns. For review again try Unknowns 4.19, 5.8, 5.9, 8.3, and 8.11.

Cyclic aliphatic alcohols. The spectrum of 2-methylcyclohexanol is shown in Figure 9F. Alpha cleavage (Reaction 4.14) is favored at the more substituted ring bond (9.19 > 9.20). Further reaction through olefin elimination gives characteristic $OE^{+\cdot}$ ions at m/z 44, 58, 72, and 86. However, this reaction is not as favorable as hydrogen abstraction, which moves the radical site to an

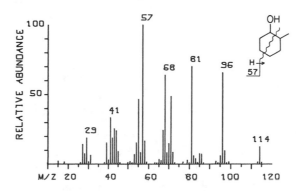

Figure 9F. Mass spectrum of 2-methylcyclohexanol.

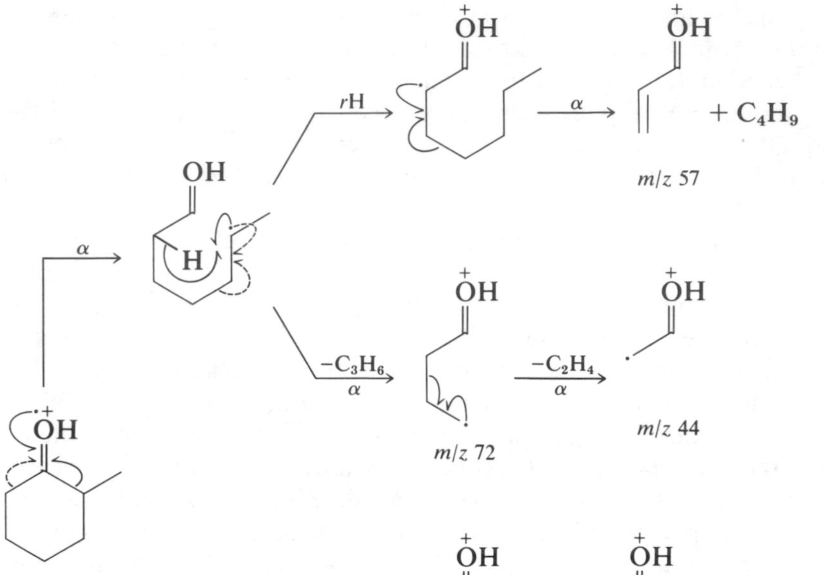

$$(9.19)$$

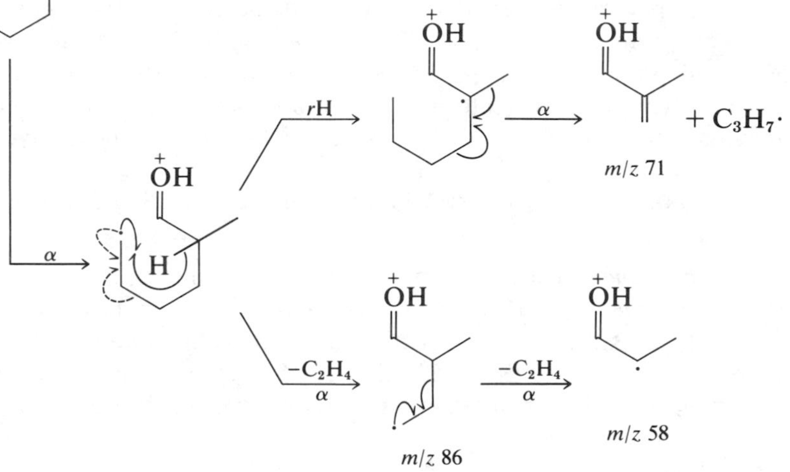

$$(9.20)$$

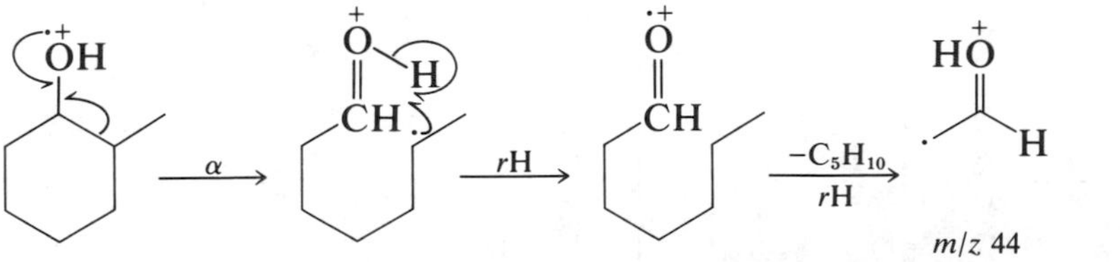

$$(9.21)$$

allylic position (Section 8.8). Elimination of an alkyl radical gives the stable $C_3H_5O^+$ (9.19) or $C_4H_7O^+$ (9.20) ions. A minor process (Reaction 9.21) indicated by labeling studies involves abstraction of the hydroxyl hydrogen (a favored type of hydrogen for rearrangement) by the initial radical product to give the heptanal molecular ion; further decomposition of this and the analogously formed 2-methylhexanal will also contribute to peaks such as m/z 44 and 58.

As in the mass spectra of acyclic alcohols, loss of H_2O (Reaction 4.38) provides the other major pathway for $M^{\ddot{+}}$ decomposition. This largely involves the hydrogen atoms attached to carbons 3, 4, and 5; transfer of 4-H (Reaction 8.9) involves *cis*-stereochemistry of the eliminated groups in the boat-form transition state. The spectrum's hydrocarbon-type ions are consistent with those expected from further decomposition of such bicyclic dehydration products; the m/z 68 and 54 $OE^{\ddot{+}}$ ions would represent elimination of a bridging group as the stable C_2H_4 or C_3H_6 (note that the "rule" of loss of the largest alkyl *radical* does not apply to the elimination of molecules; in this case the $C_5H_8^{\ddot{+}}$ ion should be more stable and thus more abundant than the alternative $C_4H_6^{\ddot{+}}$ ion).

Phenols. The spectrum of *o*-methylphenol (Figure 9G) illustrates a number of fragmentations characteristic of the hydroxyl group on an aromatic ring. The odd-electron $(M - CO)^{\ddot{+}}$ peak, which is usually accompanied by $(M - CHO)^+$, is especially useful in recognizing this functionality. The apparent mechanism (9.22) involves elimination of oxygen with its adjacent ring carbon; deuterium-labeling studies indicate that CHO loss is accompanied by substantial hydrogen scrambling.

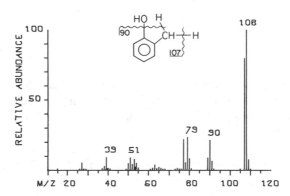

Figure 9G. Mass spectrum of *o*-methylphenol.

The characteristic $OE^{\ddot{+}}$ ion at $(M - 18)^{\ddot{+}}$ in this spectrum is an example of the "ortho effect" (9.23; see Reaction 4.41); the abundances of the $(M - 18)^{\ddot{+}}$ peak in the corresponding *meta* and *para* isomers are only 3 and 2 per cent, respectively. The larger $(M - 1)^+$ peak arises from the loss of the benzylic hydrogen atom.

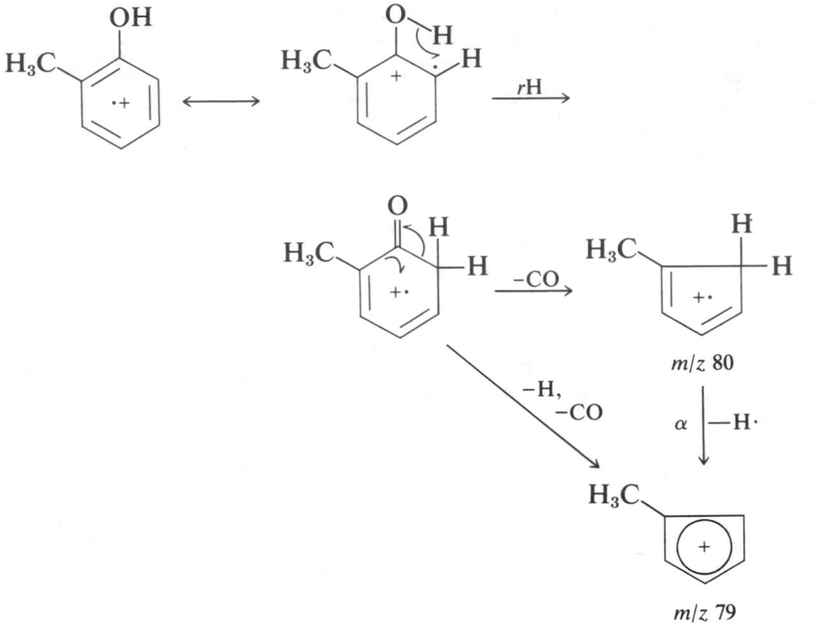

(9.22)

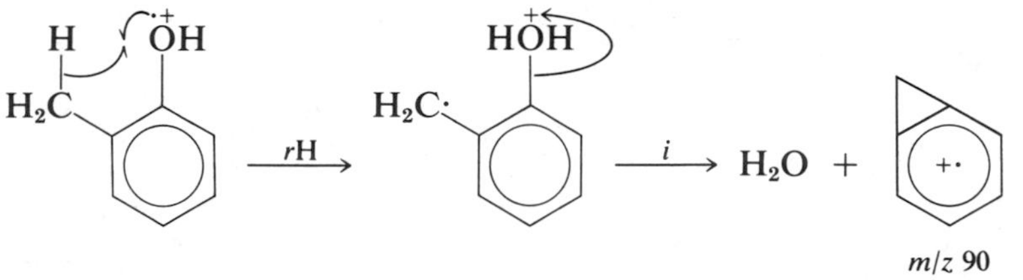

m/z 80

m/z 79

(9.23)

m/z 90

Alcohols containing other functional groups. The hydroxyl hydrogen exhibits a high tendency for transfer to a radical site. In a phenol the hydroxyl hydrogen atom can thus act as a ready donor for an ortho effect rearrangement. The base peak of methyl salicylate is due to the loss of methanol (9.24), but the corresponding abundances for the *meta* and *para* isomers are <1 per cent. Isomeric structures other than those shown are possible for the product ion of 9.23 and 9.24. Can you utilize these correlations for Unknown 9.1?

(9.24)

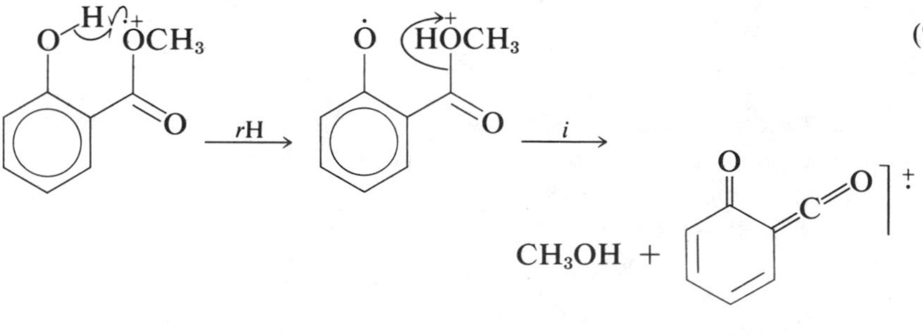

$CH_3OH\ +$

Unknown 9.1

m/z	Rel. abund.	m/z	Rel. abund.	m/z	Rel. abund.
27	3.2	77	1.6	156	0.3
28	1.7	78	0.6	157	2.8
29	5.5	91	1.4	158	0.3
38	3.2	92	1.9	172	98.
39	6.1	93	17.	173	6.7
43	0.4	94	1.1	174	100.
50	4.0	117	2.2	175	6.5
51	1.3	118	0.3	176	0.5
53	1.3	119	2.2	185	0.3
55	0.3	120	0.7	187	0.3
63	7.8	129	0.3	200	42.
64	4.5	131	0.3	201	3.7
65	17.	143	4.0	202	42.
66	1.2	144	0.7	203	3.8
74	2.1	145	4.0	204	0.2
75	3.9	146	0.7		
76	3.5	155	2.7		

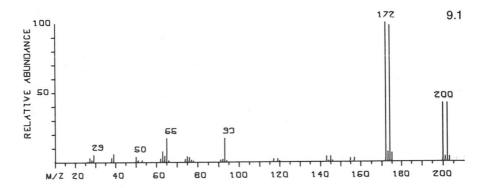

9.3 Aldehydes and ketones

Aliphatic aldehydes and ketones. The mass spectra of formaldehyde, 3-penta-none, 3-methyl-2-butanone, 4-methyl-2-pentanone, 3-methyl-2-pentanone, and 2,3-butanedione have been given as Unknowns 1.9, 4.9, 4.10, 4.16, 4.17, and 5.2; those of 2- and 6-dodecanone are Figures 3J and 3K; those of 2-ethylhexanal, 6-methyl-2-heptanone, and 6-methyl-5-heptene-2-one are shown in Figures 9H to 9J.

Addition of a carbonyl group to an alkane lowers the ionization energy substantially; typical values (Table A.3) of 9.4 to 9.8 eV are below those of the corresponding aliphatic alcohols. The molecular ion of even larger aldehydes and ketones with some degree of chain branching is usually of observable abundance.

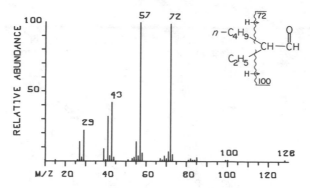

Figure 9H. Mass spectrum of 2-ethylhexanal.

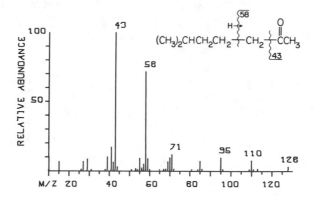

Figure 9I. Mass spectrum of 6-methyl-2-heptanone.

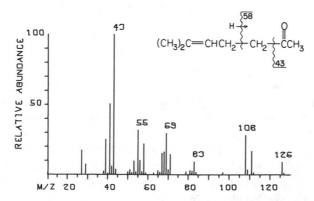

Figure 9J. Mass spectrum of 6-methyl-5-heptene-2-one.

The characteristic peaks in the spectra of carbonyl compounds provide examples of many of the major types of fragmentation pathways discussed in Chapters 4 and 8. Many of these are utilized in Reactions 9.25 and 9.26 to rationalize the formation of the characteristic peaks in 2-ethylhexanal (Figure 9H) and 6-methyl-2-heptanone (Figure 9I). The α-cleavage reaction is generally much less important for aldehydes than for ketones. The CHO^+ (m/z 29) peak is prominent only for smaller aldehydes and those with highly electronegative functionalities, such as C_3F_7CHO. The alternative α-cleavage yielding $(M - H)^+$ appears to be appreciable only when the RCO^+ ion which is formed is substantially stabilized, as for aromatic aldehydes. Alpha cleavage is much more prominent in aliphatic ketones, yielding a pair of acylium ions (m/z 113 and 43 in Figure 9I) and a pair of alkyl ions (m/z 85 and 15). The more abundant acylium ion is usually formed by the loss of the larger alkyl group, whereas the more abundant alkyl ion is usually the more stable.

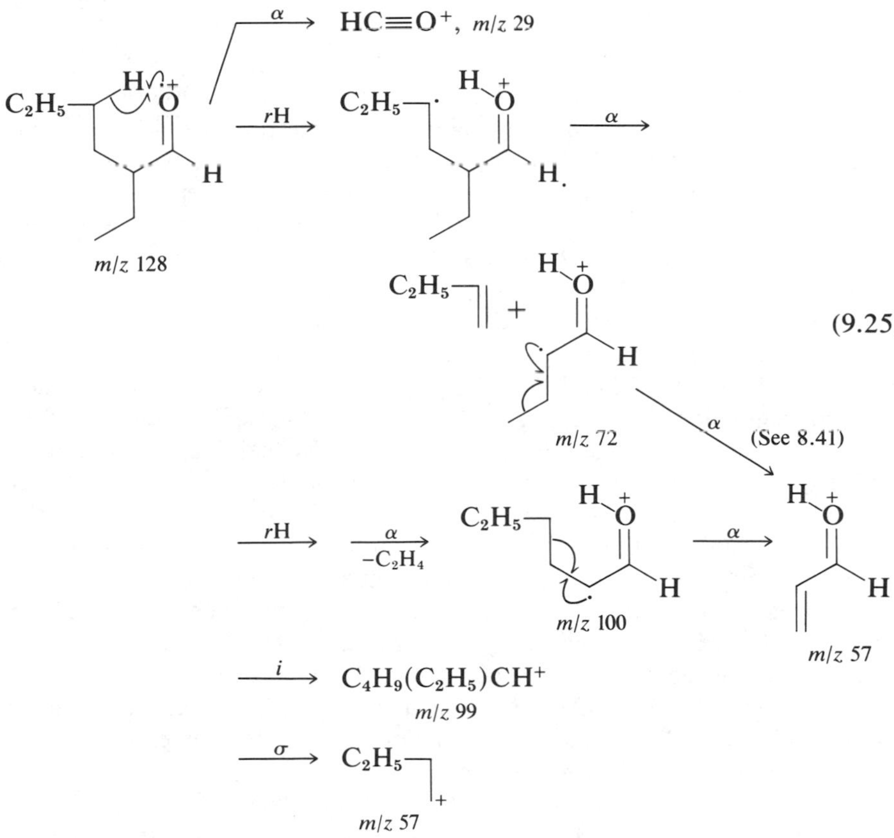

$$HC\equiv O^+, \ m/z \ 29$$

(9.25)

m/z 128

m/z 72   (See 8.41)

m/z 100

m/z 57

$$\xrightarrow{i} C_4H_9(C_2H_5)CH^+$$
$$m/z \ 99$$

m/z 57

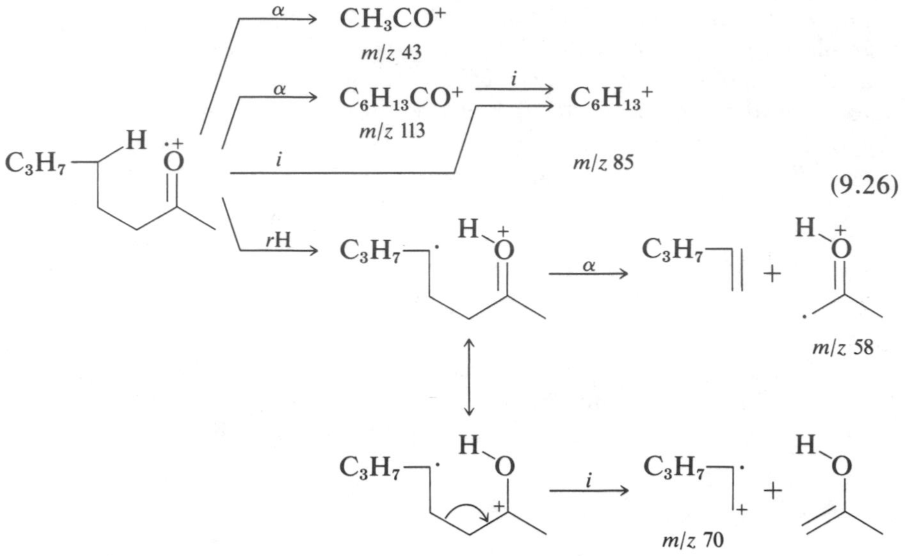

$$(9.26)$$

Ions of the general formula $C_nH_{2n+1}CO^+$ can also arise, however, through cleavages at other bonds. "Reciprocal hydrogen rearrangement" (Kraft and Spiteller 1969; Reaction 8.49) may provide an important driving force for this. Aldehydes and ketones show the characteristic ion series 15, 29, 43, 57, ... due to both $C_nH_{2n+1}CO^+$ and $C_nH_{2n+1}^+$ ions. Although these are best distinguished by high-resolution data, the absence of a particular $C_nH_{2n+1}^+$ ion can sometimes be demonstrated by a low $[(A+1)^+]/[A^+]$ value (Section 2.3).

Initial rearrangement of a γ-hydrogen atom followed by β-cleavage can give an enol ion (m/z 72 and 100 in Figure 9H and m/z 58 in Figure 9I) and its complementary alkene ion, reactions 4.33 and 4.34. The relative abundances of these competitive products are reflected by their ionization energies (Table A.3); the reaction is much less specific for aldehydes. Further α-cleavage of either $OE^{\ddagger}$ enol ion from 2-ethylhexanal will produce the resonance-stabilized $C_3H_5O^+$ ion. Comparison of the products of the α-cleavage and the β-cleavage reactions thus provides evidence of substituents at the α-position(s). If a ketone contains γ-hydrogen atoms in both chains, both hydrogen atoms can be rearranged with losses of both olefin molecules ("consecutive rearrangements," 8.52).

Ions corresponding to $(M - C_2H_4)^{\ddagger}$ have also been reported for larger n-alkyl aldehydes. Both aliphatic aldehydes and ketones can exhibit anomalous $(M - 18)^{\ddagger}$ peaks from the loss of water.

Some novel ketone decompositions have been observed for low energy, particularly metastable, ion decompositions; these can be less important for low-molecular-weight species, possibly because for these few molecular ions of appropriate internal energies are formed (Figure 7F). Products of $R'CH_2COCR_3^+$ decomposition include $CH_3C(=O^+H)CR_3$ from double-H rearrangement (Section 8.9), $(M - HCR_3)^+$ from alkane loss (Litton *et al.* 1976), and the displacement Reaction 9.27 (McLafferty *et al.* 1969).

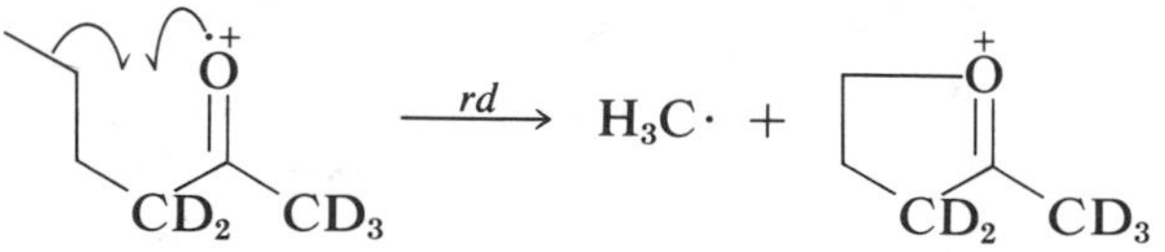

$$ \xrightarrow{rd} \quad H_3C\cdot \; + \qquad\qquad \tag{9.27}$$

Unsaturated ketones. If the γ-hydrogen is vinylic, this can drastically reduce the formation of ions from hydrogen rearrangement and subsequent β-cleavage. Thus these OE^+ peaks are of very low abundance in 4-methyl-6-hepten-3-one and 4-methyl-4-penten-2-one (9.28). However, in 6-methyl-5-hepten-2-one (Figure 9J), isomerization of the double bond to a more favorable position is possible (Reaction 9.28), and the $C_3H_6O^+$ and $C_5H_8^+$ rearrangement ions are observed in fair abundance, although $[C_3H_6O^+]$ is less than that in the saturated derivative. Paralleling the behavior of alkenes (Reaction 9.5), double-bond isomerization apparently has not occurred in all M^+; the m/z 69 ion, which would be the product of allylic cleavage, is more abundant than the α-cleavage m/z 83 ion (Dias *et al.* 1972).

In certain cases displacement reactions (Section 8.10) appear to be possible at the carbonyl group; such a reaction (9.29) accounts for the base peak corresponding to $(M - CH_3)^+$ in the spectrum of β-ionone (Unknown 4.13).

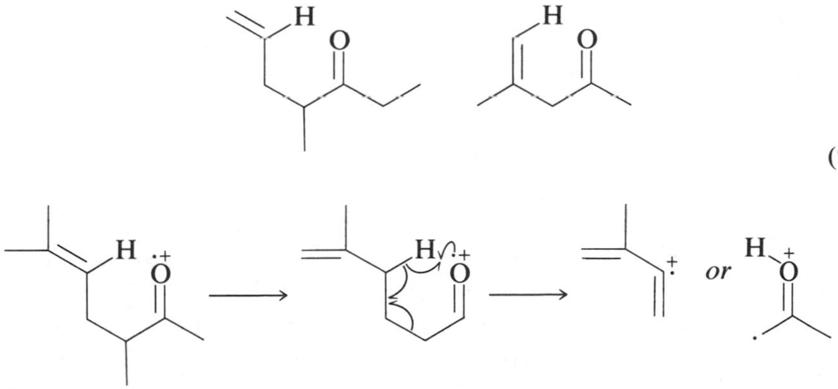

$$(9.28)$$

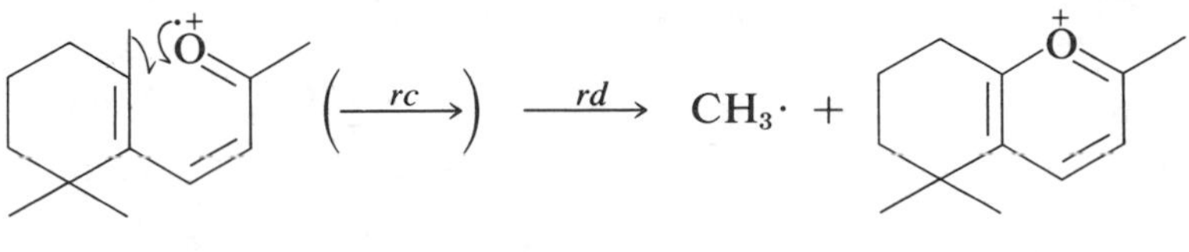

$$ \left(\xrightarrow{rc} \right) \quad \xrightarrow{rd} \quad CH_3\cdot \; + \qquad\qquad \tag{9.29}$$

Alicyclic ketones. The mass spectrum of 3,3,5-trimethylcyclohexanone (Figure 8F) illustrates major additional fragmentation pathways found for cyclic ketones (Reaction 9.30); these are similar to those discussed for cyclic alcohols and in Section 8.8.

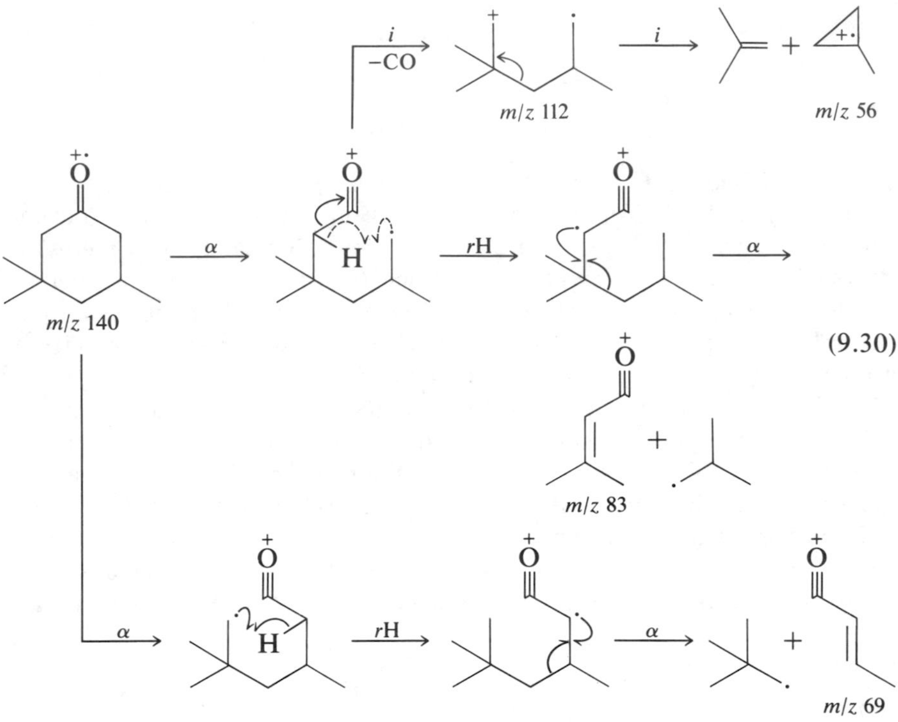

(9.30)

Aromatic aldehydes and ketones. Fragmentations in these spectra (see salicylaldehyde, Unknown 8.5; *p*-nitroacetophenone, Unknown 8.7; 5-phenyl-2-pentanone, Figure 4I; benzophenone, Unknown 5.3; and 3,6-dichlorofluorenone, Unknown 8.8) in general follow expected pathways; for example, benzaldehydes show characteristic $(M-H)^+$ and $(M-CHO)^+$ ions. Loss of CO can occur (Table 8.2), especially if the carbonyl group is endocyclic, such as in anthraquinone. The abundances of RCO^+ ions from $YC_6H_4COR^{\ddagger}$ have been shown to obey linear free energy $(\sigma\rho)$ correlations, although a number of substituent-affected factors control $[RCO^+]$ (McLafferty *et al.* 1970).

Unknowns. A variety of such compounds have been given as unknowns: 1.9, 4.9, 4.10, 4.12, 4.13, 4.16, 4.17, 5.2, 5.3, 8.5, 8.7, and 8.8.

9.4 Esters

Aliphatic methyl esters. The classic studies of Stenhagen (1972) and his co-workers in the 1950s on the mass spectra of fatty acid esters was probably the first clear demonstration of the value of mass spectrometry for deducing the structure of complex natural product molecules. The mass spectrum of methyl *n*-undecanoate is shown in Figure 3M; Figure 9K is methyl 3,7,11,15-tetramethylhexadecanoate (from butterfat). Molecular ions of appreciable abundance are observable even for methyl esters of high-molecular-weight *n*-aliphatic acids; $[M^{\ddot{+}}]$ actually increases at higher molecular weights ($>C_6$), although it decreases with chain branching.

Both the carbonyl oxygen and the saturated oxygen atom of esters can act as sites for reaction initiation to give reaction paths expected from the separate behavior of these functionalities and, in addition, new reactions apparently due to their combined effects.

Methyl esters which do not contain an additional functionality show the expected reactions of the carbonyl group—α-cleavage (Reaction 9.31) and β-cleavage with γ-hydrogen rearrangement (Reaction 9.32). For methyl *n*-alkanoates, such as Figure 3M, α-cleavage gives characteristic $(M-31)^+$ and 59^+ ions whose abundances decrease with increasing molecular weight. The $(M-59)^+$ alkyl ion and the CH_3O^+ ion are very small except for low-molecular-weight esters. Further decomposition of the alkyl moiety produces the characteristic $C_nH_{2n+1}{}^+$ and $C_nH_{2n-1}{}^+$ ion series.

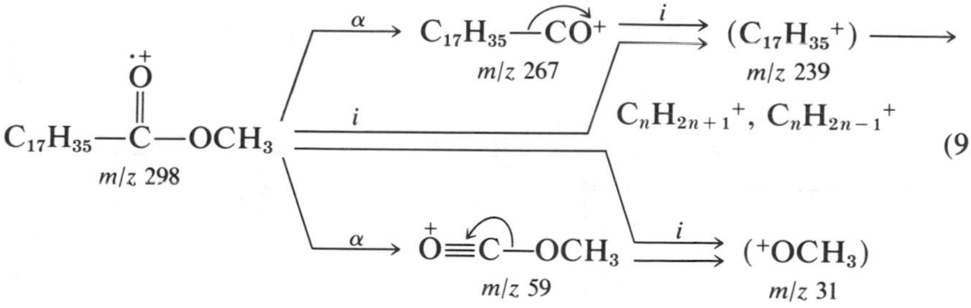

$$(9.31)$$

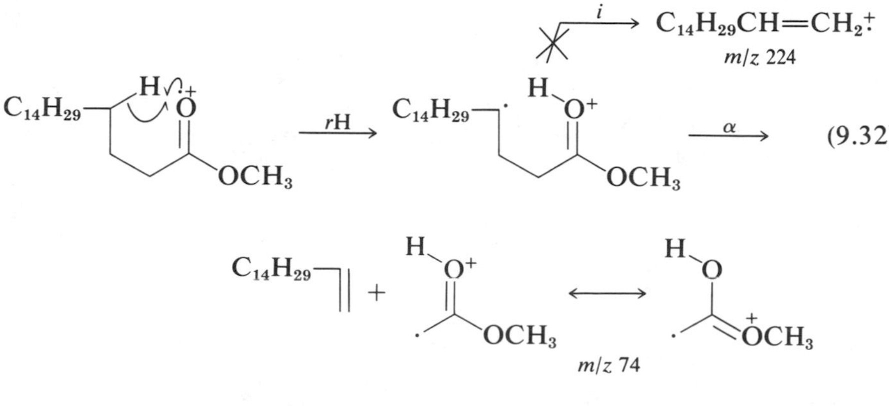

$$(9.32)$$

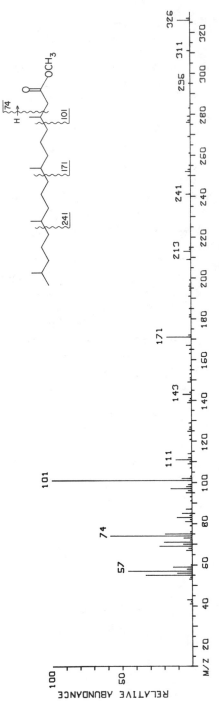

Figure 9K. Mass spectrum of 3,7,11,15-tetramethylhexadecanoate.

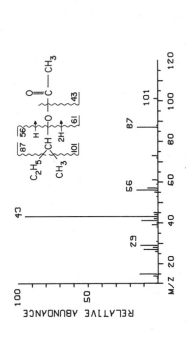

Figure 9L. Mass spectrum of *sec*-butyl acetate. The M⁺ peak at *m/z* 116 has an abundance of 0.1 per cent.

Chain branching (Figure 9K) produces characteristic changes in the $CH_3OCO(CH_2)_n^+$ ion series, m/z 59, (73), 87, 101, . . . , through increased cleavage at the tertiary carbon. However, the straight-chain alkanoates show an unusual periodicity in the $CH_3OCO(CH_2)_n^+$ ion series (quite different from the regular abundance decrease with increasing molecular weight found for the $C_nH_{2n+1}^+$ series of n-alkanes) giving unusually high abundances for m/z 87, 143, 199, 255, . . . ($n = 1, 5, 9, 13, . . .$). The first arises through initial migration of the δ- or ϵ-hydrogen, with reciprocal hydrogen transfer to cleave the bond between the β- and γ-carbon atoms (Reaction 8.49). The peaks representing other peak maxima in the series separated by four chain atoms (m/z 87 and 143 in Figure 3M; 101, 171, 241, and 311 in Figure 9K, reinforced by the chain branching) probably result from displacement reactions (Figure 8I). Another unusual rearrangement yielding $C_nH_{2n+1}CO^+$ ions identified by Stenhagen and his co-workers involves the loss of the α-, β-, and γ-carbons as $C_3H_7 \cdot$. Chain cleavage yielding the alkyl ion can be significant at fully substituted carbons; the spectrum of n-$C_{18}H_{37}C(CH_3)_2$—CH_2COOCH_3 shows a $C_{21}H_{43}^+$ peak of 60 per cent abundance (m/z 115, 100 per cent; m/z 74, 75 per cent). There is also some tendency for a double hydrogen rearrangement in methyl esters of longer-chain acids, here yielding m/z 75 (see Section 8.9). If an R_α group is ethyl or larger, secondary β-bond cleavage can give a conjugated EE^+ ion (9.33; Equation 8.51).

$$R-CH_2CR_\alpha'C(OH)-\overset{+}{O}R \xrightarrow{\quad \alpha \quad}$$

$$\text{(9.33)}$$

$$R \cdot + CH_2{=}CR_\alpha'C(OH){=}\overset{+}{O}R$$

If the size of the alcoholic alkyl group R is indicated by the $(M - OR)^+$ and $ROCO^+$ ions, the mass of the rearrangement ion $CR_\alpha R_\alpha' C(OH)OR^{+}$ thus defines the size of the α-substituents. The reciprocal hydrogen rearrangement cleaving the β—γ bond then shows the β-substituents by difference. In Figure 9K the m/z 59 and 295 ions indicate R—$COOCH_3$; m/z 74 (and 75) indicate R'—CH_2—$COOCH_3$, and m/z 101 then indicates R''—$CHCH_3$—CH_2—$COOCH_3$.

Ethyl and higher esters. The molecular ion peak of $RCOOR'$ generally becomes relatively small when R' is larger than butyl (n-decyl acetate, Figure 3N). The spectrum of *sec*-butyl acetate (Figure 9L) illustrates the additional reactions that are possible with such esters. Rearrangement of one (Reaction 4.33) and two (Reaction 4.46) hydrogen atoms is highly characteristic (Reaction 9.34). For the former reaction, alkene ion formation predominates for propyl and higher esters, consistent with the ionization energies of acids and alkenes. Although the double hydrogen rearrangement produces an even-electron ion of only moderate abundance, you should be able

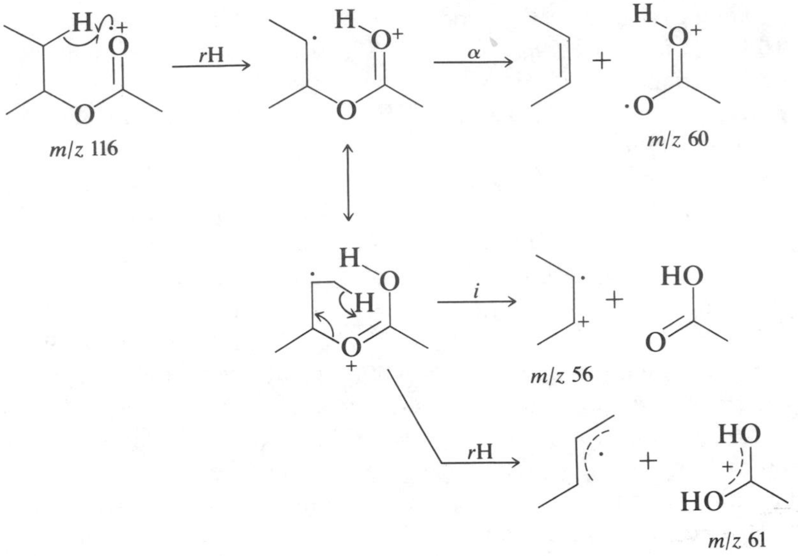

(9.34)

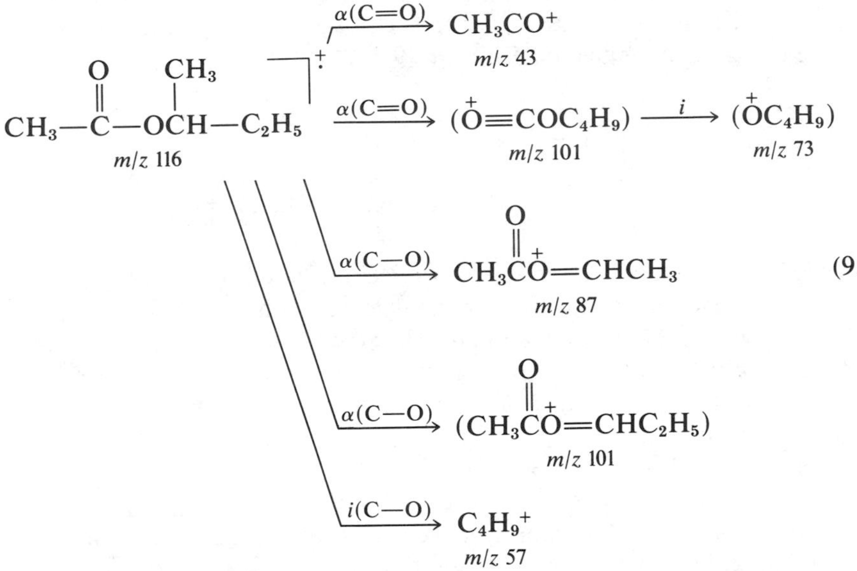

(9.35)

to recognize it from its low rings-plus-double-bonds value and because it occurs in a rather unusual ion series (47, 61, 75, . . .).

Alpha cleavage (9.35) at the carbonyl group yields the large m/z 43 ion; the alternative $(M - CH_3)^+$ peak at m/z 101 is small. This latter peak actually can also be produced by an additional α-cleavage reaction initiated by the saturated oxygen atom; a similar cleavage with loss of the largest group produces the significant m/z 87 peak, providing a means to distinguish this spectrum from that of n-butyl acetate, in which this α-cleavage gives a peak at m/z 73. Note that the ion series m/z 59, 73, 87, . . . , can be due to either $C_nH_{2n+1}O^+$ or $C_nH_{2n+1}COO^+$. A large peak in formate esters is formed by α-cleavage followed by CO loss ($HCOOCHR_2^{\ddagger} \rightarrow HCO\overset{+}{O}{=}CHR \rightarrow HOCHR^+ + CO$).

Charge-site initiation at the saturated oxygen, $RCOO{-}R'^{\ddagger} \rightarrow RCOO\cdot + R'^+$, can give important peaks when R is small and R'^+ is stable. Thus for the n-, sec-, and $tert$-butyl acetates the values of $[(M - CH_3COO)^+]/\Sigma_{ions}$ are 1, 3, and 17 per cent, respectively. A reaction like 4.38 predicts that $C_nH_{2n+1}CH_2CH_2OCOR^{\ddagger}$ should yield $C_nH_{2n}^{\ddagger}$ as well as $C_{n+2}H_{2n+4}^{\ddagger}$ ions (Figure 3N).

A rather unusual $OE^{\ddagger}$ ion can arise from the elimination of the oxygen and α-carbon of the alcohol moiety as an aldehyde molecule. (This mechanism is not important in the spectrum in Figure 9L.) Thioesters, $R{-}C_2H_4S{-}COR$, can give a significant $(M - C_2H_4S)^{\ddagger}$ ion (Table 8.2).

If both the ester and acid chains are of sufficient length (ethyl butanoate or larger), the hydrogen rearrangements 9.32 and 9.34 can occur consecutively; these characteristic ions will appear at m/z 60 and 61 if the acid's α-carbon atom is not substituted.

Esters containing other functional groups. As was mentioned above, the saturated oxygen atom of the ester group is also capable of acting as a site to which a hydrogen atom can be transferred. In reaction 4.41 of benzoate esters, the presence of an α-hydrogen on an ortho substituent gives a characteristic product ion corresponding to the loss of an alcohol molecule; the addition of a 9,10-double bond to methyl octadecanoate gives $(M-CH_3OH)^{\ddagger}$ as the largest peak in its spectrum. A labile hydrogen attached to the C-6 position of a methyl alkanoate often gives characteristic $(M - 32)^{\ddagger}$ and $(M - 76)^{\ddagger}$ peaks (Reaction 8.44). The loss of ROH from $M^{\ddagger}$ apparently is favored only if the radical site is stabilized; this $OE^{\ddagger}$ ion often then loses CO (for example, *ortho*-substituted benzoate esters) or CH_2CO. The requirement of radical site stabilization is also met in EE^+ and some $OE^{\ddagger}$ *fragment* ions of other esters that contain a labile hydrogen atom, Reaction 8.45, as discussed in Section 8.9.

The mass spectra of alkenoate esters are relatively independent of the double-bond position, which apparently migrates readily; for example, methyl alkenoates usually show the m/z 74 and 87 ions characteristic of methyl alkanoates. The spectra of pyrrolidide derivatives (Anderson *et al.* 1974) and ion-molecule reactions (Ferrer-Correia *et al.* 1975) are effective ways to differentiate between double-bond isomers. Cyclization can apparently cause displacement at the δ-position (Reaction 8.68). Unsaturated methyl and ethyl esters exhibit peaks from the elimination of CO_2 and $HCO_2\cdot$ (Table 8.2). Enol acetates and other unsaturated acetates give ions corresponding to the loss of ketene. Polyacetoxy compounds can show characteristic peaks such as $(CH_3CO)_3O^+$ (Reaction 8.75) whose formation must involve acetyl migration.

Aromatic esters. The common ester reactions cited above are found for aromatic esters (see the mass spectrum of *n*-butyl benzoate, Unknown 4.21; *n*-hexyl benzoate, Figure 3O; *n*-propyl *p*-hydroxybenzoate, Figure 10A; phenylalanine methyl ester, Unknown 8.2; and diethyl phthalate, Figure 5C. Not surprisingly, charge retention on the aryl-containing fragment is strongly favored. If the aromatic ring contains alkyl groups, these undergo the typical fragmentations of arylalkenes described above. "Ortho" effects (Reaction 4.41) are useful in identifying such ring isomerism.

A variety of new reaction pathways are made possible by the ability of the aromatic ring to reduce bond dissociation energies (Reactions 4.41 and 8.44), stabilize the ion and radical sites (Reaction 8.47), and provide an unsaturated site for rearrangement, displacement, and elimination reactions (4.36, 8.69, 8.76, and 8.77). As an example of 4.36, $(M - CH_2CO)^{+\cdot}$ is a peak characteristic of aryl acetate spectra ($C_6H_6O^{+\cdot}$ is the base peak in the spectrum of $C_6H_5OCOCH_3$).

The spectrum of diethyl phthalate was given as Figure 5C. In addition to the expected fragmentations, phthalates give a highly characteristic peak at m/z 149. Its formation is rationalized in Reaction 9.36 for di-*n*-butyl phthalate.

9.5 Acids and anhydrides

Acids generally show distinctive molecular ions which increase in relative abundance with increasing molecular weight for *n*-alkanoic acids above C_6. In polycarboxylic acids $M^{+\cdot}$ is weak or absent, but in these the molecular weight can often be determined from the $(M + 1)^+$ ion formed by an ion–molecule reaction.

Aliphatic acids. The mass spectra of aliphatic acids resemble closely those of methyl esters and primary amides; compare the spectra of *n*-dodecanoic acid (Figure 3L) and of *n*-octadecanoic acid (Figure 8I) with those of the methyl ester (Figure 3M) and amide (Figure 3S). The mass of the prominent γ-hydrogen-rearrangement peak, $CRR'C(OH)_2^{+\cdot}$, indicates the α-branching (m/z 60 for $R = R' = H$; see Section 9.4 and Equation 8.48). The $C_nH_{2n}COOH^+$ series is characteristic of chain branching and also arises through mechanisms analogous to Reactions 8.49 and 9.33. The $(M - OH)^+$ is somewhat less abundant than the $(M - OCH_3)^+$ of methyl esters, but is diagnostically useful for higher and unsaturated acids. The spectrum of

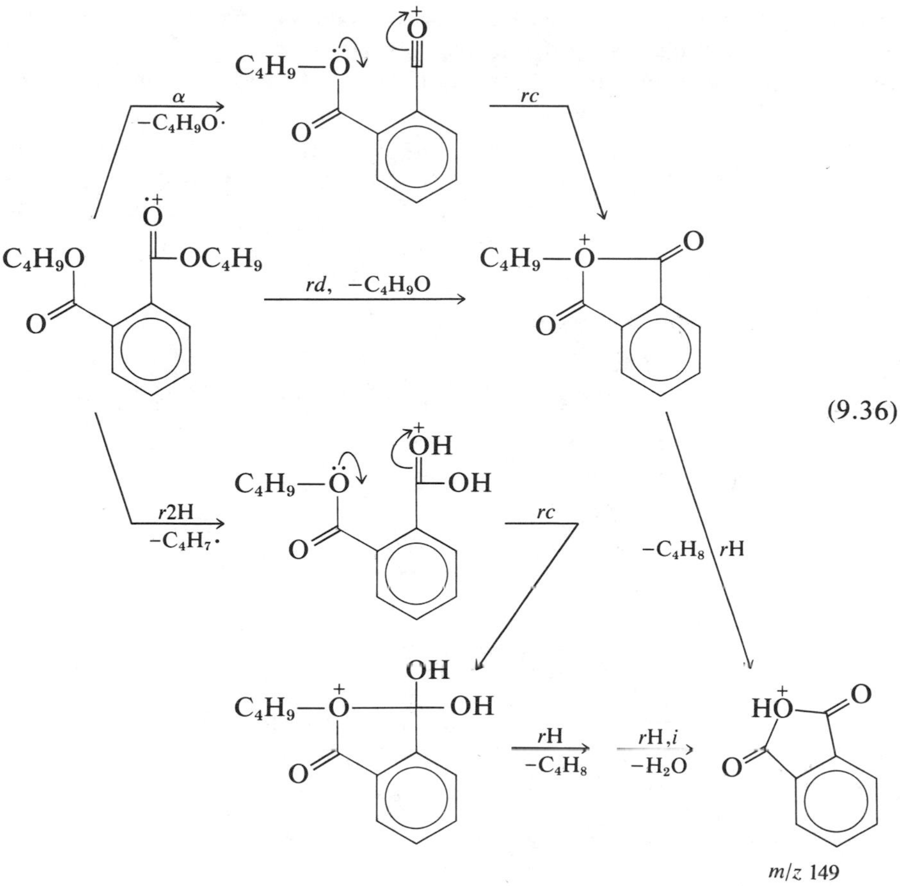

$$(9.36)$$

m/z 149

3-bromo-3-phenylpropionic acid even shows a hydroxyl group rearrangement in the decomposition of the $(M - Br)^+$ ion paralleling the methoxyl rearrangement of Reaction 8.74. Formation of $(M - H_2O)^{\ddagger}$ in acid spectra appears to be important only under the special conditions outlined for the formation of $(M - CH_3OH)^{\ddagger}$ peaks in methyl esters. The loss of CO_2 is often found in the spectra of dicarboxylic and substituted acids; however, this can also arise from thermal decarboxylation. Aliphatic dicarboxylic acids sometimes show characteristic losses of 38 (H_4O_2) and 46 (CH_2O_2).

Aromatic acids. Aromatic acids give spectra showing prominent OH loss followed by CO loss. A labile hydrogen on an *ortho* substituent causes a dominant $(M - H_2O)^{\ddagger}$ peak by the "ortho" effect, which is again followed by CO loss. Both thermal dehydration and decarboxylation are common for aromatic acids.

Anhydrides. The mass spectrum of succinic anhydride is shown in Figure 9M. The anhydrides of aliphatic acids show only a small or negligible molecular ion peak, but again the useful $(M + 1)^+$ ion can be formed at higher sample pressures. $[M^{\dot+}]$ is usually significant in unsaturated anhydrides. For mixed anhydrides, RCO—O—COR', the acylium ions RCO+ and R'CO+ are generally the largest peaks, and these lose CO as expected. A characteristic peak in the mass spectra of cyclic anhydrides of dicarboxylic acids is $(M - CO_2)^{\dot+}$, which often fragments further by the ejection of C_nH_{2n} or CO. The $(M - CO)^{\dot+}$ peak is much less common.

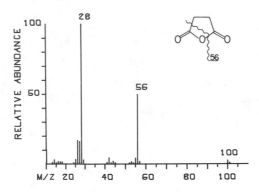

Figure 9M. Mass spectrum of succinic anhydride.

Unknown 9.2

m/z	Rel. abund.	m/z	Rel. abund.	m/z	Rel. abund.
40	1.6	73	4.5	119	3.6
41	4.5	74	20.	120	2.0
42	1.4	75	16.	121	27.
43	5.9	76	16.	122	4.8
44	7.6	77	12.	123	0.4
45	23.	78	1.7	136	3.3
46	0.3	79	0.8	137	1.0
49	4.0	80	0.3	138	2.2
50	34.	81	1.4	148	0.4
51	23.	82	0.7	149	100.
52	5.4	91	5.0	150	9.0
53	7.3	92	1.8	151	1.0
54	0.7	93	4.3	165	0.5
55	4.1	94	0.7	166	86.
63	4.7	103	3.8	167	7.9
64	1.9	104	4.9	168	1.0
65	43.	105	6.2		
66	4.6	106	0.5		

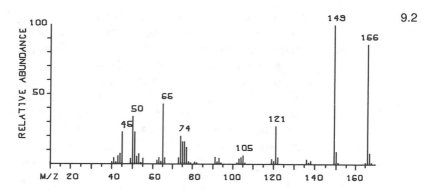

9.6 Ethers

Aliphatic ethers. The mass spectrum of diisobutyl ether is given in Unknown 4.11 and those of isopropyl *n*-pentyl ether and *n*-hexyl ether in Figures 9N and 3I.

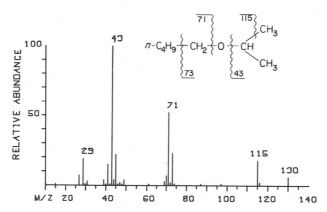

Figure 9N. Mass spectrum of isopropyl *n*-pentyl ether. (The relative abundance of *m/z* 131 increases with increasing sample pressure.)

The molecular ion peak from an aliphatic ether tends to be more abundant than that from a structurally similar alcohol of comparable molecular weight, but the M$^+$ peak is still weak to negligible. Again, ion-molecule formation of the (M + 1)$^+$ peak can be useful for molecular weight information.

The fragmentation of ethers resembles that of amines, but with a smaller propensity for α-cleavage and a much larger tendency for charge-site cleavage, as expected from the higher inductive withdrawal of electrons by oxygen. The important primary cleavages of isopropyl n-pentyl ether (Djerassi and Fenselau 1965) are due to these two types of cleavage (Reaction 9.37). For the competing α-cleavages, the loss of methyl from the tertiary carbon yielding m/z 115 is almost as favorable as the loss of the larger butyl group from the secondary carbon. Secondary product ions in the same $C_nH_{2n+1}O^+$ ion series (m/z 31, 45, 59, ...) can be formed by rearrangement loss of an olefin molecule from these ions (Reaction 8.59, but not 8.58). In contrast to the spectra of amines, the most abundant ions, $C_3H_7^+$ and $C_5H_{11}^+$, are produced by charge-site initiated reactions. Deuterium-labeling indicates that 40 per cent of the base $C_3H_7^+$ peak arises from further decomposition of $C_5H_{11}^+$.

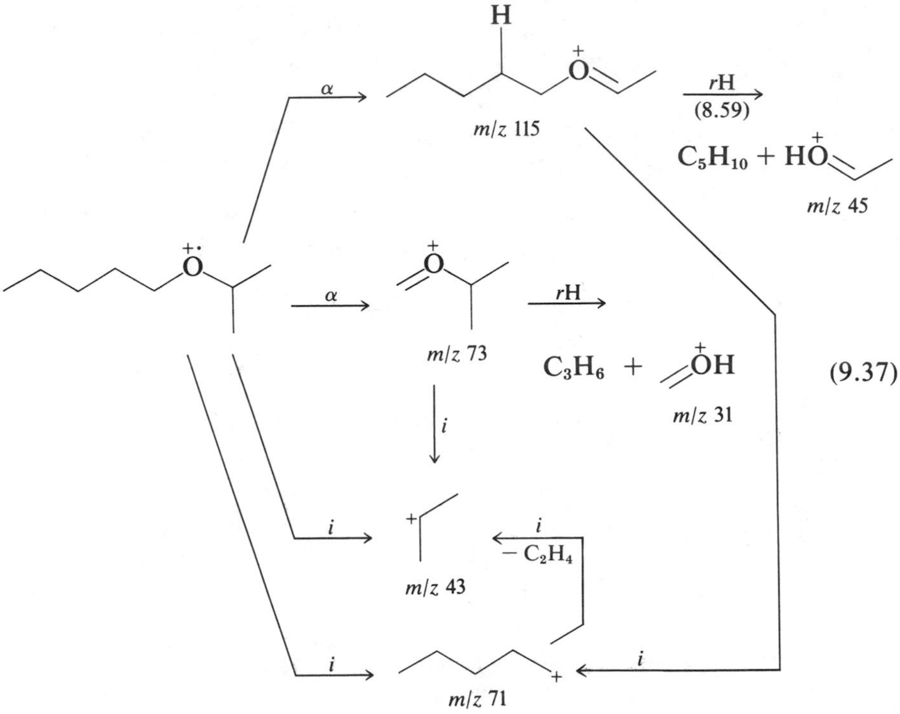

$$C_5H_{10} + HO^+ \qquad m/z\ 45$$

$$C_3H_6 + \;^+OH \qquad m/z\ 31$$

$$m/z\ 115 \qquad m/z\ 73 \qquad m/z\ 43 \qquad m/z\ 71 \qquad -C_2H_4 \tag{9.37}$$

Important odd-electron ions are notably lacking in this spectrum. Higher straight-chain ethers do show abundant $C_nH_{2n}^+$ ions; mechanism 4.38 can account for abundant ions of this type in particular cases. For example, ethyl n-hexyl ether and di-n-hexyl ether (Figure 3I) exhibit prominent m/z 56 and 84 peaks in their spectra. (Charge retention in Reaction 4.38 would give $C_nH_{2n+1}OH^{+\cdot}$, whose abundance should be low by analogy to that of the corresponding alkanol molecular ion.) Another source of $C_nH_{2n}^{+\cdot}$ ions is loss of H_2O, which yields weak $(M - H_2O)^{+\cdot}$ peaks in higher n-alkyl ethers. *n*-Hexyl and higher ethers undergo rearrangement of two hydrogen atoms, yielding $R\overset{+}{O}H_2$ ions (for example, m/z 103 from *n*-hexyl ether), which on decomposition provide another source of R^+ ions.

Alicyclic and unsaturated ethers. An illustrative example of such compounds given by Seibl (1970, p. 126) is 2,6,6-trimethyl-2-vinyltetrahydropyran (Figure 9O). A rationale of the formation of the principal peaks is given in Reaction 9.38; additional pathways are possible, such as through charge-site initiation (see Section 8.8).

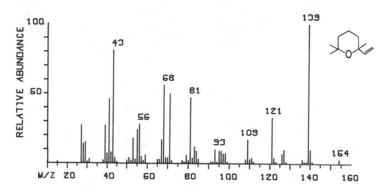

Figure 9O. Mass spectrum of 2,6,6-trimethyl-2-vinyltetrahydropyran.

Aromatic ethers. The mass spectrum of methyl phenyl ether shows characteristic peaks for $(M - CH_3)^+$, $(M - CH_3 - CO)^+$, $(M - CH_2O)^{+\cdot}$, $(M - OCH_3)^+$, and $(M - CHO)^+$ (minor — but important in some derivatives) which is typical of the mass-spectral behavior of methoxyaryl compounds. Ethyl and higher aryl ethers give a highly characteristic $OE^{+\cdot}$ ion from the elimination of C_nH_{2n}: $C_6H_5OC_nH_{2n+1}^{+\cdot} \rightarrow C_6H_5OH^{+\cdot}$ (Equation 8.47; look again at the spectrum of 1-phenoxy-2-chloroethane, Unknown 5.13). Benzyl ethers similarly give a significant $C_6H_5CH_2OH^{+\cdot}$ peak as well as a $C_6H_5CH_2^+$ peak.

Ketals and acetals. (Unknown 8.4 is methylal, $CH_3OCH_2OCH_3$.) The presence of two alkoxyl groups on the same carbon provides a powerful reaction initiating site. The activation energy for α-cleavage, $R-CH(OR')_2^{+\cdot}$ $\rightarrow R'O-CH=\overset{+}{O}R' \leftrightarrow R'\overset{+}{O}=CH-OR$, is substantially lowered by the resonance stabilization of the product ion. Unfortunately, this also causes the molecular ion for simple acetals and ketals to be negligible. Some reactions are observed which are due to the second functionality, such as the

(9.38)

elimination of H_2CO from formaldehyde acetals (Table 8.2). However, in general the ion-decomposition pathways outlined above for ethers are applicable.

An interesting example is the striking influence of the ethylene ketal moiety on steroid mass spectra, as shown in the classic work of Fétizon and Djerassi and their co-workers. The spectra of 5α-androstan 3-one and its ethylene ketal are shown in Figure 9P. Mechanisms for the formation of the two major peaks are given in 9.39; interestingly, these are directly analogous to the fragmentation of the corresponding 3-aminosteroid. Can you account for the m/z 112 peak?

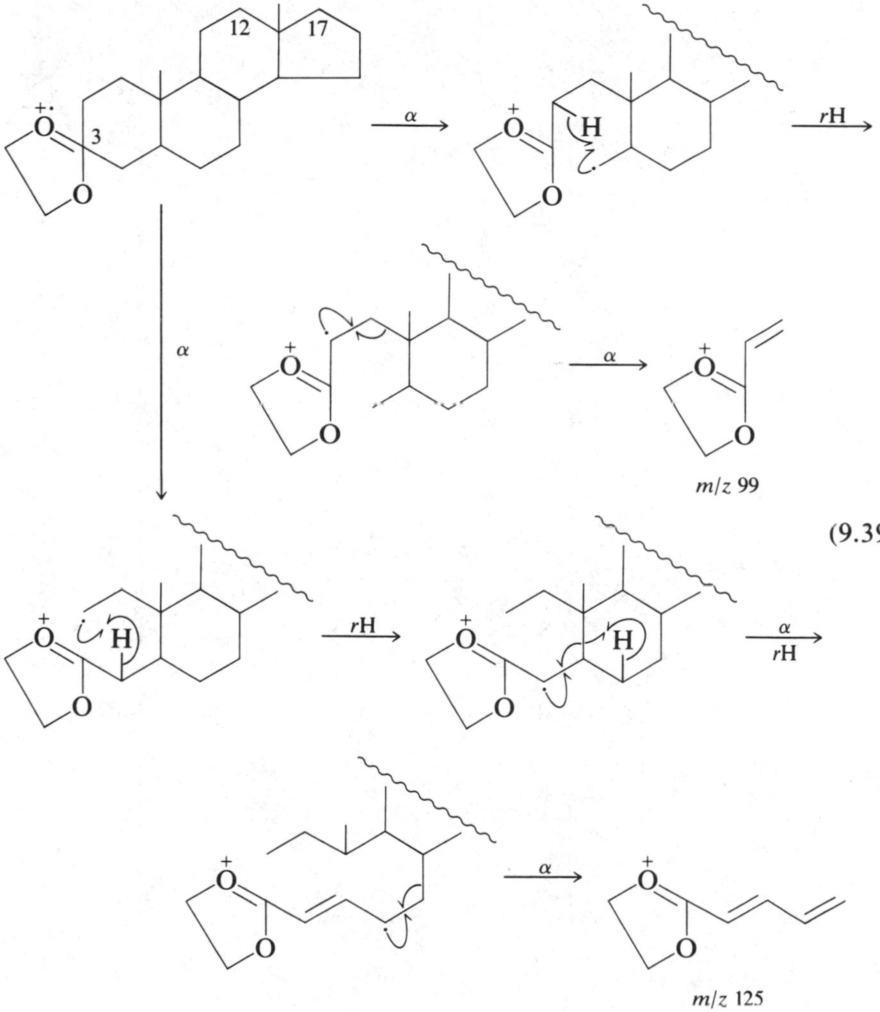

m/z 99

(9.39)

m/z 125

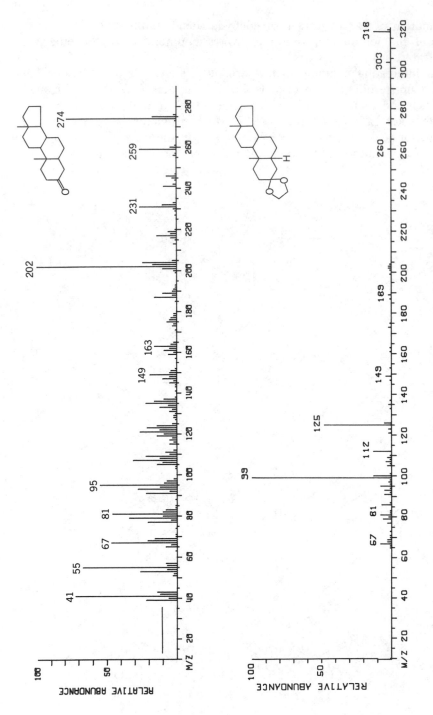

Figure 9P. Mass spectra of 5α-androstan-3-one and its ethylene ketal.

Unknown 9.3. A compound is either a 12- or 17-ketosteroid (see 9.39). The $C_5H_7O_2^+$ peak in the mass spectrum of its ethylene ketal is large. Which compound is it?

9.7 Thiols and thioethers

The mass spectra of thiols and thioesters (Sample and Djerassi 1966; van de Graaf and McLafferty 1977; Dill *et al.* 1979) resemble those of the corresponding alcohols and ethers. The spectrum of isopropyl *n*-pentyl sulfide (Figure 9Q) can be compared to that of isopropyl *n* pentyl ether (Figure 9N);

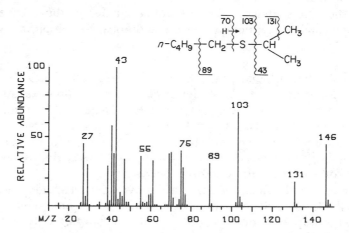

Figure 9Q. Mass spectrum of isopropyl *n*-pentyl sulfide.

Figures 3U and 3V are the spectra of *n*-dodecanethiol and *bis*-(*n*-hexyl)sulfide. A number of the most significant differences arise from the low ionization energies of sulfur compounds, which are approximately 1 eV below those of the corresponding oxygen compounds. This makes possible the formation of more lower-energy molecular ions, substantially increasing $[M^{\ddot{+}}]$. Sigma bond ionization can occur, leading to the C—S bond cleavage with charge retention on sulfur (Reactions 8.17 and 8.50) and product ion stabilization through isomerization (Reaction 9.40). Ions such as HS⁺, H_2S^+, H_3S^+, and CHS⁺ are of sufficient stability to give an ion series useful for characterization; note that few other types of compound give peaks at *m/z* 33, 34, and 35.

$$RCH_2-S-R' \xrightarrow[-R'\cdot]{\sigma} R-CH_2-S^+ \xrightarrow{rH} R-CH=SH^+ \qquad (9.40)$$

Thiols. Figure 3U is the spectrum of *n*-dodecanethiol. Alpha cleavage (see Reaction 4.13) yields the largest $C_nH_{2n+1}S^+$ peak (ion series 47, 61, 75, . . .), but cleavage at other chain positions to yield these ions is much more prevalent than in the corresponding alcohols. In straight-chain mercaptans $[89^+]$ $> [75^+]$ or $[103^+]$, indicating a favorable displacement reaction (8.65) for the formation of $C_4H_8SH^+$. The hydrocarbon ion series 27, 41, 55, . . ., and 15, 29, 43, . . ., are characteristically large, with $C_nH_{2n-1}^+ > C_nH_{2n+1}^+$ for $n > 3$. $(M - SH)^+$ is significant in secondary thiols, in contrast to the lack of $(M - OH)^+$ from alcohols. In primary thiols the $(M - SH)^+$ peak is superseded by $(M - SH_2)^{+}$; this and the prevalent peaks for $(M - SH_2 - C_nH_{2n})^{+}$, where $n \geq 2$, appear in the $C_nH_{2n}^{+}$ (28, 42, 56, . . .) ion series (Reaction 4.38).

Aromatic thiols undergo fragmentations similar to those expected for phenol, the main exceptions being the additional formation of $(M - S)^{+}$, $(M - SH)^+$, and $(M - C_2H_2)^{+}$ peaks.

Thioethers. Although the mass spectrum of $CH_3SCH_2CH_2CHO$ (the "oniony, meaty" odor from potatoes; Waller 1972, p. 716) exhibits substantial M^{+} (35 per cent) and α-cleavage ($CH_3S^+=CH_2$, 25 per cent) peaks, the base peak corresponds to CH_3SH^{+} produced by a hydrogen rearrangement similar to 4.37 and 4.39, and σ-cleavage forms CH_3S^+ as the second largest peak. Similarly, in addition to α-cleavage products, C_2H_5SR compounds typically form $C_2H_5S^+$ and $C_2H_5SH^{+}$ peaks.

As can be seen in the spectrum of isopropyl *n*-pentyl sulfide (Figure 9Q), peaks in the $C_nH_{2n+1}S^+$ ion series can now also arise from σ-ionization; this accounts for the presence of $C_5H_{11}S^+$ (m/z 103) and, in part, $C_3H_7S^+$ (m/z 75). Analogous to the behavior of isopropyl n-pentyl ether (Figure 9N and Reaction 9.37), α-cleavage yields $(CH_3)_2CH\overset{+}{S}=CH_2$ (m/z 89) and $CH_3CH=\overset{+}{S}C_5H_{11}$ (m/z 131) ions, whose decompositions by hydrogen rearrangement Reaction 8.56 should produce $H\overset{+}{S}=CH_2$ (m/z 47) and $CH_3CH=\overset{+}{S}H$ (m/z 61). Decomposition of $CH_3CH=\overset{+}{S}C_5H_{11}$ by Reaction 8.55 is an alternative mode for formation of the m/z 75 peak.

In contrast to isopropyl *n*-pentyl ether, this sulfide's mass spectrum shows characteristic OE^{+} ions at m/z 76, $(CH_3)_2CHSH^{+}$, and m/z 70, $C_5H_{10}^{+}$ (plus m/z 42, $C_3H_6^{+}$) corresponding to the competing products of Reactions 4.37 and 4.38 (formation of this charge-retained ion is a dominant pathway for disulfides). Ions resulting from the rearrangement of two hydrogen atoms, such as m/z 77 (probably $C_3H_7SH_2^+$), can also be formed in smaller abundance. On the other hand, the charge-site product $C_5H_{11}^+$ (Reaction 4.22), which is significant in the spectrum of the ether, is very small (possibly owing to the competition of 4.37), although the companion product $C_3H_7^+$ is still the base peak.

Phenyl alkyl sulfides undergo reactions typical of their counterpart ethers but show additional skeletal rearrangements that complicate structure elucidation. Elimination of SH is common in compounds whose usual cleavage pathways are less facile. Aromatic disulfides often undergo additional losses of S_2 and/or HS_2.

Unknown 9.4 is known to contain a hydroxyl group. Could you have deduced this fact from the mass spectrum? Try Unknowns 4.3, 5.14, and 8.10 again.

Unknown 9.4

m/z	Rel. abund.	m/z	Rel. abund.
27	6.0	96	3.9
29	5.8	97	40.
39	8.0	98	10.
47	1.7	99	2.3
51	2.7	100	0.4
52	1.0	108	2.3
53	11.	109	0.2
61	1.9	121	2.1
63	4.8	125	13.
64	2.3	126	52.
65	5.2	127	4.5
69	7.2	128	2.6
70	4.2	137	1.2
71	3.2	139	32.
77	2.6	140	2.7
82	3.0	141	1.6
84	5.2	153	0.8
91	4.0	154	100.
92	1.5	155	9.8
93	2.0	156	5.1
94	5.2	157	0.5
95	8.7		

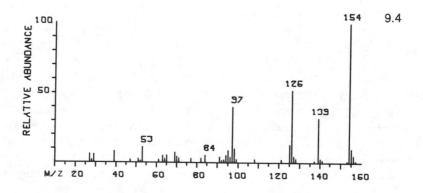

9.8 Amines

Aliphatic amines exhibit very low ionization energies; despite this, the amine group provides such a powerful driving force for reaction initiation (almost always the base peak results from α-cleavage or α-cleavage plus olefin loss rearrangement, Equation 4.44) that molecular ions are of low or negligible abundance. Fortunately, aliphatic amines have a strong tendency to undergo protonation at moderately high sample pressures to yield the characteristic $(M + H)^+$ peak. Salts of amine bases will not vaporize in the mass spectrometer; however, they often decompose on heating in the instrument to release the free amine and the acid. The presence of abundant peaks corresponding to HCl (m/z 36 and 38) or HBr (m/z 80 and 82) strongly indicates such a salt, since these ions are not formed in abundance as products of electron-impact-induced reactions.

Studies on indole alkaloid derivatives indicate that quaternary nitrogen compounds ($R \neq H$) decompose thermally on heating in the mass spectrometer by two principal paths leading to different tertiary amines (Hesse and Lenzinger 1968). In the case of the bromide or iodide salts, dealkylation with formation of the amine and the alkyl halide is favored (Reaction 9.41). For the fluoride salt, a thermal Hofmann degradation involving abstraction of a hydrogen atom β to the quaternary nitrogen is favored (Reaction 9.42).

$$R_3 - \overset{\displaystyle R_2}{\underset{\displaystyle CH_2 - CH_2}{N^+}} R_1 \quad X^- \quad \xrightarrow{\Delta} \quad R_3 - \overset{\displaystyle R_2}{\underset{\displaystyle CH_2 - CH_2}{N:}} \quad + \quad R_1X \qquad (9.41)$$

$$R_3 - \overset{\displaystyle R_2}{\underset{\displaystyle CH - CH_2}{N^+}} R_1 \quad \xrightarrow{\Delta} \quad R_3 - \overset{\displaystyle R_2}{\underset{\displaystyle CH=CH_2}{N}} - R_1 \quad + \quad HX \qquad (9.42)$$

Aliphatic amines. The mass spectra of 1-aminododecane (Figure 3P) and a number of $C_4H_{11}N$ isomeric amines (Unknowns 4.4 to 4.8; Figures 4C and 4D) have been given and features of their spectra discussed in Section 4.6. The mass spectra of diethyl-*n*-propylamine, *bis*-*n*-hexylamine, *tris*-(*n*-butyl)amine, and ephedrine were shown in Unknown 6.2 and Figures 3Q, 3R, and 4E. The mass spectra of *N*-methyl-*N*-isopropyl-*N*-*n*-butylamine

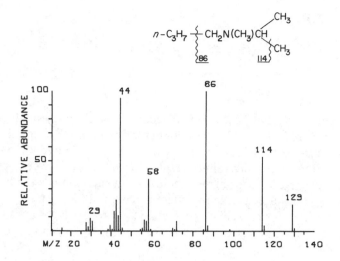

Figure 9R. Mass spectrum of *N*-methyl-*N*-isopropyl-*N*-*n*-butylamine. (Relative abundance of the *m/z* 130 peak increases with increasing sample pressure.)

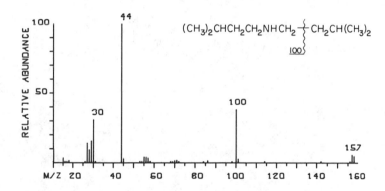

Figure 9S. Mass spectrum of *bis*(3-methylbutyl)amine. (Relative abundance of the *m/z* 158 peak increases with increasing sample pressure.)

(Unknown 4.20, see Chapter 11; Djerassi and Fenselau 1965) and *bis*(3-methylbutyl)amine are given in Figures 9R and 9S.

Alpha cleavage is a dominant reaction of amines, generally producing the base peak in *n*-alkylamines and α-substituted primary amines, with loss of the largest alkyl group favored (Equation 4.17). This cleavage accounts for the *m/z* 30 base peak in Figure 3P, the *m/z* 86 base peak and *m/z* 114 in Figure 9R, and the *m/z* 100 in Figure 9S. Peaks in this $C_nH_{2n+2}N^+$ ion series can be formed by other pathways also, however. In primary *n*-alkylamines

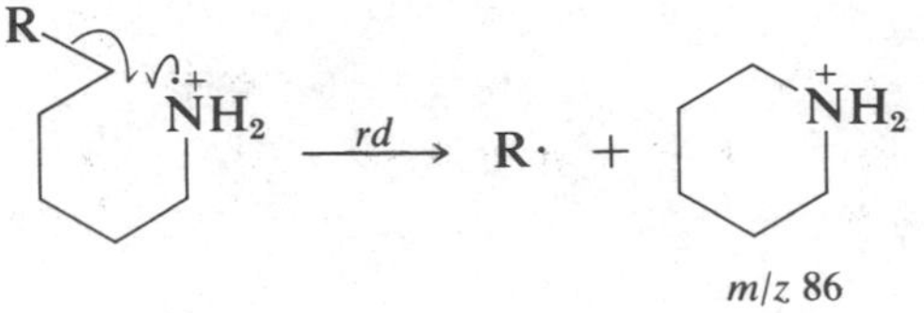

$$(9.43)$$

$m/z\ 86$

displacement reactions (9.43; also see 8.6) account for relatively small peaks with the most favored at m/z 86. For higher primary n-alkyl amines the peaks at m/z 44 (β-cleavage), and to a lesser extent m/z 58 and 72 (γ- and δ-cleavage), become increasingly important (m/z 44 = 40 per cent in the spectrum of n-$C_{16}H_{33}NH_2$). For higher-molecular-weight amines the hydrocarbon-type peaks also increase; in the spectrum of $[C_6H_{13}CH(CH_3)]_2NH$, the alkyl ion resulting from α-cleavage (m/z 85) shows 34 per cent abundance, but the immonium ion from α-cleavage (m/z 156) has an abundance of only 15 per cent.

A number of secondary rearrangement decompositions of the primary immonium ion can produce other $C_nH_{2n+2}N^+$ ions. The most common such reaction (9.44; also see 8.56) yields the important m/z 44 and 58 ions in

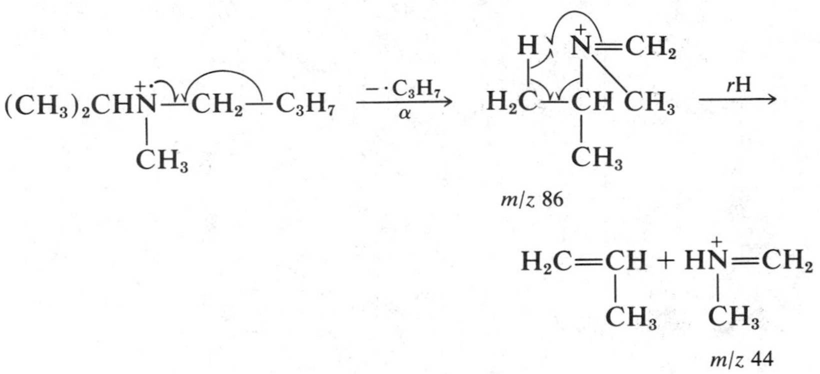

$$(9.44)$$

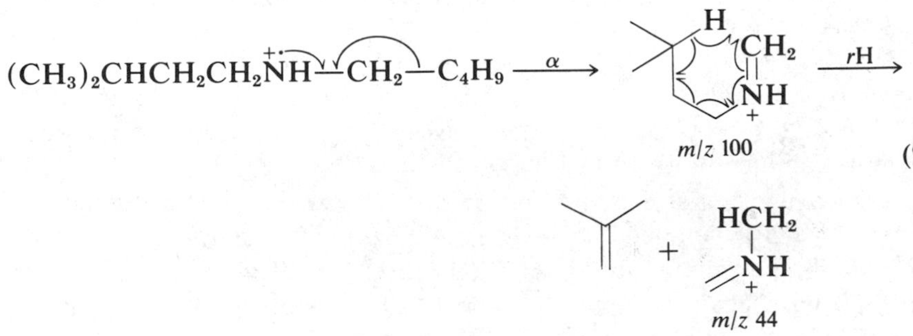

$m/z\ 86$

$m/z\ 44$

$$(9.45)$$

$m/z\ 100$

$m/z\ 44$

Figure 9R, and the less important m/z 30 peak in Figure 9S (see Section 8.9). Another possible rearrangement (9.45; see Equation 8.55) appears to account for the base peak at m/z 44 in the spectrum of *bis*(3-methylbutyl)amine (note that the hydrogen is transferred from a tertiary carbon atom). Labeling studies (Djerassi and Fenselau 1965) indicate that 65 per cent of the m/z 72 in Figure 9R arises through such a mechanism. The remaining EE^+ rearrangements, Equations 8.54 and 8.57, do not appear to be important in the aliphatic amine spectra that have been examined, although mechanism 8.54 may increase in importance for large alkyl groups. The loss of NH_3 or a neutral amine molecule (Equation 4.38) is not important except in the spectra of certain polyfunctional compounds (for example, loss of NH_3 in α-amino acids and diamines).

Cycloalkylamines (see Section 8.8). A thoroughly studied example of such spectra is that of *N*-ethylcyclopentylamine, Figure 8G. Alpha cleavage in the alkyl group leads to the $(M - CH_3)^+$ peak of moderate intensity. However, α-cleavage at the secondary carbon in the ring should be favored; this leads to an isolated radical site whose further reaction (Equation 8.37; 9.46) accounts for the m/z 84 base peak. Homologous even-electron ions at m/z 70 and 56 can be formed by a variety of pathways, as was discussed for aliphatic alcohol spectra. Thus, although the m/z 56, 70, 84, and 98 ion series is indicative of this general type of structure, the individual peaks, especially those of lower mass, are less characteristic of specific structural features. The small m/z 85 peak appears to be formed mainly by C_2H_4 loss from the ethyl group (see Reactions 4.37 and 4.39).

In the spectrum of *des-N*-methyl-α-obscurine, a similar mechanism can explain the loss of the $—CH_2—CH(CH_3)—CH_2—$ bridge plus a hydrogen atom to form the $(M - 57)^+$ ion (9.47). The $(M - C_4H_8)^{+\cdot}$ is of much lower abundance than the $(M - C_4H_9)^+$. (The mass spectra of nicotine, indole alkaloids and strychnine are shown in Figure 5A, Unknown 6.3, and Figure 3Z.)

Other types of reactions that are possible for cyclic amines are discussed in Section 8.8. A five-membered ring containing nitrogen ejects two ring carbons and their substituents as an alkene (Reaction 8.31). For larger ring compounds, such as 1,2-dimethylpiperidine (Figure 8E), this reaction is of low importance (m/z 71, 1 per cent; see Reaction 8.34). The EE^+ ion formed by α-cleavage (m/z 98, 100 per cent) can undergo a retro-Diels-Alder reaction (8.36) to yield m/z 70 in 12 per cent abundance.

Unknowns. Test your knowledge with Unknowns 4.4–4.8, 4.20, 5.10, 6.2, 8.2, 8.3, 8.9, and 8.10.

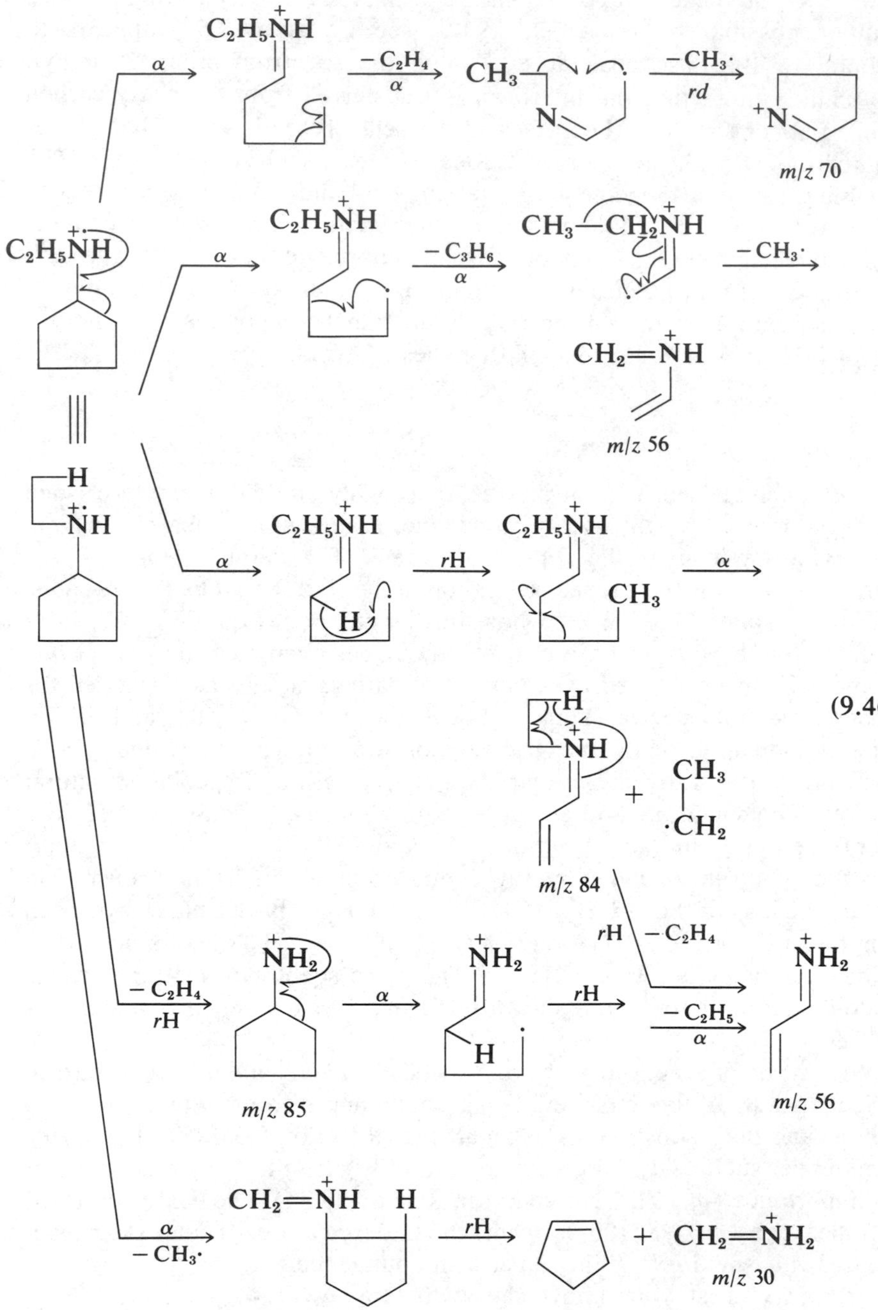

$C_2H_5\overset{+}{N}H$

α $\xrightarrow{-C_2H_4}{\alpha}$ CH_3 $\xrightarrow{-CH_3\cdot}{rd}$ $\overset{+}{N}$ $m/z\ 70$

$C_2H_5\overset{+\cdot}{N}H$

$C_2H_5\overset{+}{N}H$ $\xrightarrow{-C_3H_6}{\alpha}$ $CH_3-CH_2\overset{+}{N}H$ $\xrightarrow{-CH_3\cdot}$

$CH_2=\overset{+}{N}H$ $m/z\ 56$

$\overset{H}{\underset{\overset{+\cdot}{N}H}{|}}$

$C_2H_5\overset{+}{N}H$ $\xrightarrow{rH}$ $C_2H_5\overset{+}{N}H$ CH_3 $\xrightarrow{\alpha}$

(9.46)

$\overset{H}{\underset{\overset{+}{N}H}{}}$ $+$ CH_3 $\cdot CH_2$

$m/z\ 84$

rH $-C_2H_4$

$\overset{+}{N}H_2$ $\xrightarrow{\alpha}$ $\overset{+}{N}H_2$ $\xrightarrow{rH}$ $\xrightarrow{-C_2H_5}{\alpha}$ $\overset{+}{N}H_2$

$\xrightarrow{-C_2H_4}{rH}$

$m/z\ 85$ H $m/z\ 56$

$\xrightarrow{\alpha}{-CH_3\cdot}$ $CH_2=\overset{+}{N}H$ H $\xrightarrow{rH}$ $+$ $CH_2=\overset{+}{N}H_2$

$m/z\ 30$

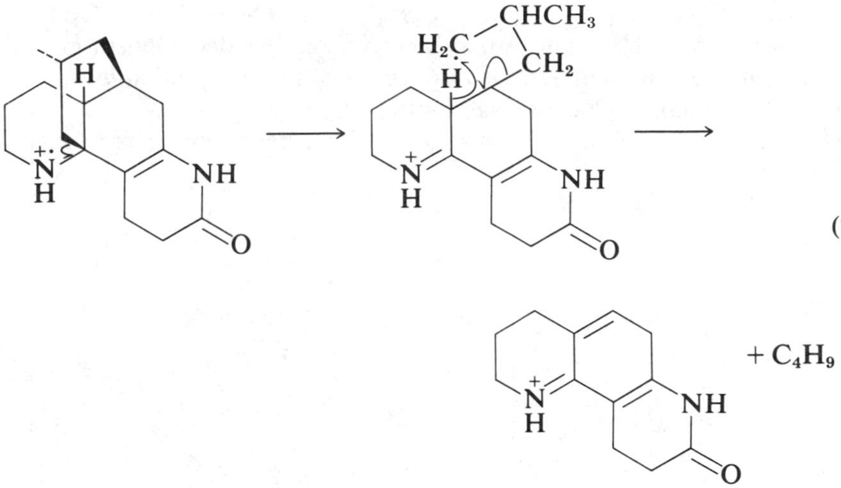

$$(9.47)$$

9.9 Amides

The mass spectral behavior of primary amides, such as n-dodecanoamide (Figure 3S), closely resembles that of the corresponding acid and methyl ester (Figures 3L and 3M). The behavior of secondary (such as sec-butyl acetamide, Figure 9T) and tertiary amides (N,N-dipentylacetamide, Figure 3T) parallels that of esters of higher alcohols, although here, in those reactions initiated by the saturated heteroatom, the more powerful directing force of nitrogen is clearly evident. Amides give molecular ion peaks that are generally more distinctive than their ester counterparts and have a strong tendency to form $(M + 1)^+$ ions by ion–molecule reactions.

The resemblance of the spectra of n-dodecanoic acid and its amide (Figures 3L and 3S) is so striking that little additional mechanistic interpretation is necessary. The nitrogen-containing peaks are shifted to lower masses by one unit; the β-cleavage–hydrogen rearrangement peak will fall in the series m/z 59, 73, 87, ... Alkyl losses produce the ion series 44, (58), 72, 86, . . . ; with longer alkyl chains maxima appear at intervals corresponding to $(CH_2)_4 - m/z$ 72, 128, 184, and 240 – although the effect is less pronounced. The numerous hydrocarbon peaks such as $C_nH_{2n+1}^+$ and $C_nH_{2n-1}^+$ are less abundant, as expected, than in the spectra of the analogous acids.

The spectra of secondary and tertiary amides show the additional expected reactions, such as β-cleavage at the N—R bond with rearrangement of one or, especially, two hydrogen atoms (m/z 60 in Figure 9T; Reaction 4.46). In tertiary amides consecutive rearrangements like reactions 4.46 and 8.52 are possible from both chains on nitrogen as well as from the acid moiety; thus N,N-diethylhexanoamide exhibits prominent peaks for $(M - C_4H_8)^{+\cdot}$,

$(M - C_4H_8 - C_2H_3)^+$, $(M - C_4H_8 - C_2H_4)^{+\cdot}$, and $(M - C_4H_8 - C_2H_3 - C_2H_4)^+$ (m/z 115, 88, 87, and 60, respectively). The decompositions of secondary and tertiary amides also reflect the strong reaction-initiating tendency of nitrogen. Alpha cleavage induced by it gives the $(M - C_2H_5)^+$ and, weakly, the $(M - CH_3)^+$ ions of Figure 9T; further hydrogen rearrange-

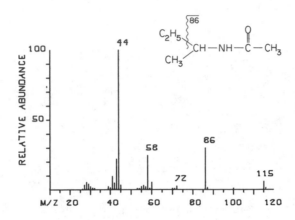

Figure 9T. Mass spectrum of *sec*-butyl acetamide.

ment decomposition of these ions by Reaction 8.56 yields $H_2\overset{+}{N}{=}CHCH_3$ and $H_2\overset{+}{N}{=}CHC_2H_5$. Cleavage of the C—N amide bond to give the alkanoyl ion (CH_3CO^+ in Figures 3T and 9T) is greatly reduced over that in esters (CH_3CO^+ is the base peak in *sec*-butyl acetate, Figure 9L). The normally unfavorable C—N bond cleavage with charge retention on N increases substantially between the secondary amide (Figure 9T, m/z 72) and the tertiary amide (Figure 3T, m/z 156; see also Equation 8.17). Some amides (for example, diethylacetamide) undergo direct ketene loss (Reaction 4.39), paralleling the behavior of unsaturated acetates; this provides a very characteristic peak in the spectra of *N*-aryl amides. Hydrogen rearrangement with charge migration away from the amide functionality (either Reaction 4.34 or Reaction 4.38) occurs only if the resulting ion $(M - RCONHR)^{+\cdot}$ is substantially stabilized and the R groups are small. The tendency for charge retention is much greater than in esters because of the greatly reduced ionization energies of amides; however, acetamides of more complex molecules, such as steroids, usually exhibit characteristic $(M - CH_3CONH_2)^{+\cdot}$ peaks.

Unknown 9.5

m/z	Rel. abund.
15	1.0
16	0.4
27	6.4
28	2.1
29	9.8
30	0.8
31	0.7
39	7.1
40	1.0
41	17.
42	6.4
43	20.
44	31.
45	1.3
46	0.9
57	11.
58	1.1
59	100.
60	2.6
61	0.2
69	3.6
73	3.4
85	1.7
86	11.
87	0.4
100	1.3
101	1.7

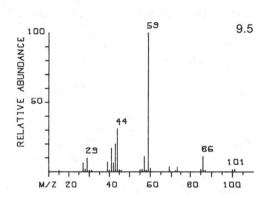

9.10 Cyanides

The ionization energies of aliphatic nitriles are unusually high, and this is reflected in the unusual degree to which these compounds undergo skeletal rearrangement on electron impact. The basic rules of ion decomposition (Chapters 4 and 8) appear to be less applicable to cyanides than to any other simple compound class. For the ionization energies of $YCH_2\cdot$, $Y = CN$ has the highest value of the common functionalities listed in Table A.3 except for $Y = H$ (Section 8.2). Further, the principal nitrogen-containing ion series found for nitriles, $C_nH_{2n-1}N^{+}$ and $C_nH_{2n-2}N^{+}$, are isobaric with the common hydrocarbon series $C_nH_{2n-1}^{+}$ (27, 41, 55, . . .) and $C_nH_{2n-2}^{+}$ (26, 40, 54, . . .).

Molecular ion peaks are weak or negligible in aliphatic nitriles, although they are usually strong in aromatic compounds. Fortunately, aliphatic nitriles have a substantial tendency to form an $(M + 1)^{+}$ peak by ion–molecule reactions.

The mass spectrum of n-dodecyl cyanide is given in Figure 3Y. Its overall appearance is unusual by mass-spectral standards; it resembles an n-alkane or alkene spectrum (Figures 3B and 3E) in its lack of specificity, but it has abundant peaks at higher masses and pairs of significant peaks separated by only one mass unit. The $(CH_2)_nCN^+$ series (40), 54, 68, 82, . . . , dominates the upper portion of the spectrum; the hydrogen-rearrangement series nominally corresponding to $C_nH_{2n-1}N^+$ (41, 55, 69, 83, . . .) is of comparable importance for $n \sim 3$ to 6. The largest peak in the former series is $(CH_2)_5CN^+$ from the displacement reaction 8.65. Chain branching does increase the abundances of the appropriate ions in this series, but such data must be used with caution for structure determination. The $C_nH_{2n-1}N^+$ odd-electron series is analogous to the $C_nH_{2n}^+$ series from 1-alkenes, 1-alkanols, and thiols (Figures 3E, 3H and 3U), replacing $-CH=CH_2$ by $-C\equiv N$. Presumably these ions are formed by hydrogen rearrangement with olefin loss; Rol (1968) has shown for 4-methylpentanonitrile that the abundant $C_3H_5N^+$ ion arises from δ-hydrogen migration followed by γ-cleavage.

A further unusual feature of the spectra of alkyl cyanides is that $[(M-1)^+]$ > $[M^+]$. Loss of HCN can occur if a labile hydrogen is available in the molecule. α-Cleavage is unfavorable, requiring formation of a quadruple bond, $C\equiv\overset{+}{N}$, to stabilize the charge on nitrogen. For n-decylcyanide, β-cleavage with hydrogen rearrangement (Reaction 4.33) to give $\cdot CH_2C\equiv\overset{+}{N}H$, is insignificant; high-resolution measurement shows that 82 per cent of the m/z 41 peak is $C_3H_5^+$ (this process is more important for smaller molecules).

This back-donation of electrons by the cyano group may also account for the absence of $(M - CN)^+$ formation, which contrasts with the behavior of electron-withdrawing groups such as chlorine and bromine (Reaction 4.23); the behavior of CN actually resembles that of fluorine much more closely in these regards. Only the lower-mass hydrocarbon ions are formed in abundance, reminiscent of the spectra of alkanes; the $C_nH_{2n-1}^+$ ions could arise through loss of $C_nH_{2n+1}\cdot$ followed by loss of HCN (Reaction 4.45), again analogous to the behavior of halogens.

Unsaturated and aromatic nitriles have much more abundant molecular ions, and those with no α-hydrogen show no significant $(M - H)^+$. Again, many cases with rather unusual breakdown pathways in their spectra have been cited.

Unknowns 5.11 and 8.6 are worth trying again.

9.11 Aliphatic halides

The following mass spectra of halogenated compounds are illustrative: HCl, Unknown 2.1; CH_3F, 1.8; CH_3Br, 2.2; CCl_4, 3.6; $CClF_3$, 2.8; 1,2-$C_2H_2Cl_2$, 3.4; C_2HCl_3, 2.14; C_2ClF_5, 3.5; CF_3CN, 5.11; 1-bromohexane, 8.1; 3-(chloromethyl)-heptane, 5.12; and 1-phenoxy-2-chloroethane, 5.13. Figures 4J, 3W, 3X, 9U, and 9V show the mass spectra of 1,1- and 1,3-dichlorobutane, 1-chlorododecane, 1-bromododecane, 3-chlorodecane, and 3-bromodecane.

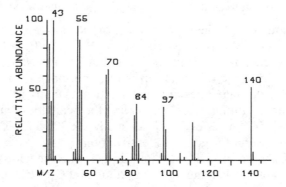

Figure 9U. Mass spectrum of 3-chlorodecane. Small (0.15 and 0.05 per cent) molecular ion peaks are present in m/z 176 and 178.

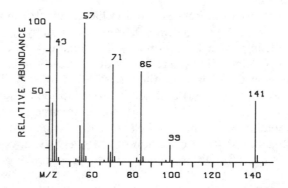

Figure 9V. Mass spectrum of 3-bromodecane. Small (0.1 per cent) molecular ion peaks are present at m/z 220 and 222.

Measurable molecular ions are obtainable from alkyl halides (except fluorides) of moderately high molecular weights or with some chain branching; $M^{\ddot{+}}$ is often negligible for perhalogenated compounds unless they are unsaturated. The presence and number of chlorine or bromine atoms in an ion is usually recognizable from its characteristic "isotopic cluster" (Section 2.2 and Table A.2).

The halogens have a relatively small influence on mass-spectral reactions. Iodine is the most influential because of the unusually low strength of the carbon–iodine bond. Also affecting carbon–halogen bond cleavage, but in the opposite order $(F > Cl > Br > I)$, is the inductive withdrawal of electrons by the halogen atom. However, the tendency for back-donation of electrons, which strengthens this bond, shows the same order (fluorine greatest), substantially compensating for this effect.

There are two main sources of $C_nH_{2n}X^+$ ions in the spectra of alkyl halides. One source is α-cleavage to form the $R_2C\overset{+}{=}X$ ion, which follows the expected order of electron-donating ability $F > Cl > Br > I$. For bromine and iodine this reaction is negligible for all but low-molecular-weight compounds, and for chlorine this reaction is of importance mainly for tertiary chlorides. Alpha cleavage does give unusually abundant $(M - H)^+$ ions for lower-molecular-weight fluorides, paralleling the behavior of alkyl cyanides. The weak $C_nH_{2n}X^+$ ion series found for many alkyl halides is quite characteristic and thus useful for general identification.

The second source of $C_nH_{2n}X^+$ ions is the displacement reaction 8.65 (Section 8-10). This leads to an unusually abundant $C_4H_8Cl^+$ or $C_4H_8Br^+$ ion in the spectra of n-alkyl chlorides or bromides containing more than five carbon atoms (Figures 3W and 3X and Unknown 8.1). The abundances of these ions are greatly lowered by chain branching and are negligible for fluorides and iodides.

There are substantial differences in alkyl halide spectra in regard to the proportions of $(M - HX)^{+\cdot}$ and $(M - X)^+$ ions formed by C—X cleavage; the former is accompanied by secondary ions characteristic of an olefin, while the $(M - X)^+$ ions characteristically decompose further by losses of C_nH_{2n}. A prominent $OE^{+\cdot}$ ion from the loss of HX is observed for fluorides and primary (lower for n-alkyl) and secondary chlorides (Figures 3W and 9U); apparently the high C—X stability can be overcome only when the tendency for donation of the unpaired electron on X is neutralized by hydrogen transfer (Reaction 4.40). Formation of $(M - HX)^{+\cdot}$, in contrast to that of $(M - X)^+$, can be strongly influenced by the presence of other functionalities.

For tertiary and activated (for example, allylic) chlorides and for bromides (Figure 9V) and iodides the formation of $(M - X)^+$ and lower $C_nH_{2n+1}^+$ ions generally is dominant. (Electron bombardment can also produce R^+ ions of unusually low appearance energy by "ion pair" processes: $R-X \rightarrow R^+ + X^-$.) Although $C_nH_{2n-1}^+$ ions are of significant abundance for low-molecular-weight ions of this type, they commonly arise by loss of HX from $C_nH_{2n}X^+$ ions (Reaction 4.45).

Perhalogenated compounds resemble hydrocarbons in their mass-spectral behavior, including a high tendency for rearrangement. The smallest perhaloalkyl ions are usually the most stable; for example, CF_3^+ (m/z 69) is the most abundant ion in perfluoroalkane spectra.

Unknowns containing halogen include 2.14, 3.4, 3.5, 3.6, 5.12, and 8.1.

9.12 Other types of compounds

A number of reviews give valuable details on the fragmentation behavior of other classes of compounds and applications in particular research areas. Only references to the literature since the second edition of this book will be given here; for earlier work check that edition, the comprehensive compilation of Budzikiewicz *et al.* (1967), Hamming and Foster (1972), or Porter and Baldas (1971). J. H. Bowie (1977, 1979) has extensively reviewed the literature on the mass-spectral behavior of particular compound classes. Several compilations of references to the original literature are also recommended: United Kingdom CIS, *Mass Spectrometry Bulletin*; Chemical Abstracts Service, *C.A. Selects Mass Spectrometry*; Johnstone (1975, 1977, 1979) and Williams (1971, 1973), *Mass Spectrometry*; the biennial reviews on mass spectrometry in *Analytical Chemistry* (Burlingame *et al.* 1978, 1980); and the indexes to *Organic Mass Spectrometry* and *Biomedical Mass Spectrometry*.

Compound Classes.
Alkaloids: Hesse and Sastry (1980).
Amino acids: Vetter (1980).
Bile acids: Elliot (1980).
Carbohydrates: Radford and De Jongh (1980).
Fatty acids, lipids: Odham and Wood (1980).
Heterocyclics: Games (1979).
Metallorganic compounds: Spaulding (1979).
Nitrogen heterocyclics: Dougherty *et al.* (1980).
Nucleic acids: Hignite (1980).
Peptides: Arpino and McLafferty (1976), Morris (1979), Biemann (1980).
Phenolics: Mabry and Vlubelen (1980).
Porphyrins: Jackson (1979).
Steroids: Brooks (1979), Budzikiewicz (1980), Zaretskii (1976).
Terpenes and terpenoids: Enzell and Wahlberg (1980).

Applications
Antibiotics: Borders and Hargreaves (1980).
Bioactive compounds: Nakanishi and Occolowitz (1979).
Drugs: Millard (1979).
Environmental: Safe (1979), Dougherty *et al.* (1980).
Extraplanetary life: Boettger (1980).
Food science: Horman (1979).
Hormones: Brooks (1980).

Medical uses: Frigerio and Castagnoli (1974), Frigerio (1980), Fenselau (1978), Caprioli *et al.* (1980).

Metabolism: Jellum (1979), Grostic and Bowman (1980).

Organic geochemistry: Eglinton *et al.* (1979), Pillinger (1979).

Pesticides and insecticides: Sphon and Brumely (1980).

Petroleum and coal: Lumpkin and Aczel (1978).

Pheromones: Anderson *et al.* (1980).

Pyrolysis mass spectrometry of biomaterials: Meuzelaar *et al.* (1980).

Stable isotopes: McCloskey and Krahmer (1977), Klein and Klein (1979), Sjovall (1980).

Vitamins and cofactors: Elliot (1980).

10

COMPUTER IDENTIFICATION
OF UNKNOWN MASS SPECTRA

Computer techniques can increase both the speed and the information gained in interpreting unknown mass spectra. This is especially true for unknowns on which very little other structural information is available, such as those produced by automated gas-chromatograph/mass-spectrometer systems in the analysis of complex organic mixtures, such as pollutants, body fluids, insect pheromones, drug metabolites, and petroleum. In general such systems are *aids to the interpreter,* and will be much more valuable to you if you have a good basic understanding of the interpretation process described in the previous chapters. However, such computer algorithms are now fast (≤ 1 minute) and readily available; you should take advantage of their help.

A number of extensive reviews of such computer algorithms have been published recently (Pesyna and McLafferty 1976; Smith 1977; Chapman 1978; Henneberg 1980; McLafferty *et al.* 1980b). These should be studied for the principles of such systems and for specific information on the variety of algorithms that have been proposed. It is useful to divide possible approaches into two types, systems for *retrieval* and those for *interpretation* of unknown mass spectra (Pesyna and McLafferty 1976). For the first type, probably the most widely used system, which has also been demonstrated to be the most efficient (Pesyna *et al.* 1976; McLafferty *et al.* 1980b), is the Probability Based Matching (PBM) system. The only interpretive system which is generally available is the "Self-Training Interpretive and Retrieval System," STIRS (Kwok *et al.* 1973). Both are made available to outside users by the Office of Computer Services, Cornell University, through the TYMNET and TELENET computer networking systems, and PBM is available from the NIH-EPA Chemical Information System (Dr. G. W. A. Milne, NIH, Bethesda, MD) on the TYMNET networking system.

10.1 Retrieval: the Probability Based Matching system

Satisfactory matching of an unknown and a reference spectrum constitutes the final proof of molecular identity (Section 5.5). However, evaluation of the reliability of such an identification must also consider the limitations of mass spectrometry; peak-abundance agreement within a very small experimental error is often observed between spectra of stereo- and geometric isomers, and is possible between spectra of other closely similar compounds. The most reliable match is obtained by running the unknown and reference mass spectra under closely identical experimental conditions on the same instrument. However, the possible structures for the unknown must be reduced to a relatively small number of available reference compounds before their comparison is feasible experimentally.

To narrow the field of possible molecular assignments for the unknown as rapidly as possible, the PBM system matches the unknown against all available reference spectra (the current file contains 41,429 spectra of 32,403 different compounds). Of necessity, these have been obtained from a wide variety of sources and under a wide variety of experimental conditions, so that large abundance differences are possible between unknown and reference spectra of the same compound. This is a major problem in the design of retrieval programs. The original PBM system (Pesyna *et al.* 1976) was shown to be superior to other available systems in identifying several hundred unknown spectra selected at random. Some more recent modifications give substantial improvements (McLafferty *et al* 1980b).

PBM incorporates two unique features: (a) data "weighting," proven valuable for document retrieval from libraries; and (b) "reverse search." The weighting involves the two principal types of data in low-resolution mass spectra: masses and abundances. The probability that particular abundances will occur follows a log-normal distribution. The probability of occurrence of most mass values also varies in a predictable manner, as discussed under rules for peak "importance" (Chapter 3). The larger molecular fragments tend to decompose to give smaller fragments; so the probability of higher m/z values decreases by a factor of two approximately every 130 mass units. The second unique feature of PBM, "reverse searching," is valuable for the identification of components in mixtures. In this, PBM ascertains whether the peaks of the reference spectrum are present in the unknown spectrum, not whether the unknown's peaks are in the reference. Thus the reverse search in effect ignores peaks in the unknown which are not in the reference spectrum, since these could be due to other components of the mixture.

PBM performance has also been evaluated statistically to find out the actual reliability of the predictions, based on the degree of match found between the unknowns and the reference spectra. Thus a reported 90 per cent reliability value indicates that in 90 per cent of the evaluation cases in which PBM retrieved a reference spectrum with this degree of similarity, the answer was correct (this measure of reliability also assumes that a spectrum of the unknown compound is present in the reference file searched). Reliability figures are reported for two degrees of structural similarity, Classes 1 and 4. For a Class 1 match, the answer is considered correct only if the retrieved spectrum is of the same compound or a stereoisomer. For Class 4 matches this definition is broadened to include also ring-position isomers, homologs, and compounds which can be made the same by moving a double bond ≤ 25 per cent of the chain length or moving one carbon atom.

Table 10.1 illustrates the PBM results for an unknown spectrum supplied by Dr. Henneberg (1980) as part of a comparative study of proposed identification algorithms. 1-Dodecene was the correct answer, and all five spectra of this compound in the data base were found by PBM as the best matches.

Table 10.1. *PBM best matches for an "unknown" mass spectrum of 1-dodecene.*

Reference	Per cent reliability	
	Class 1	Class 4
1-Dodecene	69	94
1-Dodecene	59	87
1-Dodecene	59	87
1-Dodecene	47	73
1-Dodecene	38	64
5,6-Methylenedecane	32	55
Methylundecanethiol	29	52
1-Pentadecene	23	42
Undecanol	22	41
1-Tridecene	17	32
6-Methylheptanol	14	28

Spectra of 13 other isomeric $C_{12}H_{24}$ compounds are also in the data base; one of these gave nearly as reliable a match as the most poorly matching 1-dodecene spectrum. Obviously such Class 1 results for a true unknown would leave substantial doubts about the identification (consistent with the similarity often found for spectra of isomeric hydrocarbons, Section 9.1), although the selection of the same compound for the best five matches provides useful substantiation. If a more reliable identification is necessary, the reference spectrum of 1-dodecene should be obtained under the same experimental conditions for comparison.

10.2 Interpretation: the Self-Training Interpretive and Retrieval System

If PBM cannot identify the unknown, which it could not if there is no corresponding reference spectrum in the data base, an interpretive algorithm can be used to find partial structural information. This should aid in interpretation of the spectrum and in structural identification of the unknown. STIRS combines a knowledge of mass-spectral fragmentation rules with an empirical search for correlations of reference spectra. To accomplish the former, 16 classes of mass-spectral data have been selected which are indicative of particular types of structural features; for example, low-mass ions can be characteristic of electron-donating species such as the amino group and aromatic rings, whereas the masses of neutrals lost from the molecular ion can indicate electronegative functionalities (Chapter 5). However, there are no predesignated spectra/structure correlations; instead, STIRS is "self-training," identifying particular molecular features by matching the unknown's spectral data in each of the 16 classes against the corresponding data of all reference file spectra to find those which are the most similar. If a particular substructure is found in a significant proportion of these compounds, its presence in the unknown is probable. Absence of a substructure is not predicted, since the mass-spectral features of one substructure can be made negligible by the presence of a more powerful fragmentation-directing group. The data base for the system includes information from 29,468 different organic compounds containing the common elements, H, C, N, O, F, Si, P, S, Cl, Br, and/or I, and no others.

To use the information provided by the STIRS system, one examines the compounds of the best-matching reference spectra found for each data class and identifies their common structural features. To aid the interpreter in this process, the computer examines the data for the presence of 600 common substructures (McLafferty *et al.* 1980b). The reliability of any prediction that a substructure is present in the unknown is also calculated from the statistics obtained by STIRS on 900 randomly selected unknown spectra. STIRS also predicts the molecular weight of the unknown compound producing the mass spectrum; for randomly selected unknowns, the first-choice value was correct in 91 per cent of the cases, and the first or second in 95 per cent of the cases (McLafferty *et al.* 1980b). The program also predicts the rings-plus-double-bonds value of the unknown (Dayringer and McLafferty 1977), and the number of chlorine and bromine atoms in peak clusters of the unknown spectrum, by matching against the theoretical isotopic abundance values (Mun *et al.* 1977).

The basic information available from STIRS is illustrated in Figure 10A by the best matching compounds found for three of the STIRS data classes using the unknown spectrum shown. Data class 2A uses the four most abundant even-mass and four most abundant odd-mass peaks in the range

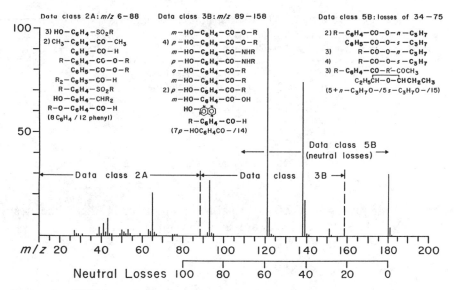

Figure 10A. STIRS results from the mass spectrum of **n**-propyl hydroxybenzoate.

m/z 6–88. The 12 reference compounds whose data in this class match best the corresponding data of the unknown all contain a phenyl group (of which 8 are disubstituted), strongly indicating that the unknown compound also contains a phenyl group. Data class 3B uses entirely different spectral data, the seven most abundant peaks in the range m/z 89–158. Again almost all the best-matching reference compounds contain phenyl, but in addition most have both a hydroxy and a carbonyl group substituted on the aromatic ring. However, note that the carbonyl group is contained in a variety of functionalities: ester, amide, ketone, acid, and aldehyde. The nature of this functionality is much better indicated by the neutrals lost from the molecular ion (Chapter 5); data class 5B uses the five most abundant peaks giving mass-loss values of 34–75. Here most of the best matching compounds are propyl esters. Adding the masses of disubstituted phenyl (76), hydroxyl (17), carbonyl (28), and propoxy (59) gives a predicted molecular weight of 180 for these substructural units, corresponding to the highest mass peak in the unknown spectrum, indicating that the molecule is a propyl ester of hydroxy-benzoic acid. The interpreter might also notice that $(M - 29)^+$ is much larger than $(M - 15)^+$, a strong indication that this is actually the n-propyl ester (an α-cleavage, Section 4.6), although STIRS has not made this differentiation. The best matching compounds of data classes 2A and 3B show no consistency for o-, m-, or p-ring positions; the interpreter would recognize that mass spectrometry is generally poor in making such a differentiation. (To reiterate, STIRS is an aid to, not a replacement for, the trained interpreter.) The computer will also search these lists of best matching com-

Table 10.2. *STIRS substructure predictions from the spectrum of* p-HO—C_6H_5—CO—O—n-C_3H_7

Substructure	Data class	Reliability (percentage)
HO—C_6H_4—CO—	3B	95
HO—aryl—	3B	95
—aryl—CO—	3B	95
aryl—CO—O—	3B	78
—CO—O—C_3H_7	5B	94

pounds for the presence of 600 substructures found to be predicted the most reliably by STIRS; these results are shown in Table 10.2.

10.3 Application of PBM and STIRS

Proficiency in the interpretation of mass spectra as described in previous chapters of this book requires practice; so does efficient use of PBM and STIRS for real unknowns. The description of these algorithms was purposely left until the end of this book, since a real understanding of their utility demands a good general knowledge of spectral interpretation. Several years of use in many laboratories on a variety of unknowns has demonstrated conclusively, in my opinion, that these programs can be a real aid to the interpreter, rapidly narrowing the field of possible compounds, and in particular cases providing structural information which could not be discerned even by experienced interpreters. Because PBM and STIRS results can now be obtained relatively easily and inexpensively, they should be used routinely for more difficult unknown spectra.

11

SOLUTIONS TO UNKNOWNS

Instructions for solutions to the unknowns

As has been emphasized in other parts of this book, the most important way to learn how to interpret mass spectra is by actually working out structures from unknown spectra. Whether this book is used in a formal course or for self-instruction, it is important to work through a variety of unknown spectra. In attempting to solve a particular unknown, use the solution here only when you think you know the answer, or when you cannot go further. When you need help, use the solution given here only to get you past your particular difficulty and keep trying to arrive at your own answer. Reference spectra should *not* be used to solve these unknowns. These unknowns are designed to illustrate principles set forth in the text. Most are common molecules; so little will be learned in locating them in a reference file of spectra that is properly indexed.

Follow *all* the applicable steps of the outline (inside the back cover) which you have covered in the text. The solutions will follow this order, although space limitations prohibit a discussion of every point for each unknown.

The spectra have been taken from the files of this laboratory or from the literature. The data have been corrected for contributions from impurities, ion-molecule reactions, and background, where such corrections appeared to be necessary. The abundance of a peak due to the contribution of natural isotopes is accurate within the larger of the following limits: ±10 per cent relative to the abundance of the peak, or ±0.2 per cent relative to the abundance of the largest peak in the spectrum. Peaks of small abundance that were thought to be inconsequential to the solution of the spectrum have been omitted. This includes unimportant peaks below 1 per cent (relative to the most abundant peak) at the low mass end of a group of peaks of unit mass separation, or at the low mass end of the spectrum.

1.1. This is the mass spectrum of water. If any peak in the spectrum corresponds to the mass of the molecule (that is, molecular weight), this peak must be the one of highest mass. This spectrum, therefore, indicates that the sample has a molecular weight of 18. This molecule must contain elements of atomic weight no greater than that of oxygen, such as hydrogen, carbon, nitrogen, or oxygen, which have atomic weights of 1, 12, 14, and 16, respectively. An obvious combination is H_2O. This is confirmed by the peaks at masses 17 and 16 which represent the molecular fragments HO and O, respectively. [Water]

1.2. For this spectrum a molecular weight of 16 is indicated, which could correspond to CH_4. The other peaks in the spectrum support this, each of them corresponding to a carbon atom to which is attached a smaller number of hydrogen atoms. [Methane]

1.3. A molecular weight of 32 corresponds to two or less oxygen atoms. O_2 is ruled out, as its only possible fragment would have a mass of 16. The elemental composition CH_4O is indicated. The only possible arrangement for these is CH_3OH, and the other peaks in the spectrum can all be justified as pieces of such a structure. [Methanol]

1.4. The knowledge that the main gaseous components of *air* are nitrogen, oxygen, and argon identifies the large 28, 32, and 40 peaks as N_2, O_2, and Ar, respectively. Masses 29 and 34 are due to heavy isotopes of nitrogen and oxygen. The significance of such naturally occurring isotopes will be discussed later. Peaks at masses 14, 16, and 20 represent monoatomic and doubly charged species (for example, m/z $Ar^{2+} = \frac{40}{2} = 20$). The peaks at m/z 17 and 18 are due to water (Unknown 1.1); mass 44 is discussed in Unknown 1.5. [Nitrogen, oxygen, argon, water]

1.5. If the isotopic peaks of masses 45 and 46 are ignored, the base (largest) peak in the spectrum indicates a molecular weight of 44. The major peaks at 12 and 16 indicate that the compound contains carbon and oxygen, and m/z 28 corresponds then to CO. Note that there are no peaks at masses 13, 14, and 15 which were indicative of CH_n groups in the spectra of CH_4 and CH_3OH. The formula of carbon dioxide does fit the observed peaks, with the m/z 22 corresponding to CO_2^{2+}. [Carbon dioxide]

1.6. Ignoring m/z 27 brings us to an indicated molecular weight of 26. The only apparent logical combination of atoms is C_2H_2, a fact which can be deduced from the presence of peaks at masses 12 (C_1) and 24 (C_2). Peaks at m/z 13 and 25 correspond to CH and C_2H, respectively, and the remaining small peaks contain heavy isotopes. [Acetylene]

1.7. [Hydrogen cyanide]

1.8. [Fluoromethane]

1.9. [Methanal]

2.1. This spectrum *could* indicate a molecule of mass 38 which could form an abundant fragment ion by the loss of two hydrogen atoms (m/z 36, $[M - 2]^+$). The hints in the text should lead to the observation in Table 2.1 of the 3/1 abundance ratio of the natural chlorine isotopes, masses 35 and 37. Thus, the spectrum is a mixture of $H^{35}Cl$ and $H^{37}Cl$, in proportions corresponding to the isotopic abundances of ^{35}Cl and ^{37}Cl. [Hydrogen chloride]

2.2. The large peaks at masses 94 and 96 should arouse your suspicions; bromine has isotopes of mass 79 and 81 and of nearly equal abundances. Because masses 94 and 96 represent the molecular ions, by difference the molecule contains 15 mass units in addition to the bromine atom; the most logical assignment for this is CH_3. Peaks at masses 95 and 97 are due to ^{13}C isotopes. [Methyl bromide]

2.3. Look first for the "A + 2" elements. In this case $[m/z\ 66]/[m/z\ 64]$ equals 5.0 per cent; the isotopic abundance table shows only two elements of such $[(A + 2)^+]/[A^+]$ values, Si (3.4 per cent) and S (4.4 per cent). The former is outside the allowed maximum relative error of 10 per cent. Thus sulfur is indicated, with the ^{33}S abundance of 0.8 agreeing with $[(A + 1)^+]$. (Although S_2 would have the correct mass, for this the abundance of the m/z 66 peak would be 8.8 per cent, since there would be twice the probability that one of the two S atoms would be ^{34}S.) With S_1 present in m/z 64, the residual 32 mass units must be assigned; O_2 is an obvious possibility. Each of these would contribute 0.2 per cent to the probability of an isotopic m/z 66 peak, or with S_1 a total of 4.8 per cent, agreeing well with the observed 5.0 per cent. The mass 32 difference between M^{+} and ^{32}S might also be assigned to CH_4O, but (as will be discussed in Section 2.3) this would cause a 1.1 per cent ^{13}C contribution to m/z 65, which would make the predicted abundance much too large. [Sulfur dioxide]

2.4. If the mass 43 peak were due to $C_2H_3O^+$, the expected 44/43 abundance ratio would be 2.2 per cent from the two ^{13}C atoms present. Thus the observed ratio corresponds more closely to that predicted for three carbon atoms, such as in the ion formula $C_3H_7^+$. However, one should always be aware of the possible contribution from "background" in the instrument, which is much more likely to give a high (*not* low; why?) indication of the number of carbon atoms. On the other hand, if an oxygen atom is present in the m/z 43 ion, there should be a 0.2 per cent contribution to the m/z 45 peak due to $C_2H_3^{18}O$; unfortunately, this is just on the limit of the absolute error assumed possible in the abundance data of these unknown spectra.

To calculate the number of carbon atoms in the large m/z 58 peak, note that the abundance ratio of masses 59/58 is 4.4 per cent. From this, Table 2.2 indicates four carbon atoms; a smaller number would not be possible unless m/z 59 contained an impurity or extraneous fragment ion contribution. Thus m/z 58 corresponds to C_4H_{10} (butane). The loss of 15 to give the base peak at m/z 43 is most logically a methyl group, confirming the prediction of the formula $C_3H_7^+$ for this fragment ion. [*n*-Butane]

2.5. The $[m/z\ 79]/[m/z\ 78]$ value of 6.8 per cent corresponds nicely to six carbon atoms (Table 2.2). By difference there should also be six hydrogen atoms in m/z 78, for a probable elemental composition of C_6H_6. The aromatic molecule benzene is a possible structure whose molecular ion should have the indicated high stability. [Benzene]

2.6. The spectral region m/z 71–74 indicates that no "A + 2" elements other than oxygen are possible; otherwise $[m/z\ 74]/[m/z\ 72] > 3$ per cent. From the $[73]/[72]$ value, Table 2.2 indicates three carbon atoms in the base peak at mass 72; however, for this, the m/z 74/72 ratio should be 0.04 per cent in comparison to the 0.5 per cent found. Check the common isotopes of Table 2.1 for an explanation. The C_3 accounts for only 36 amu of the molecular weight of 72. Both this difference and the m/z 74/72 ratio can be accounted for by assuming the presence of two oxygen atoms. For these there will be twice the probability that one of the oxygens of the molecule is an ^{18}O. Thus, from Table 2.2 the m/z 74/72 ratio should be 0.04 per cent + 2 times 0.20 per cent = 0.44 per cent, checking the 0.5 per cent found. The molecular formula $C_3H_4O_2$ is thus indicated. Similar calculations indicate $C_3H_3O^+$ for m/z 55 (loss of OH). Interferences prevent such calculations on other peaks (m/z 46/45 indicates a maximum of C_7). In interpreting the spectrum, the prominent m/z 27 must be C_2H_3 (it is far too abundant to be mass 54^{+2}). $C_3H_4O_2 - C_2H_3 = CO_2H$, suggesting (although not requiring) a carboxyl group as an assignment for the m/z 45. The only acid molecule of this composition is CH_2=CHCOOH. [Acrylic acid]

2.7. The striking feature of the isotopic ratios that you should note in this spectrum is the low $[M + 1^+]/[M^+]$ ratio and similar ion abundance ratios. Because of these ratios the elements carbon, oxygen, silicon, sulfur, and chlorine are all eliminated from consideration. Note that there are abundant peaks at $[M - 19]^+$ and $(M - 2 \times 19)^+$ as well as m/z 19 indicating the presence of fluorine. [Nitrogen trifluoride]

2.8. The presence of the "A + 2" element chlorine should be obvious from both the m/z 104/106 and 85/87 abundance ratios. The m/z 69/70 abundance ratio indicates the presence of one carbon atom, which is consistent with both m/z 85/86 and 50/51. The remaining mass of each of these ions must be provided by an "A" element, and only fluorine appears to be appropriate. [Chlorotrifluoromethane]

2.9. The $[m/z\ 134]/[m/z\ 132]$ ratio (see Table 11.1) clearly shows that no silicon, sulfur, chlorine, or bromine atoms are present. The $[m/z\ 133]/[m/z\ 132]$ value indicates C_9 (however, if the relative error limits had been slightly larger, both C_8 and C_{10} would have to be considered). This accounts

Table 11.1. *Data for Unknown 2.9.*

m/z	Rel. abund.	O ≤4	C_9	O_1
130	1.0			
131	2.0			
132	100.	100.	100.	100.
133	9.9 ±1.0		9.9	
134	0.7 ±0.2	≤ 0.8	0.44	0.20

for only 0.44 of the $[m/z\ 134]$ value of 0.7; ^{18}O nicely accounts for the difference. The remainder of the m/z mass value, which must be due to "A" elements, is accounted for by H_8. $[C_9H_8O, r + db = 6]$

2.10. The abundance of the $m/z\ 78$ peak (see Table 11.2) shows immediately that no "A + 2" elements can be present. For the "A + 1" elements, Table 2.2 tells us that the possibilities are C_3 and C_2N_3. For reasons discussed in

Table 11.2. *Data for Unknown 2.10.*

m/z	Rel. abund.	Normalize	C_3	C_2	N_3	(N_2)
73	26.					
74	19.					
75	41.					
76	80.	100.	100.	100.	100.	100.
77	2.7 ±0.27	3.4 ±0.34	3.3	2.2	1.1	0.72
78	<0.1	<0.13	0.0	0.0		

Chapter 3, the $m/z\ 76$ molecular ion cannot contain an odd number of nitrogen atoms, so that C_3 is correct. A possible combination of "A" elements is H_9P. ($C_3H_2F_2$ could not fragment to yield $m/z\ 73$.) $[C_3H_9P, r + db = 0]$

2.11. If $m/z\ 86$ and 87 are only isotope peaks of $m/z\ 85$, then the presence of one silicon atom explains their relative abundances. The absence of $m/z\ 84$ actually is evidence that there are no hydrogen atoms in $m/z\ 85$. A logical composition for the "A" elements is F_3. $[SiF_3, r + db = 0]$

Table 11.3. *Data for Unknown 2.11.*

m/z	Rel. abund.	Normalize	Si_1	$C_{<1}$[a]
84	0.0	0.0		
85	40.	100.	100.	100.
86	2.0 ±0.2	5.0 ±0.5	5.1	<1.1
87	1.3 ±0.2	3.3 ±0.5	3.4	
88	0.0 ±0.2	0.0 ±0.5		

[a]Note that *all* isotopic contributions of the "A + 2" elements (including that to A + 1) must be subtracted out before calculating the "A + 1" elements.

2.12. The most abundant peak, $m/z\ 182$ (see Table 11.4) can contain no "A + 2" elements except oxygen; the indicated abundance of the "A + 1" elements is sufficiently great that these must be approximated before a calculation for oxygen can be made. The $[m/z\ 183]/[m/z\ 182]$ value of 15.1 per cent indicates C_{13} to C_{15} (the stated error limit of ±10 per cent means that values in the range 13.6 to 16.6 per cent must be considered). The abundance

Table 11.4. *Data for Unknown 2.12.*

m/z	Rel. abund.	Normalize	C_{13}	C_{14}	C_{15}	O_1
179	2.2					
180	1.1					
181	7.4					
182	55.	100.	100.	100.	100.	100.
183	8.3 ±0.83	15.1 ±1.5	14.3	15.4	16.5	
184	0.6 ±0.2	1.1 ±0.36	0.94	1.1	1.3	0.2

values predicted for m/z 184 for C_{13}, C_{14}, and C_{15} are 0.9, 1.1, and 1.3, respectively, if no oxygen is present. Experimentally, the value of $[m/z\ 184]/$ $[m/z\ 182]$ can be 0.73 to 1.45, allowing the formulas $C_{13}O_{0-2}$, $C_{14}O_{0-1}$, and C_{15}. However, $C_{13}O_2$ (mass 188) and $C_{14}O$ (mass 184) are impossible, and C_{15} would only be possible for m/z 182 with the unlikely molecular formula $C_{15}H_2$ ($182 - 15 \times 12 = 2$). Assigning "A" elements, the remaining possibilities are $C_{13}H_{26}$, $C_{13}H_7F$, $C_{13}H_{10}O$, and $C_{14}H_{14}$. Improved experimental accuracy or composition information from other spectral peaks would be necessary to restrict this list further. $[C_{13}H_{26},\ C_{13}H_7F,\ C_{13}H_{10}O,\ C_{14}H_{14}]$

2.13. Abundant peaks formed by the loss of H_2 are unusual; the absence of an $(M - 3)^+$ would also be unusual if m/z 172 were the $(M - 2)^+$ peak. You should recognize that two peaks of nearly equal abundance separated by two mass units can be due to Br_1 (Table 11.5); this is true of m/z 173:175 as well as 172:174. Using m/z 172 as the A peak then indicates Br_1 and C_6.

Table 11.5. *Data for Unknown 2.13.*

m/z	Rel. abund.	Normalize	Br_1	C_6	O_1
171	0.1				
172	98.	100.	100.	100.	
173	6.7	6.9 ±0.7		6.6	
174	100.	102. ±10	98.	(102.)	(102.)
175	6.5	6.7 ±0.7		(6.6)	
176	0.5	0.5 ±0.2		(0.18)	(0.2)

Note that the latter can also be derived from $[175^+]/[174^+]$, since almost all ions of the m/z 174 peak have the composition $^{12}C_6{}^{81}Br$. This also means that the $[176^+]/[174^+]$ ratio can indicate the "A + 2" elements other than Br_1, so that at least one oxygen must be present. $[C_6H_5BrO,\ r + db = 4]$

2.14. In attempting to deduce elemental compositions in this group of peaks, start with the most abundant peak. The presence of one or more "A + 2" elements should be immediately obvious. Although the $[m/z\ 132]/$ $[m/z\ 130]$ value of 0.98 strongly suggests Br_1, this is ruled out by the higher mass ratios. Note that the ratios of the series $[129]/[131]/[133]/[135]$ are very similar to those of the series $[130]/[132]/[134]/[136]$, and that these correspond closely with the "isotopic cluster" for Cl_3 in Figure 2B and Table A.2. The number of "A + 1" elements is indicated from the peaks which contain only the ^{37}Cl isotope; this $[m/z\ 137]/[m/z\ 136]$ value of 0.0 ± 0.2 indicates $C_{\leq 5}$. (Bar graphs indicating these isotopic contributions are shown in Section 2.7.) The mass difference due to "A + 1" and "A" elements of 25 $(130 - 3 \times 35 = 25)$ can only be due to C_2H, with consideration of valence restrictions; this assignment is also consistent with the absence of an $(M - 2)^+$ peak. $[C_2HCl_3,\ r + db = 1]$

3.1. $C_2H_4^+$, m/z 28; $C_3H_7O^+$, m/z 59; $C_4H_9N^+$, m/z 71; $C_4H_8NO^+$, m/z 86; $C_7H_5ClBr^+$, m/z 203; $C_6H_4OS^+$, m/z 124; $C_{29}F_{59}^+$, m/z 1469; H_3O^+, m/z 19; and $C_3H_9SiO^+$, m/z 89.

3.2. $C_{10}H_{15}O^+$ is an even-electron ion, and so cannot be the molecular ion.

3.3. $C_9H_{12}^+$ is an odd-electron ion. The other ions represent losses of H, H_3 $(H + H_2$, for example), CH_3, $CH_3 + H_2$, and C_2H_5, respectively, all of which are possible losses. *n*-Propylbenzene would exhibit such a spectrum.

3.4. The presence of the large m/z 96 and 98 peaks casts doubt on the assignment of m/z 100 as the molecular ion. None of the common isotopes corresponds to this type of isotopic cluster, however. The ions of lower mass give helpful hints for elucidating this. Note particularly the m/z 63/61 ratio, the chlorine signpost. Formation of m/z 61 from m/z 96 involves the loss of 35 daltons, a hint for a second chlorine atom in the molecule. This is confirmed by checking the isotopic abundances of masses 96/98/100 against Figure 2B or Table A.2. To calculate the carbon content of M^+ from m/z 97/96, do not forget that the $^{35}Cl^{37}Cl$ isotopic ion of $(M-1)^+$ will contribute 65 per cent of the m/z 95 to the 97 peak: $(3.3\% - 0.65 \times 3.0\%)/67\% \times 1.1\% = 1.8$ or $\sim C_2$. This calculation is superfluous, however, since the balance of molecular formula of 26 amu $(M^+ - Cl_2 = 96 - 70)$ is most logically due to C_2H_2. It is difficult to distinguish between the possible isomers of $C_2H_2Cl_2$ without standards. [*trans*-1,2-Dichloroethylene]

3.5. The isotopic cluster at m/z 135–137 indicates one chlorine atom, and m/z 135 contains a maximum of two carbon atoms, $0.5/(1.1\% \times 24) = 2.0$. Similar assignments can be made for the lower mass ions: m/z 119, C_2Cl_0; m/z 85, C_1Cl_1; m/z 69, C_1Cl_0; and m/z 31, C_1Cl_0. The absence of evidence for hydrogen atoms and the differences of 19 amu between several pairs of ions containing the same numbers of chlorine atoms implicate fluorine as the missing element. Thus, the formulas CF^+, CF_2^+, CF_3^+, $CClF_2^+$, $C_2F_4^+$, $C_2F_5^+$, and $C_2ClF_4^+$ will correspond in mass to the prominent peaks in the spectrum. You should have already noted that the m/z 135 and 137 peaks failed several of the requirements for a molecular ion. Because there is no indication of ions containing more than two carbon atoms or one chlorine atom, the most probable structure to account for this spectrum is $CClF_2CF_3$. [Chloropentafluoroethane]

3.6. [Carbon tetrachloride]

4.1. The first rule emphasizes the importance of product ion stability. The highly stabilized *p*-aminobenzyl ion (11.1) produces by far the most abundant peak in this spectrum.

$$H_2N-\langle\bigcirc\rangle-CH_2^+ \longleftrightarrow H_2N^+=\langle\bigcirc\rangle=CH_2 \qquad (11.1)$$

4.2. Alpha cleavage can be initiated by the radical site on either the oxygen or the nitrogen atom (11.2, 11.3). The electron-donating ability of nitrogen

$$H_2NCH_2{\overset{\curvearrowleft}{-}}CH_2{\overset{\curvearrowleft}{-}}\overset{+\cdot}{O}H \xrightarrow{\ \alpha\ } H_2NCH_2\cdot + CH_2{=}\overset{+}{O}H \ (m/z\ 31) \qquad (11.2)$$

$$H_2\overset{+\cdot}{N}{\overset{\curvearrowleft}{-}}CH_2{-}CH_2OH \xrightarrow{\ \alpha\ } H_2\overset{+}{N}{=}CH_2 \ (m/z\ 30) + \cdot CH_2OH \qquad (11.3)$$

is superior to that of oxygen; ionization should occur more readily at nitrogen (Table A.3), and the m/z 30 ion should be more stable than m/z 31. Actually $[m/z\ 30] = 57$ per cent of total ionization, whereas $[m/z\ 31] = 2.5$ per cent.

4.3. The ratio $[m/z\ 98]/[m/z\ 97]$ is much too large for m/z 98 to be the $(A + 1)$ isotope peak of an A peak of m/z 97; thus m/z 98 should be considered first as an A peak (see Table 11.6). Looking first for "A + 2" elements,

Table 11.6. *Data for Unknown 4.3.*

m/z	Rel. abund.	Normalize	S_1	(S_1)	C_5
97	100.	179.		(179.)	
98	56.	100.	100.	(1.4)	100.
99	7.6	13.6 ±1.4	0.8	(7.9)	5.5
100	2.4	4.3 ±0.4	4.4		0.1

the value $[m/z\ 100]/[m/z\ 98] = 4.3$ is consistent with S_1. If m/z 98 is the molecular ion, m/z 97 must also contain sulfur; its ^{34}S contribution to m/z 99 must be subtracted before one calculates the "A + 1" elements. The residual abundance $(13.6 - 0.8 - 7.9 = 4.9)$ is consistent with either C_4 or C_5, from which only the elemental composition C_5H_6S $(r + db = 3)$ is logical $(C_4H_{18}S$ gives $r + db = -4)$. The low abundances of other peaks make equivocal the determination of their elemental compositions from isotopic abundances, although CHS^+ is a logical assignment for m/z 45.

The m/z 98 peak passes all tests for the molecular ion, and as such should represent a relatively stable species. The loss of H is the dominant mechanistic feature of the spectrum; its loss could originate from a radical site on either sulfur or a π-system $(r + db = 3)$, or both. Possible structures consistent with this are 2- or 3-methylthiophene or thia-2,5-cyclohexadiene. [2-Methylthiophene]

4.4 through 4.8. Although these are, as stated, isomeric C_4-amines, calculation of the elemental composition of M^+ would be too high in several cases if one were using the m/z 74/73 ratio, mainly because of the production of $(M + 1)^+$ by an ion-molecule reaction (see Section 6.1). Using the righthand ordinate scale, note the substantial differences in base-peak abundances as the percentage of total ions; this is useful in evaluating the relative importance of peaks in different spectra (I thank Dr. R. D. Grigsby for this suggestion).

In answer to the question in the text, loss of an α-methyl group is in competition with the losses of other α-substituents. The probability of α-CH_3 loss is much greater than that for α-hydrogen loss, but less than that for α-C_2H_5 loss. This helps to identify Unknown 4.8 as $C_2H_5CH(CH_3)NH_2$. Of the five α-CH_3 amines listed (see text), only $(CH_3)_2CHNHCH_3$ and $CH_3CH_2N(CH_3)_2$ remain unidentified; the first should have a larger $(M - CH_3)^+/M^+$ ratio, and so is Unknown 4.6. The assignment of the latter structure to Unknown 4.7 is borne out by the m/z 44 of this spectrum; it is formed by the elimination of C_2H_4 from the $(M - 1)^+$ ion through a rearrangement similar to the formation of m/z 30 in $C_2H_5NHC_2H_5$ (see Section 4.10).

The remaining possible isomers are $CH_3CH_2CH_2CH_2NH_2$, $(CH_3)_2CH$-CH_2NH_2, and $CH_3CH_2CH_2NHCH_3$. The first two should yield abundant ions of m/z 30, and are thus Unknowns 4.4 and 4.5. These spectra are surprisingly similar, reflecting the dominant effect of the amine function on the cleavage; however, the small absolute differences in the abundances of the alkyl ions are reproducible and significant. The abundances of the $C_3H_7^+$ and $C_4H_9^+$ ions in Unknown 4.5 are several times the corresponding abundances in Unknown 4.4 (only small contributions should be present from $C_2H_5N^+$ and $C_3H_7N^+$), compatible with n- and isobutylamine assignments for 4.4 and 4.5, respectively. No spectrum is given for n-$C_3H_7NHCH_3$; it should give a base peak at m/z 44, but an insignificant peak at $(M - CH_3)^+$. [4.4, n-Butylamine; 4.5, isobutylamine; 4.6, isopropylmethylamine; 4.7, ethyldimethylamine; and 4.8, *sec*-butylamine]

4.9, 4.10. For the symmetrical 3-pentanone, α- and i-cleavages can only give the $C_2H_5CO^+$ and $C_2H_5^+$ ions, m/z 57 and 29. On the other hand, 3-methyl-2-butanone should yield CH_3CO^+, $(CH_3)_2CHCO^+$, CH_3^+, and $C_3H_7^+$ ions at m/z 43, 71, 15, and 43, respectively. [4.9, 3-Pentanone; 4.10, 3-methyl-2-butanone]

4.11. Even if its elemental composition had not been given, mass 130 appears to be a good possibility for the molecular ion; the ion series (Section 5.2) 31, 45, 59, 73, 87, 101, and 115 strongly suggests $C_8H_{18}O$. Table A.4 shows that the abundance of $M^{\overset{+}{\cdot}}$ is indicative of an ether, not an alcohol. Facile cleavage of C_α—C_β bonds should yield the most abundant $C_nH_{2n+1}O^+$ ions; the $(M - 43)^+$ is the most abundant of these by far, and thus indicates one or two propyl groups substituted on one or both α-carbon atoms. The abundances of $(M - CH_3)^+$ and $(M - C_2H_5)^+$ are 3 and 10 per cent, respectively, of $(M - C_3H_7)^+$, making the presence of methyl or ethyl substituents on an α-carbon atom unlikely. The other common decomposition of ethers via cleavage of the C—O bond to yield the alkyl ion strongly indicates a butyl ether (m/z 57); of course $C_8H_{18}O$ must then be a dibutyl ether. The formula $C_3H_7CH_2OC_4H_9$ is required; $C_3H_7CH_2OCH(CH_3)C_2H_5$ is less probable because of the abundances of $(M - CH_3)^+$ and $(M - C_2H_5)^+$, as discussed. [Diisobutyl ether]

$$C_3H_7-CH_2OC_4H_9 \xrightarrow{-e^-} \begin{cases} \xrightarrow{\alpha} C_3H_7 \overset{+\cdot}{\frown} CH_2\overset{+\cdot}{O}C_4H_9 \longrightarrow \\ \quad C_3H_7\cdot + CH_2=\overset{+}{O}C_4H_9 \\ \\ \xrightarrow{i} C_3H_7-CH_2\overset{+}{O}C_4H_9 \longrightarrow \\ \quad C_3H_7CH_2O\cdot + C_4H_9{}^+ \end{cases} \tag{11.4}$$

4.12, 4.13. The retro-Diels-Alder reactions expected for these isomers are shown here. The low abundance of m/z 164 in the second spectrum is due to the competition of the loss of CH_3 by a displacement reaction (see Reaction 9.29). [4.12, α-Ionone; 4.13, β-Ionone]

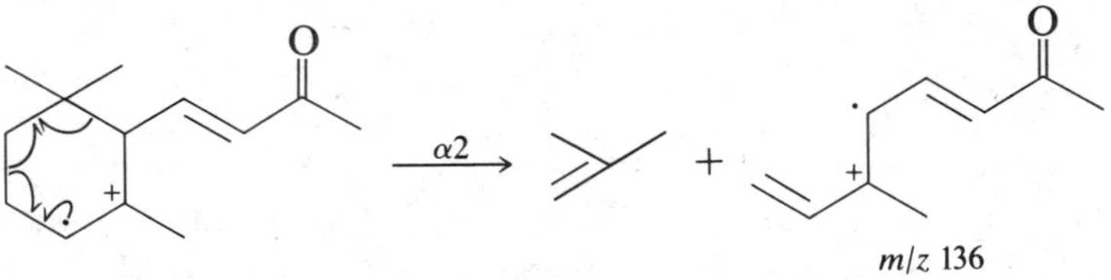

$$\tag{11.5}$$

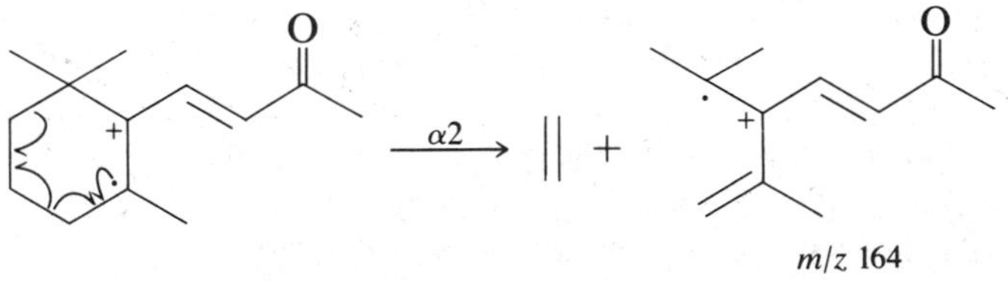

m/z 136

$$\tag{11.6}$$

m/z 164

4.14. See Reactions 11.7 and 11.8.

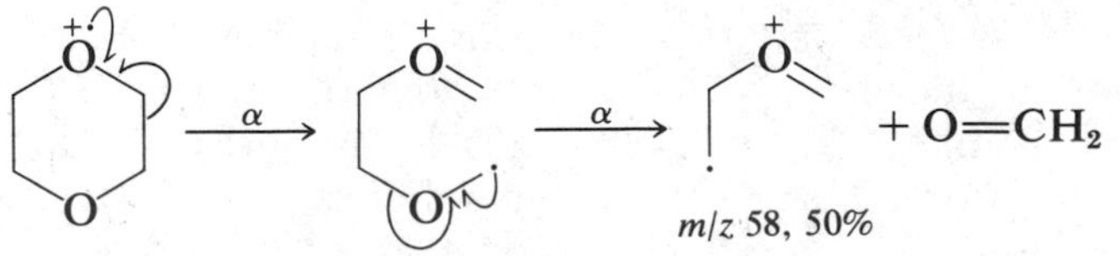

$$+ O=CH_2 \tag{11.7}$$

m/z 58, 50%

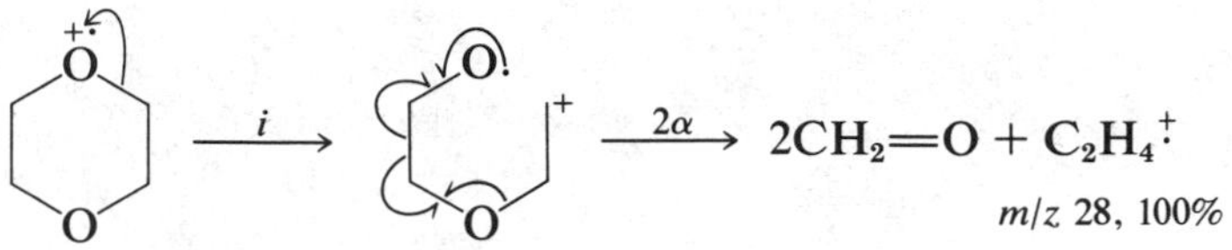

$$\xrightarrow{i} \xrightarrow{2\alpha} 2CH_2{=}O + C_2H_4^{\ddagger} \quad (11.8)$$

m/z 28, 100%

4.15. See Reactions 11.9 to 11.13.

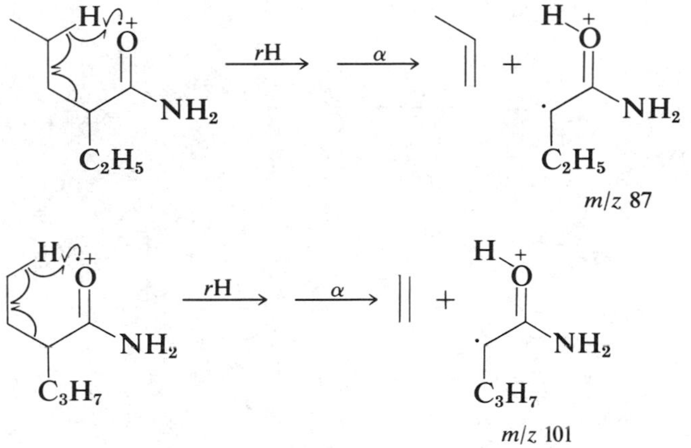

$$\xrightarrow{rH} \xrightarrow{\alpha} \quad \| \quad + \quad$$

C₂H₅   C₂H₅

m/z 87

(11.9)

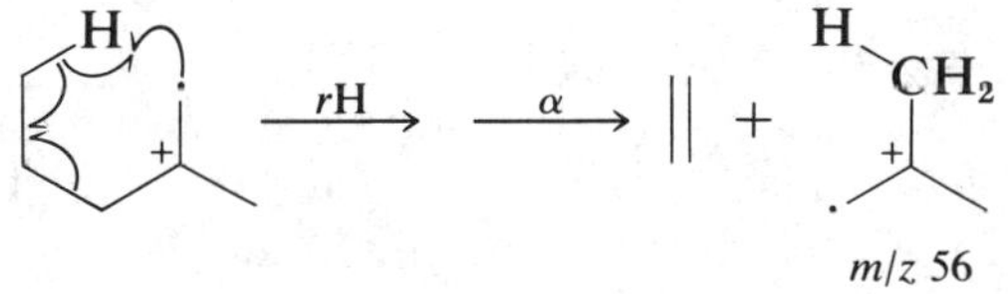

$$\xrightarrow{rH} \xrightarrow{\alpha} \quad \| \quad +$$

C₃H₇   C₃H₇

m/z 101

$$\xrightarrow{rH} \xrightarrow{\alpha} \quad \| \quad +$$

m/z 56   (11.10)

$$\xrightarrow{rH} \xrightarrow{\alpha} \quad \| \quad +$$

OC₂H₅   OC₂H₅

m/z 88

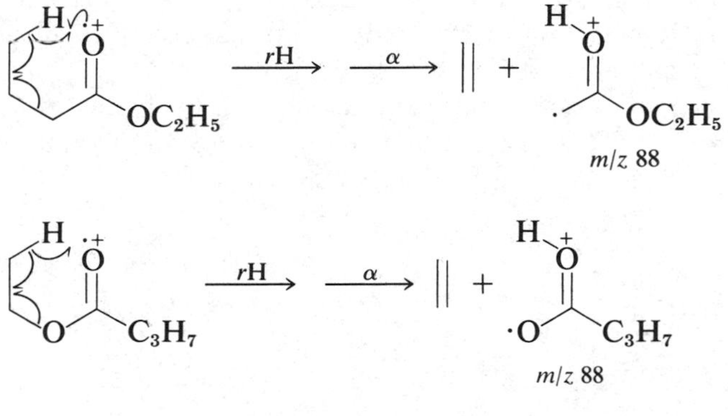

(11.11)

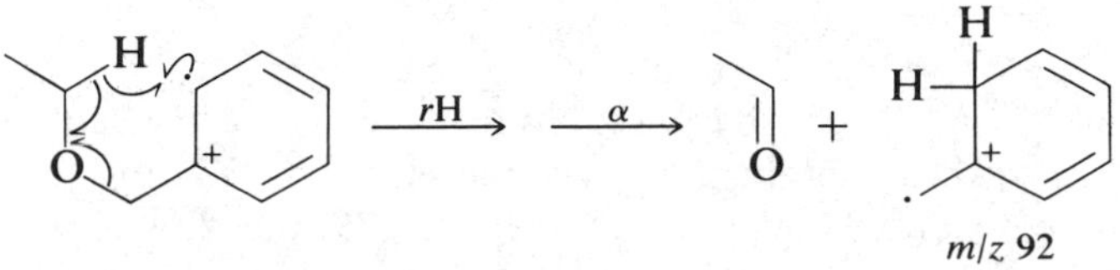

$$\xrightarrow{rH} \xrightarrow{\alpha} \quad \| \quad +$$

C₃H₇   C₃H₇

m/z 88

$$\xrightarrow{rH} \xrightarrow{\alpha} \quad + \quad$$

(11.12)

m/z 92

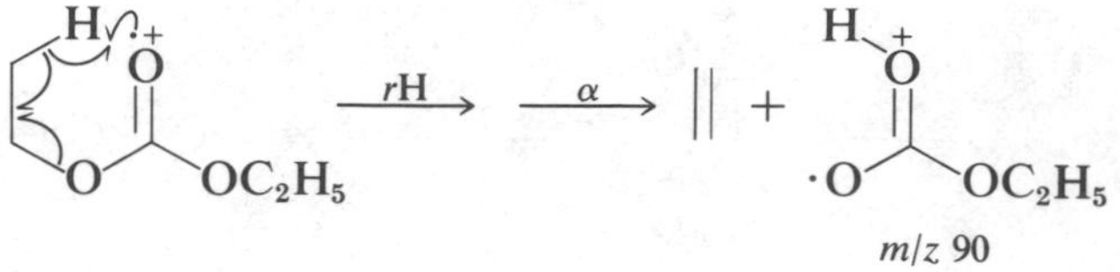

$$(11.13)$$

$m/z\ 90$

4.16, 4.17. See Reaction 11.14. [4.16, 4-Methyl-2-pentanone; 4.17, 3-methyl-2-pentanone]

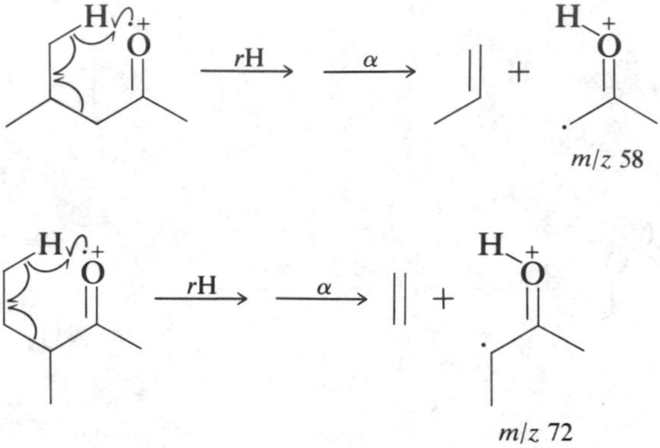

$m/z\ 58$

$$(11.14)$$

$m/z\ 72$

4.18. The apparent molecular ion at m/z 134 contains no "A + 2" elements (except oxygen) and 10 or 11 carbon atoms, consistent with $C_{10}H_{14}$. The corresponding r + db value of 4 is that expected for phenyl; this is supported by the C_7H_7 and C_7H_8 assignments for m/z 91 and 92, and by the "aromatic' ion series" (Section 5.2), suggesting that the molecule is a C_4H_9—C_6H_5 isomer. You should have marked m/z 92, C_7H_8, as an important odd-electron ion; note that it is on the high-mass side of the base peak (rule iii, Section 3.4). Both the formation of $C_7H_8^+$ by rearrangement and $C_7H_7^+$ by α-cleavage indicate the α-position is unsubstituted (see also Tables A.6 and A.7), $C_6H_5CH_2$—C_3H_7. The low abundance ratio of m/z 43/29 indicates that the alkyl chain is not branched. [*n*-Butylbenzene]

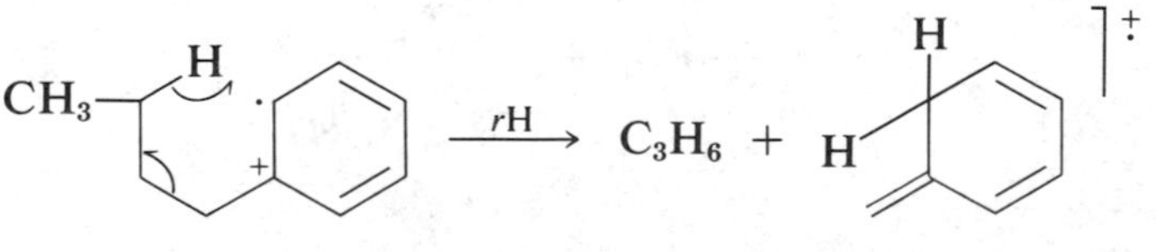

$$(11.15)$$

4.19. Masses 45, 73, and 87 probably have two, four (?), and five carbon atoms each, and thus contain one oxygen atom each. The rings-plus-double-bonds value of each of these and of the m/z 129 ions should thus equal $\frac{1}{2}$, so that they are saturated EE^+ ions. $C_8H_{17}O^+$ cannot be the molecular ion; it is an even-electron ion. The m/z 112 ion is an odd-electron ion. Mass 130 might be $M^{\ddagger}$, but its abundance is within experimental error of that expected for the heavy isotope contribution from m/z 129. The ion series m/z 31, 45, 59, 73, 87 suggests (Table A.6) that an alcohol or ether function is a dominant feature of the molecule; the lack of $M^{\ddagger}$ indicates (Table A.4) that this is not a straight-chain ether. A reasonable hypothesis is that m/z 130 is the molecular ion, and that m/z 112 is the $(M-18)^{\ddagger}$ formed by loss of H_2O from an alcohol. The formation of important $OE^{\ddagger}$ ions from ethers, R—O—R', by the loss of ROH is uncommon. Further, the loss of H_2O from primary alcohols, RCH_2OH, is so facile that $C_nH_{2n+1}O^+$ ions are generally of lower abundance. The abundant $(M-43)^+$ and $(M-57)^+$ ions indicate the identity of the octyl alcohol as C_4H_9—$CHOH$—C_3H_7; the lack of $(M-15)^+$ and $(M-29)^+$ ions indicates that there is no chain branching. (An extra clue is the metastable ion of m/z 96.5; see Section 6.4. For the transition m/z 130 → 112 a metastable would be expected at m/z 96.4; m/z 129 → 112 would appear at m/z 97.0, but would involve the unlikely loss of 17 amu.)[4-Octanol]

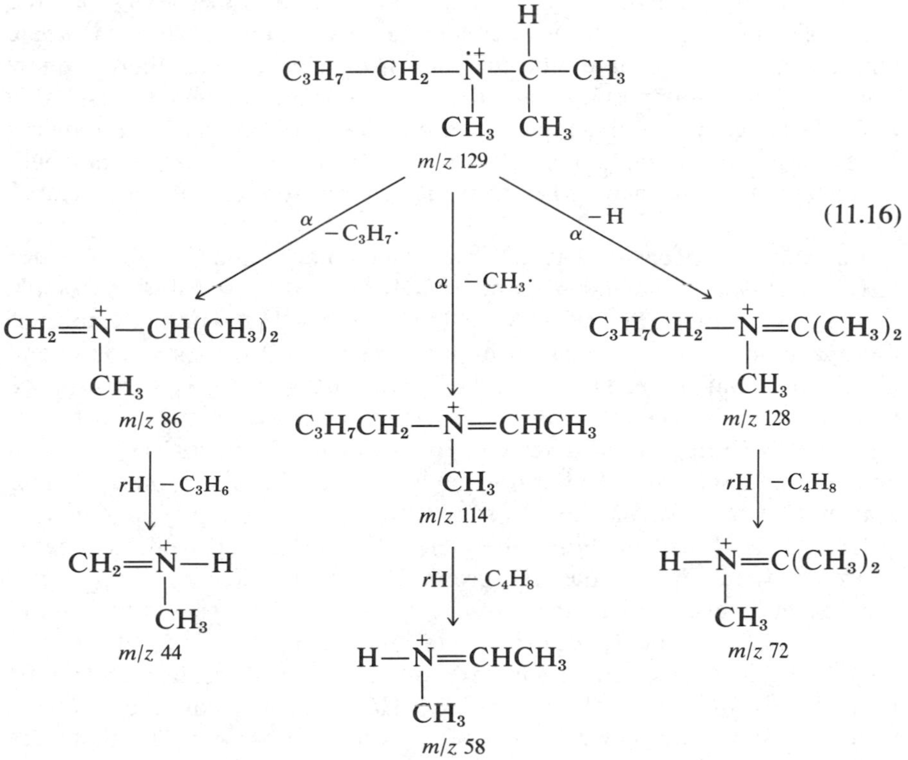

 4.20. Alpha-cleavage reactions can result in the loss (Reaction 11.16) of three different groups: $C_3H_7\cdot$, $CH_3\cdot$, and $H\cdot$. These reactions will produce peaks at m/z 86, 114, and 128, with abundances decreasing in this order as predicted by the rule for the loss of the largest alkyl group. Each of these ions can undergo loss of an olefin by cleavage of an N—C bond with hydrogen rearrangement (Reaction 4.44). See Figure 9R and Section 9.8 for a more thorough discussion of the spectrum of this compound.

4.21. Mass 178 can be postulated as the molecular ion, with roughly twelve carbon atoms. Both m/z 105 and 123 appear to contain seven carbon atoms. The m/z 122 is of questionable significance as an odd-electron ion because of the abundant m/z 123 ion; however, the mass 56 peak, $(M - 122)^{+}$, should be noted. A weak alkyl ion series through C_4 is present, as is an aromatic ion series (see Section 5.2). There are weak alkyl loss ions at $(M - 15)^+$, $(M - 29)^+$, and $(M - 43)^+$, but instead of $(M - 57)^+$ there are the major ions at $(M - 55)^+$ and $(M - 56)^{+}$. From Table A.5 a common loss of 55 is loss of C_4H_7 from esters (double hydrogen rearrangement) and the loss of 56 odd-electron ion supports this; Table A.7 indicates a corresponding C_4H_8—RY structure. Two other abundant ions provide the final clues; m/z 77 and 105 should be phenyl and benzoyl, suggesting that the compound is a butyl benzoate. This is probably not *sec-* or *tert-*butyl; for example, $(M - 15)^+$ and $(M - 29)^+$ are less abundant than $(M - 43)^+$. [*n*-Butyl benzoate]

5.1. Although the possible molecular ion at m/z 128 probably has the composition $C_{10}H_8$, a C_9 formula cannot be ruled out on the basis of isotopic abundances. The probable molecular ion is of high stability; there are few fragment peaks of importance, and these resemble the aromatic series (Table A.6). The large M^{2+} is also typical of aromatics. The formula $C_{10}H_8$ requires seven rings plus double bonds $(10 - \frac{8}{2} + 1)$. Again, reference spectra would be necessary to distinguish between the several isomers possible. [Naphthalene]

5.2. Mass 86 appears to have the elemental composition $C_4H_6O_2$ (a better check with isotopic abundances than $C_5H_{10}O$), and for m/z 43 the formula $C_2H_3O^+$ is definitely preferable. Mass 15 should be CH_3^+. The appearance of the spectrum (that is, dominated by mass 43 and 15) suggests C_2H_3O and CH_3 groups that are readily lost. If the CH_3 is part of the C_2H_3O, the balance is CO. The absence of other peaks suggests a symmetrical structure, CH_3CO—$COCH_3$. Most other compounds of the formula $C_4H_6O_2$ would be expected to give some other fragments. The $(M - CH_3)^+$ is much smaller than CH_3^+ because of the electronegative groups on the concomitant —$COCOCH_3$ fragment. When the CH_3—CO bond is cleaved, the charge is preferably retained by the CH_3 group. [2,3-Butanedione]

5.3. Note that the abundances of m/z 181 to 184 are identical to those of Unknown 2.12. Of the four possible elemental compositions for these data, only $C_{13}H_{10}O$ is consistent as a molecular ion which can produce the m/z 105 fragment $C_7H_5O^+$ as well as m/z 77 $C_6H_5^+$. The abundance of m/z 182 suggests a stable molecular ion, in keeping with r + db = 9. Notice that there is a definite resemblance between the spectra of Unknowns 5.1 and 5.3; the peaks at 39, 51, 63, and 77 suggest aromatic character. From the simplicity of the spectrum, the molecule should be generally stable, with one or a few bonds of much higher lability. The compositions of the major fragments, $C_6H_5^+$ and $C_7H_5O^+$, when combined are equivalent to the proposed molecular formula, and differ by CO, suggesting that these substructures are C_6H_5— and C_6H_5—CO—. [Benzophenone]

5.4, 5.5, 5.6. The $C_nH_{2n+1}^+$ ion series is strongly indicated, and its distribution should suggest the n-alkane $C_{16}H_{34}$ as the compound yielding spectrum 5.4. (Compare this spectrum to Figures 3B and 5B.) Several of the prominent $C_nH_{2n+1}^+$ ions of Unknowns 5.5 and 5.6 appear paired with a $C_nH_{2n}^{\overset{+}{\cdot}}$ ion; loss of (R + H) is also a common decomposition reaction for a compound with an R alkyl branch. Thus in Unknown 5.5 the ions of m/z 211, 168–169, 140–141 (weak), and 84–85 indicate chain branchings corresponding to C_1—C_{15}, C_4—C_{12}, C_6—C_{10}, and C_{10}—C_6. The grouping n-C_4H_9—CH(CH$_3$)— will explain these data. The compound is actually $CH_3(CH_2)_3CH(CH_3)(CH_2)_9CH_3$, although the compound $CH_3(CH_2)_3$-$CH(CH_3)(CH_2)_4CH(CH_3)(CH_2)_3CH_3$ also should give similar prominent alkyl ions. For Unknown 5.6 the ions of m/z 182–183, 140–141, and 85 are abundant in comparison to the spectrum of n-$C_{16}H_{34}$, indicating branches corresponding to C_3—C_{13}, C_6—C_{10}, and C_{10}—C_6. A first postulate to fit these data might be the C_3H_7—CH(C$_2$H$_5$)— group, but this is made improbable by the low abundance of $(M - C_2H_5)^+$; that is, the *negative* information that there is no appreciable loss of C_1, C_2, C_4, or C_5 is very valuable. The data can be fitted by $CH_3(CH_2)_5CH(n$-$C_3H_7)(CH_2)_5CH_3$. [n-Hexadecane, 5-methylpentadecane, and 7-n-propyltridecane]

5.7. Ion-series information greatly facilitates the solution of this unknown. Because there is no oxygen present, the peaks at m/z 15, 29, 43, 57, and 71 indicate the series $C_nH_{2n+1}^+$ for $n = 1 - 5$. The isotopic abundances of m/z 105, 92, and 91 indicate $C_8H_9^+$, $C_7H_8^{\overset{+}{\cdot}}$, and $C_7H_7^+$, respectively; this and the presence of the alkyl ion series indicates (Table A.6) that the ion series 77, 91, 105, 119, 133, 147, and 161 represents $C_6H_5(CH_2)_n$ where $n = 0 - 6$. The similar spectrum of Unknown 4.18 shows that the important OE$^{\overset{+}{\cdot}}$ ion C_7H_8 can arise by γ-H rearrangement to a phenyl—CH$_2$— moiety, and the compound is thus phenyl—CH$_2$—C$_7$H$_{15}$. Actually, the alkyl chain is unbranched; this is consistent with the abundance relationships of the $C_nH_{2n+1}^+$ ion series. The secondary maximum of m/z 133 in the $C_6H_5(CH_2)_n^+$ series is *not* due to chain branching; local maxima separated by —(CH$_2$)$_4$— are found in functionalized n-alkanes such as fatty acids (Figure 8.9). [n-Octylbenzene]

5.8, 5.9. The important OE$^{\overset{+}{\cdot}}$ series $C_nH_{2n}^{\overset{+}{\cdot}}$ in the upper part of both spectra suggests (Tables 5.1, A.7) that the unknowns are C_nH_{2n} or $C_nH_{2n+1}Y$ molecules; note these series in Figures 3E, 3H, 3N, and 3U. The only real difference is the presence of a "+4 index" series in Unknown 5.8 comprising

m/z 31, 45, 59, 73, and 87, which is indicative of a saturated alcohol or ether. [5.8, n-Hexadecanol; 5.9, 1-hexadecene]

5.10. The m/z 129 is the logical candidate to be the molecular ion, which means that the molecule contains an odd number of nitrogen atoms. The abundant ions of even mass such as 102, 98, and 78 could in this way be even-electron ions. The high abundance of this postulated molecular ion would also be consistent with the aromatic-type ion series. Again, several isomers are possible of the formula C_9H_7N (rings plus double bonds = 7), and reference spectra would give positive identification. [Isoquinoline]

5.11. There are a number of signs that this compound is quite different from the preceding ones. Masses 76, 69, 50, 31, and 26 have a maximum of two, one, one, one, and one carbon atoms, respectively. Mass 95 is either *not* the molecular ion, or else has an odd number of nitrogen atoms. Similarly, prominent ions of m/z 76, 50, and 26 are either odd-electron fragment ions, or contain an odd number of nitrogen atoms. There are few or no hydrogen atoms present (no groups of peaks in unit mass sequence, like 66, 67, 68, 69). The practiced eye may have found a valuable clue in the ion series of masses 31, 50, and 69 (see Table A.6). The lack of hydrogens even makes the low mass peaks of 12, 14, and 19 clues to the atoms present. Assuming m/z 95 to be the molecular ion gives correspondence between ions at masses 19 and M − 19, and at 26 and M − 26, suggesting the parts F—CF₂—CN. [Trifluoroacetonitrile]

5.12. Your suspicions concerning the m/z 119–121 doublet can be confirmed at masses 105–107, 91–93, and 77–79; this is the chloroalkyl ion series. Isotopic abundances indicate the $C_nH_{2n+1}^+$ ion series as another important feature of the spectrum, consistent with the elemental composition $C_8H_{17}Cl$ for m/z 148 as the molecular ion. Abundant peaks corresponding to $(M − C_2H_5)^+$ and $(M − CH_2Cl)^+$ suggest the substructure C_2H_5—C—CH_2Cl, for which a logical molecule would be $(C_2H_5)_3CCH_2Cl$. However, the formation from this of $C_4H_9^+$ as a base peak seems surprising, suggesting $C_4H_9(C_2H_5)CHCH_2Cl$ as an alternative. The loss of the largest alkyl group, C_4H_9, is actually greater than the loss of ethyl if the charge migration as well as charge retention products are considered: $([(M − C_4H_9)^+] + [C_4H_9^+]) > ([(M − C_2H_5)^+] + [C_2H_5^+])$. Stevenson's Rule is consistent with the observed abundance if the following ionization energies (I, Table A.3) are assumed: $I(n\text{-}C_4H_9\cdot) = 8.0$ eV, $I(\cdot CH(C_2H_5)CH_2Cl) = 8.2$ eV; $I(C_2H_5\cdot) = 8.4$ eV, $I(\cdot CH(C_4H_9)CH_2Cl) = 8.2$ eV (see Section 8.6). [3-(Chloromethyl)heptane]

5.13. You should have made the following probable assignments: m/z 156, M^+, C_8H_9OCl; 107, $C_7H_7O^+$; and 94, $C_6H_6O^+$, OE^+. A phenyl ring is indicated by the aromatic ion series (the high abundance of the m/z 77 is a fairly reliable indication of monosubstitution). The $C_6H_6O^+$ peak indicates that oxygen is attached to the ring. The fact that $(M − CH_2Cl)^+$ is much larger than $(M - Cl)^+$ is indicative of a labile CH_2Cl group; cleavage alpha to the C_6H_5O- would account for this, suggesting $C_6H_5OCH_2CH_2Cl$. A strong peak resulting from hydrogen rearrangement with loss of C_2H_3Cl would be expected for this compound, accounting for the base $OE^{\ddot{+}}$ ion at m/z 94. [2-Chloroethyl phenyl ether]

5.14. The mass 146–148 appearance should suggest an isotopic cluster. The most probable suspects of Table 2.1 are silicon and sulfur. One atom of sulfur fits the mass 148/146 ratio better, and this postulation leaves a residual at m/z 147 (correcting for the ^{34}S contribution from m/z 145), indicating eight or nine carbon atoms (correspondingly, Si_1 would indicate approximately C_5). Sulfur (or silicon) is also indicated in m/z 89, but not in masses 112, 98, or 84. Assuming that m/z 146 is the molecular ion and contains no nitrogen atoms, a series of abundant *odd-electron* ions is apparent; masses 42, 56, 70, 84, 98, and 112. Table A.7 suggests an olefin ion series; the abundant $(M - 34)^{\ddot{+}}$ and $(M - 62)^{\ddot{+}}$ indicate a thiol. This is supported by the alkyl ion series and by the m/z 47, 61, 75, 89, 103 ion series. Mass 89, the most abundant ion of these, corresponds to the cyclic ion formed in *n*-alkyl mercaptans (Equation 4.42). For a dramatic confirmation of this evidence, compare this spectrum to those of n-$C_{12}H_{25}SH$ and n-$C_{12}H_{25}OH$, Figures 3U and 3H [n-Octyl mercaptan]

6.1. Metastable peak assignments are m/z 8.0, 45 → 19; 16.2, 59 → 31 or 45 → 27; 18.7 (flat-topped) 45 → 29; 24.1, 28 → 26; 28.5, 59 → 41; 37.1, 41 → 39; 39.1, 43 → 41; 40.1, 44 → 42; and 42.1, 44 → 43.

6.2. The isotopic abundances of the large peaks indicate the elemental compositions CH_4N, C_3H_8N, $C_5H_{12}N$, and $C_6H_{14}N$. These are consistent with the saturated amine ion series, and a molecular composition of $C_7H_{17}N$. The largest and second largest peaks correspond to the losses of C_2H_5 and CH_3, respectively; there should be ethyl and methyl, but no larger alkyl, groups on the carbon atom(s) alpha to the nitrogen atom. However, many isomeric $C_7H_{17}N$ amines fit this description.

The three diffuse peaks correspond to the metastable decompositions $58^+ \rightarrow 30^+$, $100^+ \rightarrow 58^+$, and $86^+ \rightarrow 58^+$. The most common pathway for C_nH_{2n} loss from such even-electron ions is the charge-site rearrangement reaction 4.44 involving cleavage of the C—N bond. Thus the $(M - CH_3)^+$ → $(M - CH_3 - C_3H_6)^+$ reaction indicates the structure CH_3—C—N—C_3H_7, and the pathway $(M - C_2H_5)^+$ → $(M - C_2H_5 - C_2H_4)^+$ indicates the presence of C_2H_5—C—N—C_2H_5. The only $C_7H_{17}N$ isomer containing both of these substructures is $(C_2H_5)_2NCH_2CH_2CH_3$. [Diethyl-*n*-propylamine]

6.3. (A) $R_1 = R_2 = R_3 = H$, $Y = H_2$;
    (B) $R_1 = H$, $R_2 = CH_3O$, $R_3 = CH_3$, $Y = H_2$;
    (C) $R_1 = CH_3O$, $R_2 = H$, $R_3 = CH_3$, $Y = O$; see Biemann 1962, p. 309.

The peaks at $(M - 28)^{+\cdot}$ in A and B, and $(M - 42)^{+\cdot}$ in C can be explained by 11.17.

$$\xrightarrow{\alpha} \quad \text{loss of} \quad \underset{H_2C}{\overset{}{\diagup}} C{=}Y \qquad (11.17)$$

8.1. A relatively abundant $(M - 2)^{+\cdot}$ ion is unusual in mass spectra, so the 164/166 should be checked out as a possible isotopic cluster. One bromine atom is indicated by a number of such ion pairs. The ion series m/z 15, 29, 43, 57, 71, and 85 indicates alkyl to C_6, or possibly carbonyl to C_5. The bromine-containing ion series m/z 79–81, 93–95, 107–109, 121–123, and 135–137 gives strong support to the presence of the alkyl chain, and to the formula $C_6H_{13}Br$ for the m/z 164–166 ions. The loss of the small ethyl radical should not be favored over $C_nH_{2n}Br\cdot$ loss from the same carbon; a logical explanation for the large m/z 135–137 ions is that they arise through the displacement reaction 4.42 [1-Bromohexane]

8.2. The pressure-dependent m/z 180 peak is probably the $(M + 1)^+$ ion and is approximately C_{11}; $M^{+\cdot}$ (m/z 179) then contains an odd number of nitrogen atoms. The m/z 120 and 88 peaks have values of 8 and 3, respectively, for their maximum number of carbon atoms. Low-mass peaks indicate the presence of aromatic and saturated amine (m/z 30) and oxygenated (m/z 31, 45) functionalities. A methyl ester, $-CO-OCH_3$, is a probable cause (Table A.5) of the peaks representing $(M - 31)^+$, $(M - 32)^{+\cdot}$, and $(M - 59)^+$. Two of the largest peaks in the spectrum, m/z 88 and 91, could arise from cleavage of the same bond ($88 + 91 = 179$); the m/z 91 could be $C_6H_5CH_2^+$. The greater charge retention by the m/z 88 suggests stabilization by the amine moiety; $-CH(=\overset{+}{N}H_2)COOCH_3$ appears to be the only possible group containing the carboxymethyl functionality; this also is consistent with the high abundance of m/z 120. The m/z 131 peak probably arises from the combined losses of CH_3O and NH_3. [Phenylalanine methyl ester]

8.3. The ion of m/z 103 should be C_5, based on the C_4 composition of m/z 88. The abundant 30, 44, 58, 72 ion series is indicative of a saturated amine; the m/z 103 composition thus should be $C_5H_{13}NO$, and it passes the $M^{+\cdot}$ tests. Alpha cleavages initiated by the amine should be dominant reactions; the $(M - CH_3)^+$ and $(M - CH_3O)^+$ peaks thus indicate the moieties $N-C-CH_3$ and $N-C-CH_2OH$, respectively, given that a hydroxyl is present. The possible molecules are (a) $HOCH_2CH(CH_3)NHC_2H_5$, (b) $HOCH_2CH(CH_3)N(CH_3)_2$, (c) $HOCH_2C(CH_3)_2NHCH_3$, (d) $HOCH_2$-$CH_2NHCH(CH_3)_2$, and (e) $HOCH_2CH_2N(CH_3)C_2H_5$; no other α-loss, except $(M - H)^+$, is possible from any of these. The base m/z 30 peak should arise by the further decomposition of one of the α-cleavage products (m/z 88 or 72) through hydrogen rearrangement (Reaction 4.44); this would not be possible with the tertiary amines b and e. Reaction sequences for the remainder should be:

$$a \quad \xrightarrow{\alpha} HOCH_2CH(CH_3)\overset{+}{N}H=CH_2 \xrightarrow{rH} H_2\overset{+}{N}=CH_2 \; (m/z \; 30)$$

$$\xrightarrow{\alpha} HOCH_2CH=\overset{+}{N}HC_2H_5 \xrightarrow{rH}$$
$$HOCH_2CH=\overset{+}{N}H_2 \; (m/z \; 60) \tag{11.18}$$

$$\xrightarrow{\alpha} CH_3CH=\overset{+}{N}HC_2H_5 \xrightarrow{rH} CH_3CH=\overset{+}{N}H_2 \; (m/z \; 44)$$

$$c \quad \xrightarrow{\alpha} HOCH_2C(CH_3)=NHCH_3 \xrightarrow{\times}$$
$$\xrightarrow{\alpha} (CH_3)_2C=\overset{+}{N}HCH_3 \xrightarrow{\times} \tag{11.19}$$

$$d \quad \xrightarrow{\alpha} HOCH_2CH_2\overset{+}{N}H=CHCH_3 \xrightarrow{rH}$$
$$H_2\overset{+}{N}=CHCH_3 \; (m/z \; 44) \tag{11.20}$$

$$\xrightarrow{\alpha} H_2C=\overset{+}{N}HCH(CH_3)_2 \xrightarrow{rH} H_2C=\overset{+}{N}H_2 \; (m/z \; 30)$$

Thus c should give weak m/z 30 and 44 peaks. The m/z 88 ion of a should arise mainly by loss of CH_3 from the more highly branched α-carbon, so that there should be a significant m/z 60 ion. Structure d correctly predicts the abundant m/z 30 and 44 peaks; note that the significant m/z 70 ion is consistent with the stable structure $CH_2=CH-\overset{+}{N}H=CHCH_3$. [$N$-(2-hydroxyethyl)-$N$-isopropylamine]

8.4. Have you assigned the formulas $C_3H_7O_2$ to m/z 75 and C_2H_5O to m/z 45? If not, rework the problem assuming these data were supplied by high-resolution mass spectrometry.

This spectrum is difficult because of the variety of isobaric formulas which can be applied to each major ion, and because of the absence of a molecular ion. Yet many students are able to arrive at the correct structure in a relatively short time, mainly because the data fit this structure so well. Without reference spectra it is difficult to eliminate many other possible structures unequivocally. However, such a "best guess" can often be very valuable.

For such low-molecular-weight compounds, the lack of $M^{\ddagger}$ suggests a saturated structure (Table A.4); CH_3O^+, $C_2H_5O^+$, and $C_3H_7O_2^+$ are the most favorable postulates for the abundant m/z 31, 45, and 75 ions, respectively. Thus m/z 75 must contain two functional groups; there is no odd-electron $(M - 18)^{\ddagger}$ ion to indicate that either is a hydroxy group. The abundant CH_3^+ suggests a CH_3O moiety; relative ion stabilities predict a low abundance for CH_3^+ unless it is attached to an electron-attracting group. The CH_3O group is probably attached to $-CH_2-$, since it cannot be part of a $CH_3OCH(CH_3)-$ or $CH_3OC(CH_3)_2-$ group. If these groups were present, the abundance of the m/z 59 or 73 ions, respectively, would be much greater than is observed. The few abundant ions in the spectrum suggest a simple molecular structure; so one of the two oxygen-containing functional groups in the m/z 75 ion may well be the same as the one in the m/z 45 ion. If there is no hydroxyl group, the most logical structure for the m/z 75 ion is $(CH_3O)_2CH-$, and for the molecule it is $CH_3OCH_2OCH_3$. It should be reemphasized that this was not a rigorous proof of structure by mass spectrometry, but rather a way to hypothesize a probable structure which can then be checked against reference spectra or other techniques. [Methylal]

$$\begin{array}{ccc} \underset{CH_3O \quad OCH_3}{\overset{H \quad H}{C^{+}}} & \xrightarrow{\alpha} & H\cdot \;+\; \underset{CH_3O \quad OCH_3}{\overset{H}{C^{+}}} \quad m/z\ 75 \end{array}$$

(11.21)

$$\begin{array}{ccc} \underset{CH_3O \quad OCH_3}{\overset{H \quad H}{C^{+}}} & \xrightarrow{\alpha} & CH_3O\cdot \;+\; \underset{OCH_3}{\overset{}{H_2C^{+}}} \quad m/z\ 45 \end{array}$$

8.5. The mass 122 ion appears to have seven carbon atoms and two oxygen atoms, so it could be $C_7H_6O_2$, rings-plus-double-bonds value $= 5$. The m/z 104 ion possibly has seven carbon atoms, and it should be noted as an odd-electron ion along with masses 94 and 76. Mass 122 seems to be a prime candidate for the molecular ion, and its abundance indicates a highly stable molecule. In the low-mass region the predominant ion series indicates an aromatic hydrocarbon; the rings-plus-double-bonds value of 5 corresponds to a phenyl ring plus an additional ring or double bond. The striking $(M - 1)^+$ peak indicates the presence of a labile hydrogen atom, and $(M - 18)^{\ddagger}$ indicates a ready loss of H_2O (Table A.5 suggests that this is due to a hydroxyl group). The $(M - 28)^{\ddagger}$ and $(M - 29)^+$ ions are probably $C_6H_6O^{\ddagger}$ and $C_6H_5O^+$, respectively, due to loss of CO, and COH, as losses of C_2H_4 and C_2H_5 would yield unusually unsaturated structures. The $(M - 29)^+$ suggests that the $-CHO$ group is present, which would account for the extra double bond predicted by the rings-plus-double-bonds calculation above. From these indications of a labile hydrogen, phenyl, $-OH$, and $-CHO$ an obvious structure postulation is $HO-C_6H_4-CHO$, a structure which also is consistent with the m/z 76 ion, $C_6H_4^{+\cdot}$. The odd-electron ion m/z 104 indicates that this is the ortho isomer; an $(M - CO)^{+\cdot}$ peak is also commonly found in the spectra of phenols. [Salicylaldehyde]

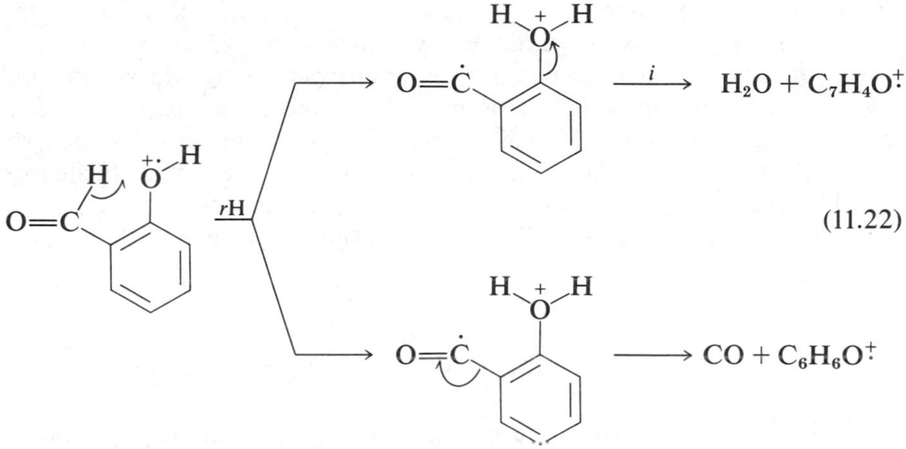

$$(11.22)$$

8.6. The m/z 99/98 ratio is too large for m/z 98 to be the molecular ion; the m/z 84 makes m/z 99 a much more likely candidate. This suggests that the molecule contains an odd number of nitrogen atoms, consistent with the C_3H_4N assignment of m/z 54 (EE$^+$, r + db = 2). Because m/z 59 contains oxygen (EE', r + db = 0), the most logical assignment for m/z 84 is C_4H_6NO (EE$^+$, r + db = 2), and thus C_5H_9NO for M$^{+\cdot}$. The significant mass 31, 45, and 59 ions suggest (Table A.6) an alcohol or ether functionality. For m/z 54 Table A.6 suggests $-C_2H_4CN$, and the presence of a nitrile group would also explain the unusual $(M - 1)^+$ and $(M - 27)^+$ ions. The m/z 59 ion is most likely $HOC(CH_3)_2-$, $CH_3OCH(CH_3)-$, or $C_2H_5OCH_2-$, in light of $(M - 15)^+ \gg (M - 29)^+$. The only one of these compatible with the base peak at mass 31 is $C_2H_5OCH_2-$. [β-Cyanoethyl ethyl ether]

$$CH_3\dot{C}H_2\overset{+\cdot}{O}CH_2CH_2CN \xrightarrow{\alpha}$$

$$CH_3\cdot + CH_2=\overset{+}{O}CH_2CH_2CN \quad (m/z\ 84)$$

$$(11.23)$$

$$CH_3CH_2\overset{+\cdot}{O}CH_2CH_2CN \xrightarrow{i}$$

$$CH_3CH_2O\cdot + \overset{+}{C}H_2CH_2CN \quad (m/z\ 54)$$

$$(11.24)$$

$$CH_3CH_2\overset{+\cdot}{O}CH_2CH_2CN \xrightarrow[\alpha]{-\cdot CH_2CN}$$

$$\underset{\underset{(m/z\ 59)}{CH_2CH_2}}{\overset{H}{\underset{|}{\overset{+}{O}}}=CH_2} \xrightarrow{rH} C_2H_4 + H\overset{+}{O}=CH_2 \qquad (11.25)$$

$$(m/z\ 31)$$

8.7. The m/z 150 ion appears to have C_7 and O_3 components; this is consistent with C_8 and O_3 components in m/z 165. The mass 104 peak has a maximum of seven carbon atoms. The m/z 43 ion isotope ratios point to the composition C_2H_3O. None of the tests eliminates m/z 165 as the molecular ion; it must then contain an odd number of nitrogen atoms. Masses 104 and 76 *may* be significant odd-electron ions. The molecule appears to be generally stable, giving a few abundant ions. The low aromatic series suggests electronegative substituents. The abundant $(M-15)^+$ indicates facile loss of CH_3, and the m/z 43 and $(M-43)^+$ make the acetyl group, $—COCH_3$, a possible source of this loss. If the acetyl group is attached to the phenyl,

$$Y—C_6H_4—CO—CH_3\overset{+}{\cdot} \xrightarrow{\alpha}$$

$$Y—C_6H_4—CO^+ \xrightarrow{i} Y\cdot + C_6H_4CO\overset{+}{\cdot} \qquad (11.26)$$

then 11.26 offers an explanation for the abundant m/z 104. This decomposition of an even-electron ion to yield an abundant odd-electron ion is unusual, and requires Y to be strongly electron-attracting. Additional evidence comes from m/z 76; this is commonly formed from disubstituted aromatic compounds (Table A.7). If so, this second functional group contains $165 - (76 + 43) = 46$ amu, and is probably NO_2, C_7H_4NO (m/z 150) $-$ C_7H_4O (m/z 104). Loss of NO is typical of nitroaromatics (Table A.5); m/z 120 may be $(M-CH_3-NO)^+$, and m/z 92 may be $(M-CH_3CO-NO)\overset{+}{\cdot}$. [*p*-Nitroacetophenone]

8.8. The molecular composition of the m/z 248 peak could be $C_{13}H_6OCl_2$ from the isotopic abundances. This corresponds to the replacement of the elements of CH_2Cl_2 in DDE by an oxygen atom, and an increase in the rings-plus-double-bonds value from 9 to 10. The m/z 248 peak passes the tests for the molecular ion; there are no other peaks containing more than two chlorine atoms. Strong additional evidence for the presence of oxygen is provided by the abundant $OE\overset{+}{\cdot}$ ion at m/z 220; this should represent the loss of CO, as C_2H_4 loss would be unlikely from a highly aromatic molecule. Note that the probable elemental composition of the $OE\overset{+}{\cdot}$ ion of m/z 150 is $C_{12}H_6$; it can contain no chlorine *or oxygen*, so it must be $(M-Cl_2-CO)\overset{+}{\cdot}$. This is also evidence that the $OE\overset{+}{\cdot}$ ion of m/z 220 is $(M-CO)\overset{+}{\cdot}$, not

$(M - C_2H_4)^{\ddagger}$. A large CO loss peak suggests a phenol or endocyclic carbonyl group; the latter would be consistent with the chemically logical molecule. [3,6-Dichlorofluorenone]

(11.27)

8.9. The m/z 303 peak corresponds in mass and isotopic abundances to the $M^{\ddagger}$ of cocaine, and the relative abundance of this peak is consistent with the stability of this cyclic molecule. There is an aromatic ion series; the prominent (68), 82, and 96 ions are consistent with an EE^+ bicycloalkylamine series (Table A.6) and/or an $OE^{\ddagger}$ cycloalkyl series (Table A.7); these are also supported by the abundances of m/z 42, 55, 83, and 97. The intense high mass peaks at $(M - 31)^+$, $(M - 105)^+$, and $(M - 121)^+$ give strong evidence for the presence of the methyl ester, benzoyl, and benzoyloxy groups, respectively. The corresponding ions at m/z 59, 105, and 122 ($OE^{\ddagger}$, Reaction 4.33) provide additional evidence for the methoxycarbonyl, benzoyl, and benzoyloxy groups, respectively. High-resolution information on elemental compositions would make all this evidence more positive, however; for example, the large m/z 122 peak may also be due to the loss of benzoyloxy and acetic acid to give the stable EE^+ ion, $C_8H_{12}N^+$. The m/z 94 could correspond to the loss of C_2H_4 from this ion (Reaction 4.31). [Cocaine]

(11.28)

m/z 122   *m/z* 94

8.10. The presence of at least two oxygen atoms and one nitrogen atom is required by the fact that this is the ethyl ester of an amino acid. The m/z 177/179 ratio suggests an isotopic cluster, and the abundances are consistent with one sulfur and around eight carbon atoms for m/z 177. There is probably a sulfur atom in the peaks at m/z 148, 131, 104, and 61. To calculate the elemental composition, mass $177 - 32$ (S_1) $- 32$ (O_2) $- 14$ (N) $= 99$, which would be C_8H_3 (impossible for an ethyl ester) or, more probably, C_7H_{15}.

None of the tests eliminates m/z 177 as the molecular ion, $C_7H_{15}NO_2S$. Other odd-electron ions are difficult to identify because of the presence of the nitrogen atom.

The large number of functional groups present makes it more difficult to judge the significance of the low mass ion series. In the alkyl amine series m/z 30 appears to be the only ion of importance; this fact plus an appreciable $(M - 17)^+$ indicate a free amine group. Of the ion series of Table A.6 containing sulfur, the alkythiol or sulfide series m/z 47, 61, 75 offer a structure assignment for the sulfur atom in the molecule. The $(M - 15)^+$, $(M - 29)^+$, and $(M - 46)^{\ddot{+}}$ are roughly as expected for an ethyl ester. From Table A.5 the $(M - 48)^{\ddot{+}}$ ion could be due to a methyl sulfide, and this supports the assignment of $CH_3SCH_2{}^+$ as the structure of the base peak at m/z 61 and for the group lost in m/z 116. The $(M - 73)^+$ is expected for R—$COOC_2H_5$

$$R—\overset{\overset{\displaystyle \widehat{NH_2}^{+\cdot}}{\big|}}{C}H—COOC_2H_5 \xrightarrow{\ \alpha\ } R—CH{=}\overset{+}{N}H_2 + \cdot COOC_2H_5 \qquad (11.29)$$

when R^+ is stabilized, such as occurs for α-amino acids (Table A.5). The stable R^+ also accounts for the lack of the expected $(M - C_2H_4)^{\ddot{+}}$ and $(M - C_2H_3)^+$. R—$CHNH_2COOC_2H_5$ must have a mass of $177 - 102 = 75$, and must contain CH_3SCH_2—, so that the molecular structure CH_3-$SCH_2CH_2CHNH_2COOC_2H_5$ is indicated. The formation of m/z 56 is rationalized in Equation 8.60. [Methionine ethyl ester]

8.11. The maximum values for elemental compositions are not particularly meaningful. The m/z 75 peak cannot be $M^{\ddot{+}}$; m/z 61 would then be $(M - 14)^+$. The scarcity of even-mass peaks indicates that the true $M^{\ddot{+}}$ is at an even mass number. A saturated oxygen functionality is indicated by the 31, 45, (59, 73) ion series; a second one is indicated by m/z 75 and 61 ($C_2H_5O_2{}^+$; $C_3H_9O^+$ should not occur in this abundance without ion-molecule reactions; also the m/z 62 cannot be $C_3H_{10}O^{\ddot{+}}$). The metastable at m/z 30.4 indicates that m/z 61 can lose H_2O ($30.4 = 43^2/61$); the most logical structures for it are $—CH_2OCH_2OH$ and $—CH(OH)CH_2OH$. There are significant $OE^{\ddot{+}}$ ions at m/z 74, 62, and 44 (and 32? Did you mark these?). The most likely process for m/z 74 formation is the loss of H_2O (or ROH); so m/z 92 ($74 + 18$), $C_3H_8O_3$, or a higher homolog, is a possible candidate for the $M^{\ddot{+}}$. Compounds such as $HOCH_2CH_2OCH_2OH$ and $CH_3OCH_2OCH_2OH$ (if stable) should give significant m/z 45 and $(M - 17)^+$, $RCH_2O{=}CH_2$. Glycerol, $HOCH_2CH(OH)CH_2OH$, is a logical alternative; hydrogen rearrangement can explain the m/z 62 and 44 peaks. [Glycerol]

$$(11.30)$$

$$m/z\ 62$$

$$m/z\ 44$$

8.12. The m/z 141 could be $M^{\ddot{+}}$ if it contains an odd number of nitrogen atoms; it appears to contain C_4NO (or C_3N_3O), and m/z 126 corresponds to C_3NO. The presence of "A" elements F and/or P appears to be required. Calculated maxima for the carbon number of other ions are: m/z 114, 97, and 78, C_2; and m/z 72, C_3. The latter $(M-69)^+$ peak has thus lost only C_1; a logical assignment is $(M-CF_3)^+$, which also accounts for the abundant m/z 69 as CF_3^+. Thus $M^{\ddot{+}}$ must be $C_4H_6NOF_3$, $r+db=1$. The unusual $(M-27)^+$ ion must be due to the loss of $C_2H_3\cdot$, not HCN, as a cyanide would require $r+db>1$; Table A.5 suggests that an ethyl amide could account for $(M-C_2H_3)^+$, or a molecular structure $CF_3CONHC_2H_5$. This should show an abundant $(M-CH_3)^+$ and m/z 72 by cleavages alpha to the nitrogen function, and at m/z 97 and 72 (again) by cleavages alpha to the carbonyl function. Note that the hydrogen rearrangement decomposition of m/z 72 by Reaction 4.44 yields m/z 44: $H-CH_2CH_2-NH=C=O \rightarrow C_2H_4 + H_2N=C=O$. [Ethyl trifluoroacetamide]

9.1. This was a "trick" question; the compound is not phenolic, which you might have assumed if you incorrectly identified the mass 172 $OE^{\ddot{+}}$ peak as $(M-CO)$. Its elemental composition is C_6H_5OBr, and that of the probable $M^{\ddot{+}}$ ion at m/z 200 is C_8H_9OBr. Don't skip the steps in Table A.1! The rearrangement loss of C_2H_4 cannot occur if the C_2H_4 moiety is directly attached to the ring; it is likely, however, if it is attached to another atom (such as oxygen) which is attached to the ring. [p-Bromophenyl ethyl ether]

9.2. Calculations on isotopic abundances suggest elemental compositions of $C_8H_6O_4$ and $C_8H_5O_3$ for the peaks at m/z 166, and 149, respectively (rings-plus-double-bonds value for each equals 6). The $OE^{\ddot{+}}$ m/z 166 ion is a logical candidate for the molecular ion. There are characteristic low-mass aromatic ions, and the intense peaks at m/z 149 [$(M-17)$ '], m/z 121 [$(M-17-28)$ '], and m/z 45 (CHO_2^+), provide strong evidence for a carboxyl group. These C_6H_5- and $-COOH$ moieties leave only CO_2 unaccounted for in the molecular formula, so that a benzenedicarboxylic acid is a logical postulate. Although the intensity of the peak at m/z 122 is slightly greater than that expected from the isotopic contributions, in this case an "ortho effect" would give an m/z 122 (and 104) peak of much higher abundance. [Terephthalic acid]

9.3. Alpha cleavage of the 12-isomer leads to two possible radical species, neither of which can abstract a hydrogen to give m/z 99 ($C_5H_7O_2^+$) by mechanisms similar to Reaction 9.39. For the 17-keto derivative, however, α-cleavage of the weakest bond leads to m/z 99.

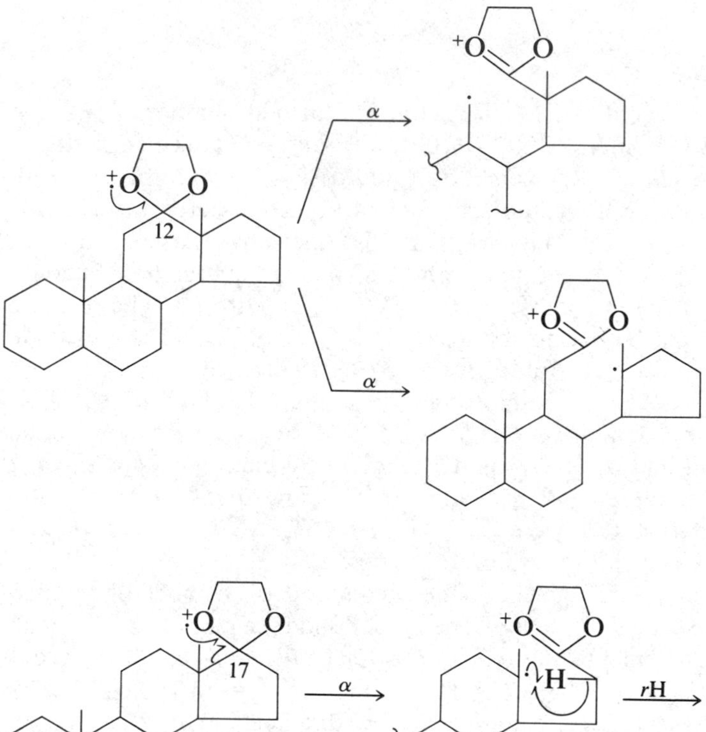

(11.31)

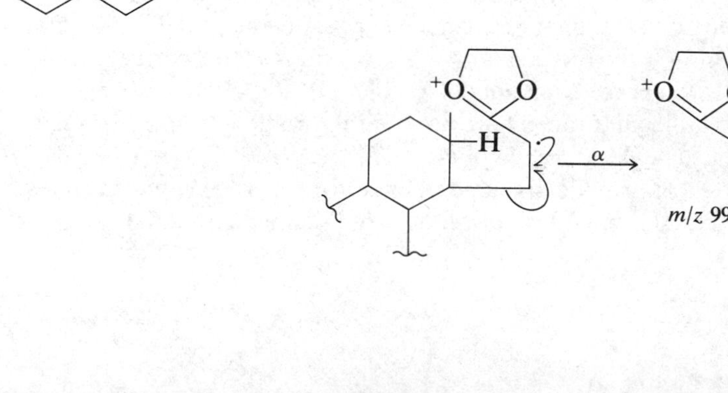

(11.32)

9.4. The isotopic abundances of the m/z 154 indicate the composition $C_8H_{10}OS$, r + db = 4; this peak passes the M^{+} tests also. There is an aromatic ion series, and the peaks at 69, 82, 95–7, and 108 are evidence for sulfur attached to an aromatic ring (Table A.6), although a thiophene substituted with an unsaturated or cyclic group is also possible. The m/z 47 and 61 peaks indicate that an alkyl moiety is also attached to sulfur. The OE^{+} m/z 126 peak has the composition C_6H_6OS (don't forget that the m/z 125 isotopes contribute to the m/z 127 peak), or $(M - C_2H_4)^{+}$; possible mechanisms for its formation include Reaction 8.47 with *a*, and Reaction 4.31 with *b*. The presence of a hydroxyl group, possibly in an ortho position, is

(11.33)

indicated by the peak at m/z 121, $(M - CH_3 - H_2O)^+$. The m/z 98 and, presumably, 97 peaks still contain sulfur, and thus apparently are $(M - C_2H_4 - CO)^{+}$ and $(M - C_2H_4 - CHO)^+$, consistent with a phenolic-type hydroxyl group. [Ethyl *o*-hydroxyphenyl sulfide]

9.5. If mass 101 is the molecular ion, it must contain an odd number of nitrogen atoms. The $(M - CH_3)^+$ ion appears to have no more than four carbon atoms, and m/z 59 to have two carbon atoms and no more than one oxygen atom. Any even-electron ion series of Table A.6 that includes m/z 59 is ruled out by the isotopic abundances. In fact, the identity of this ion fragment is a major key to the identity of Unknown 9.5. With a maximum of C_2 and O_1, and with the indicated presence of nitrogen, the formula C_2H_5NO is a logical postulate for m/z 59, and is an *odd-electron ion*, $(M - 42)^{+}$. Table A.7 suggests $H_2NCOCH_2—Z—H$ or $HON=CHCH_2—Z—H$; for Z, 42 amu, Table A.5 suggests C_3H_6 or CH_2CO. Corresponding to these, the prominent $(M - 15)^+$ and m/z 43 peaks suggest either $—CH(CH_3)_2$ or $—COCH_3$ as terminal groups. Differentiation between the possible amide and oxime structures also requires study of reference spectra of similar compounds. [3-Methylbutyramide]

(11.34)

BIBLIOGRAPHY

Abbott, S. J., S. R. Jones, S. A. Weinman, F. M. Bockhoff, F. W. McLafferty, and J. R. Knowles. 1979. *J. Am. Chem. Soc.* 101:4323.

Andersson, B. A., W. H. Heimermann, and R. T. Holman. 1974. *Lipids.* 9:443.

Andersson, B. A., L. Lundgren, and G. Stenhagen. 1980. In Waller 1980. 1:855–894.

Arnett, E. M., J. C. Sanda, J. M. Bollinger, and M. Barber. 1967. *J. Am. Chem. Soc.* 89:5389.

Arpino, P. J., and G. Guiochon. 1979. *Anal. Chem.* 51:682A.

Arpino, P. J., and F. W. McLafferty. 1976. In Nachod *et al.* 1976. 6:1–89.

Arsenault, G. P. 1980. In Waller 1980.

Audier, H. E. 1969. *Org. Mass Spectrom.* 2:283.

Audier, H. E., M. Fetizon, and J.-C. Tabet. 1975. *Org. Mass Spectrom.* 10:639 and references cited therein.

Audier, H. E., A. Milliet, C. Perret, J.-C. Tabet, and P. Varenne. 1977. *Nouv. J. Chim.* 1:269.

Baer, T. 1979. In Bowers 1979. 1:153.

Baldwin, M. A. 1979. *Org. Mass Spectrom.* 14:601.

Baldwin, M. A., and F. W. McLafferty. 1973. *Org. Mass Spectrom.* 7:1353.

Beckey, H. D. 1977. *Principles of Field Ionization and Field Desorption Mass Spectrometry.* Pergamon Press. 335 pp.

Benninghoven, A., and W. K. Sichtermann. 1978. *Anal. Chem.* 50:1180.

Benoit, F. W., and A. G. Harrison. 1976. *Org. Mass Spectrom.* 11:599.

———. 1977. *J. Am. Chem. Soc.* 99:3980.

Benoit, F. W., A. G. Harrison, and F. P. Lossing. 1977. *Org. Mass Spectrom.* 12:78.

Bente, P. F., III, F. W. McLafferty, D. J. McAdoo, and C. Lifshitz. 1975. *J. Phys. Chem.* 79:713.

Benz, R. C., and R. C. Dunbar. 1979. *J. Am. Chem. Soc.* 101:6363.

Beynon, J. H. 1960. *Mass Spectrometry and Its Applications to Organic Chemistry.* Elsevier. 640 pp.

Biemann, K. 1962. *Mass Spectrometry: Organic Chemical Applications.* McGraw-Hill. 370 pp.

_____. 1980. In Waller 1980. 1:469–526.

Boettger, H. 1980. In Waller 1980.

Borchers, F., K. Levsen, H. Schwarz, C. Wesdemoitis, and R. Wolfschutz. 1977. *J. Am. Chem. Soc.* 99:1716.

Borchers, F., K. Levsen, H. Schwarz, C. Wesdemiotis, and H. U. Winkler. 1977. *J. Am. Chem. Soc.* 99:6359.

Borders, D. B., and R. Hargreaves. 1980. In Waller 1980. 1:567–610.

Bosshardt, H., and M. Hesse. 1974. *Angew. Chem. Intern. Ed. Engl.* 13:252.

Bowen, R. D., B. J. Stapleton, and D. H. Williams. 1978. *Chem. Commun.* p. 24.

Bowen, R. D., and D. H. Williams. 1977. *Org. Mass Spectrom.* 12:475.

Bowen, R. D., D. H. Williams, and H. Schwarz. 1979. *Angew. Chem. Intern. Ed. Engl.* 18:451.

Bowers, M. T., ed. 1979. *Gas Phase Ion Chemistry.* Academic Press, 2 vols. 345 and 346 pp.

Bowie, J. H. 1973. *J. Am. Chem. Soc.* 95:2547.

_____. 1977 and 1979. In Johnstone 1977, pp. 217–241; 1979, pp. 262–284.

Bowie, J. H., and S. Janposri. 1976. *Org. Mass Spectrom.* 11:1290.

Boyd, R. K., and J. H. Beynon. 1977. *Org. Mass Spectrom.* 12:163.

Brion, C. E., and L. D. Hall. 1966. *J. Am. Chem. Soc.* 88:3661.

Brooks, C. J. W. 1979. *Phil. Trans. R. Soc., Lond. A.* 253:53.

_____. 1980. In Waller 1980. 1:611–660.

Broxton, T. J., Y. T. Pang, J. F. Smith, and F. W. McLafferty. 1977. *Org. Mass Spectrom.* 12:180.

Budde, W. L., and J. W. Eichelberger. 1979. *Organic Analysis Using GC/MS.* Ann Arbor Science. 241 pp.

Budzikiewicz, H. 1980. In Waller 1980.

Budzikiewicz, H., C. Djerassi, and D. H. Williams. 1967. *Mass Spectrometry of Organic Compounds.* Holden-Day. 690 pp.

Burlingame, A. L. 1970. *Topics in Organic Mass Spectrometry.* Wiley-Interscience. 471 pp.

Burlingame, A. L., T. A. Baillie, P. J. Derrick, and O.S. Chizhov. 1980. *Anal. Chem.* 52:214R.

——. 1980. *Anal. Chem.* 52:000R.

Burlingame, A. L., C. H. L. Shackleton, I. Howe, and O. S. Chizhov. 1978. *Anal. Chem.* 50:346R.

Bursey, J. T., M. M. Bursey, and D. G. I. Kingston. 1973. *Chem. Rev.* 73:191–233.

Bursey, M. M., D. J. Harvan, C. E. Parker, L. G. Pederson, and J. R. Hass. 1979. *J. Am. Chem. Soc.* 101:5489.

Caprioli, R.M., J. Liehr, and W. Seifert. 1980. In Waller 1980. 1:007–1084.

Chapman, J. R. 1978. *Computers in Mass Spectrometry.* Academic Press.

Chemical Abstracts Service, *C. A. Selects Mass Spectrometry.* Columbus, Ohio.

Comisarow, M. 1979. *27th Conf. Mass Spectrom. Allied Topics.* Seattle.

Cooks, R. G., ed. 1978. *Collision Spectroscopy.* Plenum Press. 458 pp.

Cooks, R. G., J. H. Beynon, R. M. Caprioli, and G. R. Lester. 1973. *Metastable Ions.* Elsevier. 296 pp.

Dawkins, B. G., and F. W. McLafferty. 1978. In K. Tsuji and W. Morozowich, eds., *GLC and HPLC Determination of Therapeutic Agents* (Dekker), pp 259–275.

Dawson, P. H. 1976. *Quadrupole Mass Spectrometry and Its Applications.* Elsevier.

Dayringer, H. E., and F. W. McLafferty. 1977. *Org. Mass Spectrom.* 12:53–54.

DeKock, R. L., and M. R. Barbachyn. 1979. *J. Am. Chem. Soc.* 101:6515.

Dell, A., D. H. Williams, H. R. Morris, G. A. Smith, J. Feeney, and G. C. K. Roberts. 1975. *J. Am. Chem. Soc.* 97:2497.

Dias, J. R., Y. M. Sheikh, and C. Djerassi. 1972. *J. Am. Chem. Soc.* 94:473.

Dill, J. D., C. L. Fisher, and F. W. McLafferty. 1979. *J. Am. Chem. Soc.* 101:6526 and 6531.

Dillard, J. 1980. In Waller 1980. 1:927–950.

Dinh-Nguyen, N., R. Ryhage, S. Stallberg-Stenhagen, and E. Stenhagen. 1961. *Ark. Kemi.* 18:393; 28:289.

Djerassi, C., and C. Fenselau. 1965. *J. Am. Chem. Soc.* 87:5747, 5752.

Djerassi, C., and L. Tokes. 1966. *J. Am. Chem. Soc.* 88:536.

Dole, M., H. L. Cox, Jr., and J. Gieniec. 1971. *Adv. Chem. Ser.* 125:73.

Dougherty, R. C., and D. Schuetzle. 1980. In Waller 1980. 1:951–1006.

Dromey, R. G. 1976. *Anal. Chem.* 48:1464.

Eadon, G. 1977. *Org. Mass Spectrom.* 12:671.

Eglinton, G., *et al.* 1979. *Phil. Trans. R. Soc., Lond. A.* 293:69.

Elliot, W. H. 1980. In Waller 1980. 1:229–254.

Enzell, C. R., and I. Wahlberg. 1980. In Waller 1980. 1:311–438.

Fales, H. 1971. In Milne 1971, p. 179.

Fenselau, C. 1978. *Physical Methods in Modern Chemical Analysis.* Academic Press. pp. 103–186.

Ferrer-Correia, A. J. V., K. R. Jennings, and D. K. Sensharma. 1975. *Chem. Commun.* p. 973.

Field, F. H. 1968. *Accounts Chem. Res.* 1:42.

————. 1972. In Maccoll 1972, p. 133.

Friedman, L. 1979. Personal communication.

Frigerio, A. 1980. *Adv. Mass Spectrom.* 8:1076.

Frigerio, A., and N. Castagnoli, Jr. 1974. *Mass Spectrometry in Biochemistry and Medicine.* Raven Press. 389 pp. See also subsequent volumes in this series.

Gaffney, J. S., R. C. Pierce, and L. Friedman. 1977. *J. Am. Chem. Soc.* 99:4293.

Games, D. E. 1979. In Johnstone 1979, pp. 285–311.

Green, M. M. 1976. In N. L. Allinger and E. L. Eliel, eds., *Topics in Stereochemistry* (Interscience), pp. 35–110.

Green, M. M., R. J. Cook, J. M. Schwab, and R. B. Roy. 1970. *J. Am. Chem. Soc.* 92:3076.

Green, M. M., R. J. Giguere, and J. R. P. Nicholson. 1978. *J. Am. Chem. Soc.* 100:8020.

Gross, M. L. 1978. *High-Performance Mass Spectrometry: Chemical Applications.* American Chemical Society. 358 pp.

Grostic, M. F., and P. B. Bowman. 1980. In Waller 1980. 1:661–692.

Grubb, H. M., and S. Meyerson. 1963. In McLafferty 1963, p. 453.

Hamming, M. G., and N. G. Foster. 1972. *Interpretation of Mass Spectra of Organic Compounds.* Academic Press. 694 pp.

Happ, G. P., and D. W. Stewart. 1952. *J. Am. Chem. Soc.* 74:4404.

Harrison, A. G., C. D. Finney, and J. A. Sherk. 1971. *Org. Mass Spectrom.* 5:1313.

Harrison, A. G., and F. I. Onuska. 1978. *Org. Mass Spectrom.* 13:35.

Hartman, K. N., S. Lias, P. Ausloos, H. M. Rosenstock, S. S. Schroyer, C. Schmidt, D. Martinsen, and G. W. A. Milne. 1979. *A Compendium of Gas Phase Basicity and Proton Affinity Measurements.* U.S. National Bureau of Standards.

Heller, S. R., and G. W. A. Milne. 1978. *EPA/NIH Mass Spectral Data Base.* NSRDS-NBS 63. U.S. Government Printing Office. 3975 pp.

Henneberg, D. 1980. *Adv. Mass Spectrom.* 8:1511.

Henneberg, D., and G. Schomburg. 1968. *Adv. Mass Spectrom.* 4:333.

Hesse, M., and F. Lenzinger. 1968. *Adv. Mass Spectrom.* 4:163.

Hesse, M., S. D. Sastry, and K. M. Madyastha. 1980. In Waller 1980. 1: 751–820.

Hickling, R. D., and K. R. Jennings. 1970. *Org. Mass Spectrom.* 3:1499.

Hignite, C. 1980. In Waller 1980. 1:527–566.

Hoffman, M. K., and J. C. Wallace. 1973. *J. Am. Chem. Soc.* 95:5064.

Horman, I. 1979. In Johnstone 1979, pp. 211–233.

Houle, F. A., and J. L. Beauchamp. 1979. *J. Am. Chem. Soc.* 101:4067.

Houriet, R., G. Parisod, and T. Gaumann. 1977. *J. Am. Chem. Soc.* 99:3599.

Howe, I., and D. H. Williams. 1968. *J. Chem. Soc. (C).* p. 202.

———. 1973. *Organic Mass Spectrometry.* McGraw-Hill. 245 pp.

Hunt, D. F., C. N. McEwen, and T. M. Harvey. 1975. *Anal. Chem.* 47:1730.

Hunt, D. F., C. N. McEwen, and R. A. Upham. 1972. *Anal. Chem.* 44:1292.

Hunt, D. F., J. Shabanowitz, F. K. Botz, and D. A. Brent. 1977. *Anal. Chem.* 49:1160

Hunt, D. F., G. C. Stafford, Jr., F. W. Crow, and J. W. Russell. 1976. *Anal. Chem.* 48:2098.

Jackson, A. H. 1979. *Phil. Trans. R. Soc., Lond. A.* 293:21.

Jellum, E. 1979. *Phil. Trans. R. Soc., Lond. A.* 293:13.

Jennings, K. R. 1979a. In Bowers 1979. 2:124.

———. 1979b. *Phil. Trans. R. Soc., Lond. A.* 293:125.

Johnstone, R. A. W. 1975, 1977, 1979. *Mass Spectrometry,* Vols 3, 4, and 5. The Chemical Society, London.

Karni, M., and A. Mandelbaum. 1980. *Org. Mass Spectrom.* 15:53.

Kilburn, K. B., P. H. Lewis, J. G. Underwood, S. Evans, J. Holmes, and M. Dean. 1979. *Anal. Chem.* 51:1420.

Kingston, D. G. I., J. T. Bursey, and M. M. Bursey. 1974. *Chem. Rev.* 74:215.

Kingston, D. G. I., B. W. Hobrock, M. M. Bursey, and J. T. Bursey. 1975. *Chem. Rev.* 75:693–730.

Klein, E. R., and P. D. Klein. 1979. *Biomed. Mass Spectrom.* 6:515.

Koppel, C., and F. W. McLafferty. 1976. *J. Am. Chem. Soc.* 98:8293.

Kraft, M., and G. Spiteller. 1967. *Chem. Commun.* p. 943.

———. 1969. *Org. Mass Spectrom.* 2:541.

Kuster, T., and J. Seibl. 1976. *Org. Mass Spectrom.* 11:644.

Kwok, K.-S., R. Venkataraghavan, and F. W. McLafferty. 1973. *J. Am. Chem. Soc.* 95:4185.

Land, D. G., and H. E. Nursten. 1979. *Progress in Flavour Research.* Applied Science.

Lavanchy, A., R. Houriet, and T. Gaumann. 1979. *Org. Mass Spectrom.* 14:79.

Ledford, E. B., Jr., S. Ghaderi, R. L. White, R. B. Spencer, P. S. Kulkarni, C. L. Wilkins, and M. L. Gross. 1980. *Anal. Chem.* 52:463.

Lehman, T. A., and M. M. Bursey. 1976. *Ion Cyclotron Resonance Spectrometry.* Wiley. 230 pp.

Leung, H.-W., and A. G. Harrison. 1977. *Org. Mass Spectrom.* 12:582.

Levsen, K. 1978. *Fundamental Aspects of Organic Mass Spectrometry.* Verlag Chemie, Weinheim. 312 pp.

Levsen, K., G. E. Gerendsen, N. M. M. Nibbering, and H. Schwarz. 1977. *Org. Mass Spectrom.* 12:125.

Levsen, K., R. Weber, F. Borchers, H. Heimbach, and H. D. Beckey. 1978. *Anal. Chem.* 50:1655.

Lifshitz, C. 1977. *Adv. Mass Spectrom.* 7:3.

Ligon, W. V., Jr. 1979. *Science.* 205:151.

Lin, Y. Y., and L. L. Smith. 1979. *Biomed. Mass Spectrom.* 6:15.

Litton, J. F., T. L. Kruger, and R. G. Cooks. 1976. *J. Am. Chem. Soc.* 98:2011.

Longevialle, P., and R. Botter. 1980. Euchem Conference, Lunteren, Holland.

Longevialle, P., J.-P. Girard, J.-C. Rossi, and M. Tichy. 1979. *Org. Mass Spectrom.* 14:414.

Lory, E. R., and F. W. McLafferty. 1980. *Adv. Mass Spectrom.* 8:954.

Lumpkin, E., and T. Aczel. 1978. In Gross 1978, pp. 261–273.

Mabry, T., and A. Ulubelen. 1980. In Waller 1980. 1:1131–1158.

Maccoll, A., ed. 1972. *Mass Spectrometry.* MTP Int. Rev. Sci. Butterworths. 300 pp.

———. 1980. *Org. Mass Spectrom.* 15:109.

Macfarlane, R. D. 1980. In Waller 1980. 1:1209–1220.

Macfarlane, R. D., and D. F. Torgerson. 1976. *Science.* 191:930.

Mandelbaum, A. 1977. In H. B. Kagan, ed., *Stereochemistry* (G. Thieme, Stuttgart), 1:137–180.

Mass Spectrometry Data Centre. 1975. *Eight-Peak Index of Mass Spectra,* AWRE, Aldermaston, Berkshire, England.

McAdoo, D. J., P. F. Bente III, M. L. Gross, and F. W. McLafferty. 1974. *Org. Mass Spectrom.* 9:525.

McAdoo, D. J., D. M. Witiak, F. W. McLafferty, and J. D. Dill. 1978. *J. Am. Chem. Soc.* 100:6639.

McCloskey, J. A., and V. I. Krahmer. 1977. *Adv. Mass Spectrom.* 7:1483.

McDowell, C. A., ed. 1963. *Mass Spectrometry.* McGraw-Hill.

McFadden, W. H. 1973. *Techniques of Combined Gas Chromotography/ Mass Spectrometry.* Wiley. 463 pp.

McLafferty, F. W. 1956. *Anal. Chem.* 28:306.

————. 1957. *Anal. Chem.* 29:1782.

————, ed. 1963. *Mass Spectrometry of Organic Ions.* Academic Press. 730 pp.

————. 1979. *Phil. Trans. R. Soc., Lond. A.* 293:93.

————. 1980a. *Accounts Chem. Res.* 13:33.

————. 1980b. *Org. Mass Spectrom.* 15:114.

McLafferty, F. W., B. L. Atwater (Fell), K. S. Haraki, K. Hosakawa, I. K. Mun, and R. Venkataraghavan. 1980b. *Adv. Mass Spectrom.* 8:1564.

McLafferty, F. W., P. F. Bente III, R. Kornfeld, S.-C. Tsai, and I. Howe. 1973a. *J. Am. Chem. Soc.* 95:2120.

McLafferty, F. W., and F. M. Bockoff. 1979. *J. Am. Chem. Soc.* 101:1783.

McLafferty, F. W., R. Kornfeld, W. F. Haddon, K. Levsen, I. Sakai, P. F. Bente III, S.-C. Tsai, and H. D. R. Schuddemage. 1973b. *J. Am. Chem. Soc.* 95:3886.

McLafferty, F. W., D. J. McAdoo, and J. S. Smith. 1969. *J. Am. Chem. Soc.* 91:5400.

McLafferty, F. W., and I. Sakai. 1973. *Org. Mass Spectrom.* 7:791.

McLafferty, F. W., P. J. Todd, D. C. McGilvery, and M. A. Baldwin. 1980a. *J. Am. Chem. Soc.* 102:3360.

McLafferty, F. W., and R. Venkataraghavan. 1982. *Mass Spectral Correlations*, 2d ed. American Chemical Society.

McLafferty, F. W., T. Wachs, C. Lifshitz, G. Innorta, and P. Irving. 1970. *J. Am. Chem. Soc.* 92:6867.

McLoughlin, R. G., and J. C. Traeger. 1979. *Org. Mass Spectrom.* 14:434.

Meisels, G. G., C. T. Chen, B. G. Giessner, and R. H. Emmel. 1972. *J. Chem. Phys.* 56:793.

Meuzelaar, H. L. C., J. Haverkamp, and S. D. Hileman. 1980. *Pyrolysis Mass Spectrometry of Biomaterials.* Elsevier.

Meyerson, S., and L. C. Leitch. 1966. *J. Am. Chem. Soc.* 88:56.

Millard, B. J. 1978. *Quantitative Mass Spectrometry.* Heyden, London.

_____. 1979. In Johnstone 1979, pp. 186–210.

Milne, G. W. A., ed. 1971. *Mass Spectrometry: Techniques and Applications*. Wiley-Interscience. 521 pp.

Morris, H. R. 1979. *Phil. Trans. R. Soc., Lond. A.* 293:39.

Mun, I. K., R. Venkataraghaven, and F. W. McLafferty. 1977. *Anal. Chem.* 49:1723–6.

Munson, M. S. B., and F. H. Field. 1966. *J. Am. Chem. Soc.* 88:2621.

Nachod, F. C., J. J. Zuckerman, and E. W. Randall, eds. 1976. *Determination of Organic Structures by Physical Methods*, Vol. 6. Academic Press.

Nakanishi, K., and J. L. Occolowitz. 1979. *Phil. Trans. R. Soc., Lond. A.* 293:3.

Nibbering, N. M. M. 1979. *Phil. Trans. R. Soc., Lond. A.* 293:103.

Nicholson, A. J. C. 1954. *Trans. Faraday Soc.* 50:1067.

Nishishita, T., F. M. Bockhoff, and F. W. McLafferty. 1977. *Org. Mass Spectrom.* 12:16.

Odham, G., and G. Wood. 1980. In Waller 1980. 1:153–210.

Pesheck, C. V., and S. E. Buttrill, Jr. 1974. *J. Am. Chem. Soc.* 96:6027.

Pesyna, G. M., and F. W. McLafferty. 1976. In Nachod *et al.* 1976, 6:91–155.

Pesyna, G. M., R. Venkataraghaven, H. E. Dayringer, and F. W. McLafferty. 1976. *Anal. Chem.* 48:1362–8.

Pillinger, L. T. 1979. In Johnstone 1979, pp. 250–261.

Porter, Q. N., and J. Baldas. 1971. *Mass Spectrometry of Heterocyclic Compounds*. Wiley-Interscience. 564 pp.

Posthumus, M. A., P. G. Kistemaker, H. L. C. Meuzelaar, and M. C. Ten Noever de Brauw. 1978. *Anal. Chem.* 50:985.

Radford, T., and D. C. DeJongh. 1980. In Waller 1980. 1:255–310.

Remberg, G., and G. Spiteller. 1970. *Chem. Ber.* 103:3640.

Resink, J. J., A. Venema, and N. M. M. Nibbering. 1974. *Org. Mass Spectrom.* 9:1055.

Respondek, J., H. Schwarz, F. Van Gaever, and C. C. Van de Sande. 1978. *Org. Mass Spectrom.* 13:618.

Richter, W. J., and H. Schwarz. 1978. *Angew. Chem. Int. Ed. Engl.* 17:424–439.

Robinson, P. J., and K. A. Holbrook. 1972. *Unimolecular Reactions*. Wiley.

Rol, N. C. 1968. *Rec. Trav. Chim.* 87:321.

Rosenstock, H. M., K. Draxl, B. W. Steiner, and J. T. Herron. 1977. *J. Phys. Chem. Ref. Data* 6, Suppl. 1. 783 pp.

Rosenstock, H. M., M. B. Wallenstein, A. L. Wahrhaftig, and H. Eyring. 1952. *Proc. Nat. Acad. Sci.* 38:667.

Safe, S. 1979. In Johnstone 1979, pp. 234–249.

Sample, S. D., and C. Djerassi. 1966. *J. Am. Chem. Soc.* 88:1937.

Sanders, J. H., and A. H. Wapstra. 1976. *Atomic Masses and Fundamental Constants.* 5:655. Plenum Press.

Schulten, H.-R. 1979. *Int. J. Mass Spectrom. Ion Phys.* 32:97–283.

Schwarz, H. 1978. *Topics Current Chem.* 73:231–263.

Schwarz, H., C. Wesdemiotis, K. Levsen, H. Heimbach, and W. Wagner. 1979. *Org. Mass Spectrom.* 14:244.

Seibl, J. 1970. *Massenspektrometrie.* Akademische Verlagsgesellschaft, Frankfurt/Main.

Seibl, J., and T. Gaumann. 1963. *Helv. Chim. Acta.* 46:2857.

Seifert, W. K., E. J. Gallegos, and R. M. Teeter. 1972. *J. Am. Chem. Soc.* 94:5880.

Seiler, K., and M. Hesse. 1968. *Helv. Chim. Acta.* 51:1817.

Sjovall, J. 1980. *Adv. Mass Spectrom.* 8:1069.

Smith, D. H., ed. 1977. *Computer-Assisted Structure Elucidation.* Amer. Chem. Soc. 151 pp.

Spaulding, T. R. 1979. In Johnstone 1979, pp. 312–346.

Speck, D. D., R. Venkataraghavan, and F. W. McLafferty. 1978. *Org. Mass Spectrom.* 13:209.

Sphon, J. A., and W. C. Brumley. 1980. In Waller 1980. 1:713–750.

Spiteller, G. 1966. *Massenspektrometrische Strukturanalyse Organischer Verbindungen.* Verlag Chemie, Weinheim. 354 pp.

Stenhagen, E. 1972. In Waller 1972, p. 11.

Stenhagen, E., S. Abrahamsson, and F. W. McLafferty. 1974. *Registry of Mass Spectral Data.* Wiley. 3358 pp.

Stevenson, D. P. 1951. *Disc. Faraday Soc.* 10:35.

Svec, H. J., and G. A. Junk. 1967. *J. Am. Chem. Soc.* 89:790.

Thomson, J. J. 1913. *Rays of Positive Electricity.* Longmans, Green.

Tokes, L., G. Jones, and C. Djerassi. 1968. *J. Am. Chem. Soc.* 90:5465.

Tsang, C. W., and A. G. Harrison. 1971. *Org. Mass Spectrom.* 5:877.

Turecek, F., and V. Hanus. 1980. *Org. Mass Spectrom.* 15:4.

Uccella, N. A., I. Howe, and D. H. Williams. 1971. *J. Chem. Soc. (B)* 1933.

United Kingdom Chemical Information Service. *The Mass Spectrometry Bulletin.* Nottingham University, NG7 2RD, England.

Van de Graaf, B., and F. W. McLafferty. 1977. *J. Am. Chem. Soc.* 99:6806, 6810.

Van Gaever, F., J. Monstrey, and C. C. Van de Sande. 1977. *Org. Mass Spectrom.* 12:200.

Vetter, W. 1980. In Waller 1980. 1:439–468.

Wahrhaftig, A. L. 1962. Personal communication.

Waller, G. R., ed. 1972. *Biochemical Applications of Mass Spectrometry.* Wiley-Interscience. 872 pp.

Waller, G. R., and O. C. Dermer, eds. 1980. *Biochemical Applications of Mass Spectrometry, Suppl. Vol.* Wiley. 1279 pp.

Wapstra, A. H., and N. B. Gove. 1971. *Nuclear Data Tables.* 9:267.

Watson, J. T. 1976. *Introduction to Mass Spectrometry.* Raven Press. 239 pp.

Wiersig, J. R., A. N. H. Yeo, and C. Djerassi. 1977. *J. Am. Chem. Soc.* 99:532.

Williams, D. H., ed. 1971 and 1973. *Mass Spectrometry,* Vols. 1 and 2. The Chemical Society, London.

――――. 1977. *Accounts Chem. Res.* 10:280.

――――. 1979. *Phil. Trans. R. Soc., Lond. A.* 293:117.

Winkler, F. J., and H. F. Grutzmacher. 1970. *Org. Mass Spectrom.* 3:1117.

Winkler, F. J., and F. W. McLafferty. 1974. *Tetrahedron.* 30:2971.

Winkler, F. J., and D. Stahl. 1979. *J. Am. Chem. Soc.* 101:3685.

Wolf, J. F., R. H. Staley, I. Koppel, M. Taagepera, R. T. McIver, Jr., J. L. Beauchamp, and R. W. Taft. 1977. *J. Am. Chem. Soc.* 99:5417.

Woodgate, P. D., K. K. Mayer, and C. Djerassi. 1972. *J. Am. Chem. Soc.* 94:3115.

Yamdagni, R., and P. Kebarle. 1976. *J. Am. Chem. Soc.* 98:1320.

Yost, R. A., and C. G. Enke. 1979. *Anal. Chem.* 51:1251A.

Zahorszky, U. I. 1979. *Org. Mass Spectrom.* 14:66, and references cited therein.

Zaretskii, Z. V. 1976. *Mass Spectrometry of Steroids.* Wiley. 182 pp.

APPENDIX

Table A.1. *Nuclidic Masses and Isotopic Abundances*

Mass[a]	Element	Isotope (mass), per cent abundance[b]
1.00782506	H	^{2}H (2.0140), 0.015
4.00260336	He	^{3}He (3.0160), 0.00013
7.0160048	Li	^{6}Li (6.0151), 8.0
9.0121828	Be	
11.00930533	B	^{10}B (10.0129), 24.6
12.00000000	C	^{13}C (13.0034), 1.1
14.00307407	N	^{15}N (15.0001), 0.37
15.99491475	O	^{17}O, 0.04; ^{18}O (17.9992), 0.20
18.9984046	F	
19.9924405	Ne	^{21}Ne, 0.28; ^{22}Ne (21.9914), 9.7
22.9897703	Na	
23.9850443	Mg	^{25}Mg, 12.9; ^{26}Mg, 14.2
26.9815406	Al	
27.9769286	Si	^{29}Si (28.9765), 5.1; ^{30}Si (29.9738), 3.4
30.9737633	P	
31.9720728	S	^{33}S (32.9715), 0.80; ^{34}S (33.9679), 4.4; ^{36}S, 0.015
34.9688530	Cl	^{37}Cl (36.9659), 32.5
39.9623842	Ar	^{36}Ar, 0.34; ^{38}Ar, 0.06
38.9637089	K	^{40}K, 0.01; ^{41}K, 7.4
39.9625921	Ca	^{42}Ca, 0.67; ^{43}Ca, 0.14; ^{44}Ca, 2.2; ^{46}Ca, 0.19; ^{48}Ca, 0.19
48.9478721	Ti	^{46}Ti, 10.7; ^{47}Ti, 9.9; ^{49}Ti, 7.5; ^{50}Ti, 7.2
50.9439644	V	^{50}V, 0.24
54.9380464	Mn	
55.9349339	Fe	^{54}Fe, 6.4; ^{57}Fe, 2.4; ^{58}Fe, 0.36
58.9331879	Co	
57.9353358	Ni	^{60}Ni, 38.2; ^{61}Ni, 1.7; ^{62}Ni, 5.3; ^{64}Ni, 1.3
62.9295898	Cu	^{65}Cu, 44.7
63.9291400	Zn	^{66}Zn, 56.9; ^{67}Zn, 8.4; ^{68}Zn, 38.0; ^{70}Zn, 1.3

Table A.1. *Nuclidic Masses and Isotopic Abundances* (*continued*)

Mass[a]	Element	Isotope (mass), per cent abundance[b]
74.9216003	As	
79.9165253	Se	^{74}Se, 1.8; ^{76}Se, 18.1; ^{77}Se, 15.2; ^{78}Se, 47.2; ^{82}Se, 18.5
78.9183320	Br	^{81}Br (80.9163), 98.0
83.9115053	Kr	^{78}Kr, 0.62; ^{80}Kr, 4.0; ^{82}Kr, 20.3; ^{83}Kr, 20.3; ^{86}Kr, 30.5
106.905091	Ag	^{108}Ag, 93.0
119.9022073	Sn	^{112}Sn, 3.0; ^{114}Sn, 2.0; ^{115}Sn, 1.1; ^{116}Sn, 43.5; ^{117}Sn, 23.2; ^{118}Sn, 73.2; ^{119}Sn, 26.1; ^{122}Sn, 14.4; ^{124}Sn, 18.1
126.9044755	I	
131.9041568	Xe	^{124}Xe, 0.37; ^{126}Xe, 0.33; ^{128}Xe, 7.1; ^{129}Xe, 98.3; ^{130}Xe, 15.2; ^{131}Xe, 78.8; ^{134}Xe, 38.8; ^{136}Xe, 33.0
132.905436	Cs	
196.966548	Au	
201.970643	Hg	^{196}Hg, 0.50; ^{198}Hg, 33.7; ^{199}Hg, 56.5; ^{200}Hg, 77.6; ^{201}Hg, 44.4; ^{204}Hg, 23.0
207.976658	Pb	^{204}Pb, 2.8; ^{206}Pb, 45.1; ^{207}Pb, 43.2
238.050816	U	^{235}U, ~0.8

[a]Mass of the most abundant isotope
[b]Based on the most abundant isotope as 100 per cent

Table A.2. *Natural abundances of combinations of chlorine, bromine, silicon, and sulfur*

Number of chlorine atoms	Mass	Number of bromine atoms				
		0	1	2	3	4
0	A		100.0	51.0	34.0	17.4
	A + 2		98.0	100.0	100.0	68.0
	A + 4			49.0	98.0	100.0
	A + 6				32.0	65.3
	A + 8					16.0
1	A	100.0	76.6	43.8	26.1	14.2
	A + 2	32.5	100.0	100.0	85.1	60.3
	A + 4		24.4	69.9	100.0	100.0
	A + 6			13.7	48.9	80.1
	A + 8				8.0	30.5
	A + 10					4.3
2	A	100.0	61.4	38.3	20.4	11.9
	A + 2	65.0	100.0	100.0	73.3	54.3
	A + 4	10.6	45.6	89.7	100.0	100.0
	A + 6		6.6	31.9	63.8	94.2
	A + 8			3.9	18.7	47.3
	A + 10				2.0	11.8
	A + 12					1.2
3	A	100.0	51.2	31.3	16.4	
	A + 2	97.5	100.0	92.0	64.5	
	A + 4	31.7	65.0	100.0	100.0	
	A + 6	3.4	17.6	49.9	77.7	
	A + 8		1.7	11.6	31.8	
	A + 10			1.0	6.5	
	A + 12				0.5	

Table A.2. *Natural abundances of combinations of chlorine, bromine, silicon, and sulfur* (*continued*)

Number of chlorine atoms	Mass	Number of bromine atoms				
		0	1	2	3	4
	A + 14					
	A	76.9	43.8	24.1	13.6	
	A + 2	100.0	100.0	78.7	57.7	
	A + 4	48.7	83.6	100.0	100.0	
4	A + 6	10.5	33.2	63.4	91.1	
	A + 8	0.9	6.3	21.4	47.2	
	A + 10		0.5	3.7	13.9	
	A + 12			0.3	2.2	
	A + 14				0.1	
	A	61.5	37.7	19.2		
	A + 2	100.0	98.3	68.9		
	A + 4	65.0	100.0	100.0		
5	A + 6	21.1	52.0	76.4		
	A + 8	3.4	14.8	33.5		
	A + 10	0.2	2.2	8.5		
	A + 12		0.1	1.1		
	A + 14			0.1		

Mass	Number of chlorine atoms				
	6	7	8	9	10
A	51.2	43.9	33.8	26.3	21.0
A + 2	100.0	100.0	87.9	76.9	68.3
A + 4	81.2	97.5	100.0	100.0	100.0
A + 6	35.2	52.8	65.0	75.8	86.6
A + 8	8.5	17.1	26.4	36.9	49.2
A + 10	1.1	3.3	6.8	12.0	19.2
A + 12	0.06	0.36	1.1	2.6	5.2
A + 14		0.02	0.10	0.36	0.97
A + 16				0.03	0.12
A + 18					0.01

Mass	Number of silicon atoms				
	1	2	3	4	5
A	100.0	100.0	100.00	100.00	100.00
A + 1	5.1	10.20	15.30	20.40	25.50
A + 2	3.4	7.1	11.00	15.16	19.60
A + 3		0.35	1.05	2.13	3.60
A + 4		0.12	0.37	0.80	1.42
A + 5			0.02	0.07	0.19
A + 6				0.02	0.05

Mass	Number of sulfur atoms				
	1	2	3	4	5
A	100.0	100.0	100.0	100.0	100.0
A + 1	0.80	1.6	2.4	3.2	4.0
A + 2	4.4	8.8	13.2	17.6	22.0
A + 3		0.07	0.21	0.42	0.70
A + 4		0.19	0.58	1.2	1.9
A + 5				0.02	0.05
A + 6				0.03	0.09

Table A.3. *Ionization energy and proton affinity values* (1 eV = 96.49 kJ/mol = 23.06 Kcal/mol) [a]

H, C		O, S, Se		N, P, He, Ar		Halogen, etc.		Radicals	
H_2	15.4			He	24.6 (2.1)	HF	16.0 (5.2)	F·	17.4
				Ar	15.8 (4)			NC·	14.1
				N_2	15.6 (5.8)			H·	13.6
CH_4	12.5 (5.9)	CO	14.0 (6.5)					HO·	~13.0
		CO_2	13.8 (5.7)	HCN	13.6	HCl	12.7 (6.4)	Cl·	13.0
		H_2O	12.6 (7.7)	NO_2	12.9	CH_3F	12.5 (6.9)	Br·	11.8
		SO_2	12.3					$H_2N·$	11.2
C_2H_6	11.5 (6.2)	O_2	12.1 (4.6)	CH_3CN	12.2 (8.4)	HBr	11.7 (6.4)	$NCCH_2·$	10.8
C_2H_2	11.4			$n\text{-}C_3H_7CN$	11.7 (8.6)	Cl_2	11.5	I·	10.5
C_3H_8	11.0 (6.7)	HCOOH	11.3 (8.1)			CH_3Cl	11.3	$CH_3·$	9.8
$n\text{-}C_4H_{10}$	10.6 (7.5)	CH_2O	10.9 (7.9)	$CH_2{=}CHCN$	10.9 (8.5)	$n\text{-}C_4H_9Cl$	10.7	$CH_2{=}CH·$	9.8
$i\text{-}C_4H_{10}$	10.5	CH_3OH	10.8 (8.2)	$n\text{-}C_3H_7NO_2$	10.8	Br_2	10.5	HCO·	9.8
$CH_2{=}CH_2$	10.5 (7.3)	CH_2CH_2O	10.6 (8.3)					$CH_3O·$	9.8[c]
$CH_3C{\equiv}CH$	10.4 (7.9)	C_2H_5OH	10.5 (8.4)			$CH_2{=}CHF$	10.4 (7.9)	$ClCH_2·$	9.3
		$CH_2{=}C(OH)_2$	10.5			HI	10.4 (6.6)	$C_3H_3·$	8.7
		H_2S	10.4 (7.8)					$BrCH_2·$	~8.6[d]
		CH_3COOH	10.4 (8.5)					HOOC·	~8.6[d]
		$n\text{-}C_3H_7COOH$	10.2 (8.6)					$CH_3OOC·$	~8.6[d]
$n\text{-}C_6H_{14}$	10.2	CH_3CHO	10.2 (8.3)	NH_3	10.2 (9.1)	$n\text{-}C_4H_9Br$	10.1	$C_2H_5·$	8.2
		$C_2H_5COOCH_3$	10.2 (8.9)	PH_3	10.0 (8.4)	$CH_2{=}CHCl$	10.0	$CH_3S·$	8.1
$((CH_3)_2CH)_2$	10.0	$n\text{-}C_4H_9OH$	10.1 (8.5)					$CH_2{=}CHCH_2·$	8.1
		CS_2	10.1 (7.6)					$C_6H_5·$	8.1
Cyclohexane	9.9	H_2Se	9.9 (7.9)	$C_6H_5NO_2$	9.9 (8.6)			$n\text{-}C_3H_7·$	8.1
$CH_2{=}C{=}CH_2$	9.7 (7.9)	$n\text{-}C_3H_7CHO$	9.9 (8.5)					$n\text{-},i\text{-}C_4H_9·$	8.0
$CH_3CH{=}CH_2$	9.7 (8.2)	C_6H_5COOH	9.7	CH_3CONH_2	9.8 (9.2)				
		CH_3COCH_3	9.7 (8.7)						

Table A.3. Ionization energy and proton affinity values (1 eV = 96.49 kJ/mol = 23.06 Kcal/mol) [a] (continued)

H, C		O, S, Se		N, P, He, Ar		Halogen, etc.		Radicals	
$CH_3C{\equiv}CCH_3$	9.6	$CH_3CH(OCH_3)_2$	9.7	C_6H_5CN	9.7 (8.8)	$(CH_3)_4Si$	9.5	$CH_3CO\cdot$	7.9
$C_2H_5CH{=}CH_2$	9.6 (8.3)	CH_2CO	9.6 (8.7)					$C_2H_5CO\cdot$	7.7
Benzyne	9.5 (9.5)	$(C_2H_5)_2O$	9.6 (8.9)	NO	9.3 (5.5)			Cyclohexyl$\cdot$	7.7
$(CH_3)_2C{=}CH_2$	9.2 (8.6)	$C_2H_5COCH_3$	9.5 (8.8)	Pyridine	9.3 (9.8)	$n\text{-}C_4H_9I$	9.2	$s\text{-}C_4H_9\cdot$	7.4
Benzene	9.2 (8.2)	$C_6H_5COCH_3$	9.3 (9.1)	$CH_5CH{=}NCH_3$	9.1	C_6H_5F	9.2 (8.2)	$HOCH_2\cdot$	7.4
$CH_3CH{=}CHCH_3$	9.1 (8.2)	$CH_2{=}C(OH)CH_3$	9.1			C_6H_5Cl	9.1 (8.2)	$HSCH_2\cdot$	7.3
$(CH_2{=}CH{-})_2$	9.1	$n\text{-}C_4H_9SH$	9.1			C_6H_5Br	9.0	$C_6H_5CH_2\cdot$	7.3
Cyclohexene	8.9	Thiophene	8.9	$C_2H_5NH_2$	8.9 (9.6)			$CH_3OCH_2\cdot$	6.9
$C_6H_5CH_3$	8.8 (8.5)	Furan	8.9 (8.5)	$CH_3CON(CH_3)_2$	8.8 (9.2)			$CH_3SCH_2\cdot$	~6.9[d]
				$n\text{-}C_4H_9NH_2$	8.7 (9.7)			$CH_3(HO)CH\cdot$	6.7
$C_6H_5CH{=}CH_2$	8.4 (9.0)	C_6H_5OH	8.5 (8.8)	$(C_2H_5)_2PH$	8.5	$(C_2H_5)_2Hg$	8.5	$t\text{-}C_4H_9\cdot$	6.7
Biphenyl	8.2	$C_2H_5SC_2H_5$	8.4 (9.1)	Pyrrole	8.2 (9.3)	$(CH_3)_4Sn$	8.2	$(CH_2O)_2{=}CH\cdot$[e]	~6.3[d]
Naphthalene	8.1	$C_6H_5OCH_3$	8.2 (8.9)	$(CH_3NH)_2$	8.2			$H_2NCH_2\cdot$	6.2
				$(C_2H_5)_2NH$	8.0 (10.0)			$(CH_3)_2NCH_2\cdot$	~5.8[d]
Anthracene	7.5			$C_6H_5NH_2$	7.7 (9.4)				
				$(C_6H_5)_3N$	6.9				

[a] Figures not in parentheses are ionization energy, and those in parentheses are proton affinity, both in eV. Most ionization energy values are taken from Rosenstock et al. 1977, and proton affinity values (based on 9.1 eV for NH_3; McLoughlin and Traeger 1979) from Hartman et al. 1979. Other references are Yamdagni and Kebarle 1976; Benoit et al. 1977; Wolf et al. 1977; Dill et al. 1979; McLoughlin and Traeger 1979; and Houle and Beauchamp 1979.
[b] Harrison et al. 1971.
[c] Dill et al. 1979; value for the C_{2v} complex of H_2 and HCO^+, 8.2 eV.
[d] Estimated from mass spectral abundances by the method of Harrison et al. 1971.
[e] Ethylene ketal.

Table A.4. *Molecular ion abundances vs. compound type* (C_n indicates a n-alkyl chain of *n* carbon atoms).

Compound type	Intensity of molecular ion relative to most intense ion					M.W. for [M+] <0.1%
	M.W. ~ 75		M.W. ~ 130		M.W. ~ 185	
Aromatic	[benzene]	100	[naphthalene]	100	 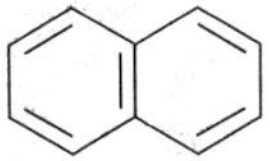 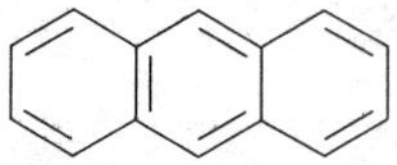 [anthracene] 100	>500
Heterocyclic 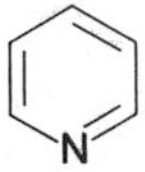 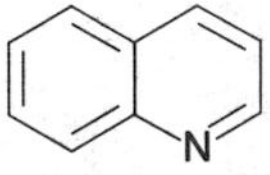 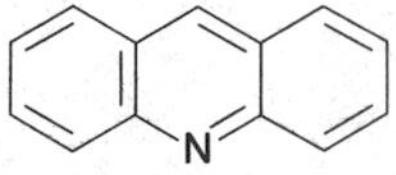	[pyridine]	100	[quinoline]	100	[acridine] 100	>500
	[thiophene]	100	[benzothiophene]	100	 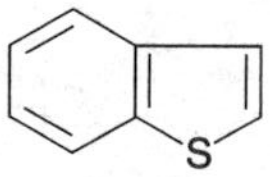 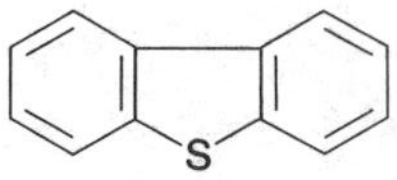 [dibenzothiophene] 100	>500
Cycloalkane	[cyclohexane]	70	[decalin]	90	 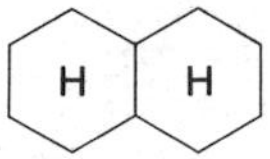 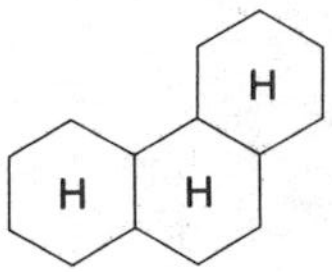 [perhydroanthracene] 90	>500
Mercaptan	C_3SH	100	C_7SH	40	$C_{10}SH$ 46	≥200
Sulfide	C_1SC_2	65	C_1SC_6	45	C_5SC_5 13	≥200
Conjugated olefin	Hexatriene	55	*allo*-Ocimene	40	–	>500
Olefin	$C_2C=CC_2$	35	$C_3C=CC_4$ 20 $C_6C=CC$ 7		$C_{11}C=C$ 3	>500
Amide	C_2CONH_2 55 $HCON(C_1)_2$ 100		C_6CONH_2 1 $C_1CON(C_2)_2$ 4		$C_{11}CONH_2$ 1 $C_1CON(C_4)_2$ 5	–
Acid	C_2COOH	80	C_6COOH	0.5	C_9COOH 9	—[b]

Table A.4. *Molecular ion abundances vs. compound type (Cₙ indicates a n-alkyl chain of n carbon atoms) (continued).*

Compound type	Intensity of molecular ion relative to most intense ion			M.W. for $[M^{+}] < 0.1\%$
	M.W. ~ 75	M.W. ~ 130	M.W. ~ 185	
Ketone[a]	C_1COC_2 25	C_2COC_5 8	C_6COC_5 8	>500
		C_1COC_6 3	C_1COC_9 10	
Aldehyde[a]	C_3CHO 45	C_7CHO 2	$C_{12}CHO$ 5	>500
Alkane	C_5 9	C_9 6	C_{13} 5	>500
Amine[a]	C_4NH_2 10	C_8NH_2 0.5	$C_{12}NH_2$ 2	–[b]
	$(C_2)_2NH$ 30	$(C_4)_2NH$ 11	$(C_7)_2NH$ 4	–
		$(C_2)_3N$ 20	$(C_4)_3N$ 7	–
Ether[a]	C_2OC_2 30	C_4OC_4 2	C_6OC_6 0.05	180
Ester[a]	C_1COOC 20	C_1COOC_5 0.1	C_1COOC_8 0.1	–[b]
		C_5COOC_1 0.3	C_7COOC_1 3	–[b]
Halide	C_4F 0.1	C_7F 0.1		RF >120
	C_3Cl 4	C_7Cl 0.1	$C_{11}Cl$ 0.3	RCl >300[a]
		C_3Br 45	C_7Br 2	RBr 300
		C_1I 100	C_4I 6	RI 320
Branched alkane	C—C—C—C / with C and C—C—C branches 6 ; 0.01	$(C_2)_2CC_4$ 6 ; branched structure 0.05	$(C_4)_3CH$ 1 ; branched structure 0.03	~400 ; 70
Nitrile[a]	C_4CN 0.3	C_8CN 0.4	$C_{11}CN$ 0.8	–[b]
Alcohol[a]	C_4OH 1	C_8OH 0.1	$C_{12}OH$ 0.0	90
Acetal[a]	$C(OC)_2$ 0.00	$C_2(OC_3)_2$ 0.00	$C_7(OC_2)_2$ 0.0	All

[a] $(M + 1)^{+}$ possible at higher sample pressures (see Section 6.1).
[b] $[M^{+}]$ increases with increasing molecular weight at higher molecular weights.

Table A.5. *Common neutral fragments.*

Δ^a	Mass	Formula	Example^{b,c}
−4	79	Br	R⫽Br
	51 65	C_3HN, C_4H_3N	Some nitrogen heterocyclic compounds
−3	38	H_6O_2	Some polycarboxylic acids
−2	39 53 67	C_nH_{2n-3}	Allyl esters and some cyclic carbonates—specific rearrangement loss of $(C_nH_{2n-1} - 2H)$; some propargyl and allenic derivatives^c
−1	26 40 54 68	C_nH_{2n-2}	Aromatics; alkenyl aryl ethers; $M^{\ddagger} - 69 - (68)_n$ in polyisoprenes^c
	54	C_3H_2O	Cyclic —CO—CH=CH—
0	26 40	$C_nH_{2n}CN$	R⫽CN, R⫽CH$_2$CN (stable R$^+$ only)
	27 41 55 69, etc.	C_nH_{2n-1}	RCOOR'-specific rearrangement loss of $(R - 2H)$ or $(R' - 2H) + (R - H)$, also from carbonates, amides, larger ketones, etc.; loss of activated C_nH_{2n-1} groups^c
+1	27	HCN	Nitrogen heterocyclic compounds, cyanides, aryl-NH$_2$, enamines, imines
	(±14: Homologous impurity)		
	28 42 56	C_nH_{2n}	RCH$_2$COCH$_2$R-specific rearrangement loss of $(R - H)$ or $(R - H)_2$, also from many unsaturated functional groups; retro-Diels-Alder^c
	14	N	Aryl—NO
	28	N_2	Aryl—N=N—Aryl, $>$C=N$_2$, cyclic —N=N—
	28	CO	Aromatic oxygen compounds (carbonyls, phenols), cyclic ketones, R⫽C≡O^{+c}
	42 56 70	$C_nH_{2n}CO$	Unsaturated alkanoates, acetamides; di-, cyclic, and complex ketones; specific H rearrangement loss of —CR$_2$—CO—
+2	29	CH_3N	Some unsaturated-, aryl—N(CH$_3$)$_2$
	43 57 71	HNCO	Loss of —NR—CO— from carbamates, cyclic amides, uracils
1	1	H	Labile H; alkyl cyanides, lower fluorides and aldehydes (stable RCO$^+$), cyclopropyl compounds
	15 29 43 57 71, etc.	C_nH_{2n+1}	Alkyl loss (α-cleavage or branched site favored); elimination from cycloalkyl group with hydrogen rearrangement^b
	29 43 57	$C_nH_{2n+1}CO$	$C_nH_{2n+1}CO$⫽R (stable R$^+$ only)

Table A.5. *Common neutral fragments (continued).*

Δ^a	Mass	Formula	Example[b,c]
+2	127 etc.	I	R⊣I
+3	2 16 30 44 58 72, etc.	C_nH_{2n+2}	Loss of RH from branched site; loss of H_2 or CH_4, mainly from even-electron ion
	16	NH_2	Aromatic and other amines; uncommon to be abundant
	30	NO	Nitroaromatics, nitroesters, N-nitroso compounds
	44	$CONH_2$	R⊣$CONH_2$ (stable R⁺ only)
	16	O	N-oxides, some sulfoxides; smaller for epoxides, nitro compounds, quinones, alicyclic ketoximes, diketoenamines
	30	CH_2O	$ROCH_2OR$, cyclic ethers, methoxy aromatics[c]
	44 58 72, etc.	$C_nH_{2n+2}CO$	$RCOCH_2R'$-specific hydrogen rearrangement $\longrightarrow$ (R′ − H)⁺ (stable ion only)[c]
	44	CO_2	Carbonates, cyclic anhydrides,[c] lactones, CH_3—N— and aryl—N—phthalimides
	44	CS	Thiophenols, aryl—S—aryl
	2,3,4	H_n	Boranes, silanes, phosphines.
+4	17	NH_3	Amines: uncommon unless other group to stabilize charge
	17	OH	Acids, oximes; rearrangement (e.g., o-$NO_2C_6H_4CH_3$)
	31 45 59	$C_nH_{2n+1}O$ ($n = 0, 1, 2$)	R⊣OR′; RCO⊣OR′
	59 73	$C_nH_{2n+1}CO_2$	R⊣OCOR′, R⊣COOR′ (stable R⁺, small R′ only)[c]
	73	C_3H_9Si	$(CH_3)_3Si$—
+5	18	H_2O	Alcohols (primary favored); higher mol. wt. aldehydes, ketones, ethers[c]
	18 32 46 60 74	$C_nH_{2n+1}OH$ ($n = 0, 1, 2$), $C_nH_{2n+1}OH + C_nH_{2n}$	Loss of ROH from $R'CH_2OR$; from R′COOR with labile hydrogen; these with further loss of C_nH_{2n} ($n > 1$)
	46 60	$CO + C_nH_{2n+1}OH$	Loss of CO + HOR from R′COOR with labile H; cyclic —OCHRO—

Table A.5. *Common neutral fragments (continued).*

Δ[a]	Mass	Formula	Example[b,c]
+5	46	NO₂	Nitroaromatics, nitroesters
	60 74	C_nH_{2n+1}COOH	R'COO—R—H (stable R⁺; small R')
	60	COS	Thiocarbonates
	32	S	Sulfides, polysulfides,[c] aryl thiols
	46	CH₂S	Methyl aryl sulfides, some cyclothioalkanes
+6	33	CH₃ + H₂O	Some alcohols
	19 33 47	C_nH_{2n}F	Fluoroalkanes
	33 47 61	C_nH_{2n+1}S	R—SR', R⁺ more stable than R'⁺; (M − 33)⁺ in aryl thiols, isothiocyanates
+7	20	HF	R—HF (primary favored)
	34 48 62 76	H₂S, H₂S + C_nH_{2n}	Thiols (primary favored); methyl sulfides; these with further loss of C_nH_{2n} (n >1)
	48 62 76 90	C_nH_{2n+3}SiOH	Silyl ethers with a labile H, e.g., mass 90 from (CH₃)₃SiOR
+8	77 91	C_nH_{2n-7}	R'—C₆H₄—R (stable R⁺)
	35 49	HF + C_nH_{2n+1}	C_nH_{2n+1}—R—HF
	35 49 63	C_nH_{2n}Cl	Chlorides (labile bond cleaved)
+9	36 50 64	(C_nH_{2n+1}OH)₂	Dialcohols, methoxyalcohols, etc.
	36	HCl	R—HCl (distinctive ³⁷Cl, primary favored)
	64	SO₂	RSO₂R, Aryl—SO₂OR[c]

[a] Δ = mass − 14n + 1 (Dromey 1976).

[b] Specific cleavages giving a major peak are usually indicative of a particular structural moiety (Chapter 4). Lower mass even-electron ions which are formed through secondary decompositions involving randomizing rearrangements ("ion series") are often of significant abundance, so that such ions are generally useful to indicate compound types, not specific structural moieties.

[c] Also see elimination rearrangement in Table 8.2.

Table A.6. Common fragment ions (mainly even-electron)

Δ^a	m/z	Formula	Compound type[b]
X	31, 50, 69, 100, 119, 131, 169, 181, 193	C_nF_m	Perfluoroalkanes
X	38, 39, 50, 51, 63, 64, 74, 75, 76	Aromatic series—low	More abundant for aromatic compounds with electronegative substituents
X	39, 40, 51, 52, 65, 66, 77, 78, 79	Aromatic series—high	More abundant with electron-donating substituents, N and O heterocyclic compounds
X	45, 57, 58, 59, 69, 70, 71, 85	Endo-sulfur aromatic series	Thiophenes
X	69, 81–84, 95–97, 107–110	Exo-sulfur aromatic series	Sulfur attached to an aromatic ring
X	73, 147, 207, 221, 281	Dimethylsiloxanes	Dimethylsiloxanes
−6	77 91 105 119 133	$C_6H_5C_nH_{2n}$	Phenylalkyl (specific cleavage, also rearrangement)
	105 119 133	$C_nH_{2n+1}C_6H_4CO$	Benzoyl (specific); 119 and above, unsaturated or cyclic phenoxy
	63 77 91	$C_nH_{2n+1}O_3$	ROCOOR-specific rearrangement loss of $(R-2H)$ or $(R-2H)+(R-H)$
	49 63 77 91 105	$C_nH_{2n}Cl$	Chloroalkyl; m/z 91 largest for $R(CH_2)_5Cl$
−5	78 92 106 120 134	$C_5H_4NC_nH_{2n}$	Pyridyl, aminoaromatic (specific cleavage, also rearrangement)
−4	79 93 107 121	C_nH_{2n-5}	Terpenes and derivatives (rearrangement; other m/z also common)
	107 121 135	$C_nH_{2n+1}C_6H_4O$	$HOC_6H_4CHR\!\!\not\!-\!Y$, $RC_6H_4OCH_2\!\!\not\!-\!Y$, $C_6H_5CH\underbrace{\;}OR$, etc.
−2	93 107 121 135	$C_nH_{2n}Br$	Bromoalkyl; m/z 135 largest for $R(CH_2)_5Br$
	95 109	C_nH_{2n-3}	Dienes, alkynes, cycloalkenes[c]
	39 53 67 81 95 109	$C_nH_{2n-1}O$	Furylalkyl (specific); polyunsaturated or cyclic oxygen compounds (alcohols, ethers, aldehydes)

285

Table A.6. *Common fragment ions (mainly even-electron)* *(continued)*

Δ[a]	m/z										Formula	Compound type[b]
−1				54	68	82	96	110	124	138	$C_nH_{2n}CN$	Alkyl cyanides; cycloalkenyl, bicycloamines
0						83	97	111	125		$C_4H_3SC_nH_{2n}$	Thiophenes (specific), thiaphenyl
		27	41	55	69	83	97	111			C_nH_{2n-1}	Alkenes, cycloalkanes[c]
				55	69	83	97				$C_nH_{2n-1}CO$	Alkenyl-, cycloalkyl carbonyl (specific); diunsaturated, cyclic alcohols, ethers[c]
+1				56	70	84	98				$C_nH_{2n}N$	Alkenyl-, cycloalkylamines, cyclic amines[c]
				56	70	84	98	112	126		$C_nH_{2n}NCO$	Alkyl isocyanates
+2	15	29	43	57	71	85	99	113			C_nH_{2n+1}	Alkyl
		29	43	57	71	85	99				$C_nH_{2n+1}CO$	Saturated carbonyl (specific); m/z > 43: cyclic ethers, cycloalkanols[c]
+3		30	44	58	72	86	100	114	128		$C_nH_{2n+2}N$	Alkyl amines (specific; also secondary rearrangement reactions)
			44	58	72	86	100				$C_nH_{2n+2}NCO$	Amides, ureas, carbamates (some specific)
					72	86	100	114			$C_nH_{2n}NCS$	Alkyl isothiocyanates
+4		31	45	59	73	87	101				$C_nH_{2n+1}O$	Aliphatic alcohols, ethers (some specific), double H rearrangement of higher ketones
			45	59	73	87	101				$C_nH_{2n-1}O_2$	Acids, esters; cyclic acetals, ketals (some specific)
						87	101	115	129 etc.		$C_nH_{2n-3}O_3$	ROCO—R'—CO$\div$Y (specific; Y = R", OR", etc.)
		31	45	59	73	87	101				$C_nH_{2n+3}Si$	Alkyl silanes (rearrangement common); m/z 73 in Me$_3$SiOR
			45	59	73	87	101				$C_nH_{2n-1}S$	Thiacycloalkanes; unsaturated, substituted sulfur compounds
				60	74	88	102				$C_nH_{2n+2}NO$	RCONHR', RCONR$_2$'-specific rearragement loss of (R' − 2H) or (R' − 2H) + (R − H) or (R' − 2H) + (R' − H)

286

Table A.6. Common fragment ions (mainly even-electron) (continued)

Δ^a m/z						Formula	Compound type[b]
+4	60	74	88	102		$C_nH_{2n}NO_2$	$R\!\!+\!\!CR_2\!-\!ONO$ (specific)
	46					NO_2	Nitrates
+5	19 33	47 61	75 89	103	117	$C_nH_{2n+3}O$	Alcohols, polyols, $ROR' \rightarrow ROH_2^+$ ($R' > n\text{-}C_5$)
	33	47 61	75 89	103	117	$C_nH_{2n+1}O_2$	RCOOR'-specific rearrangement loss of $(R' - 2H)$ or $(R' - 2H) + (R - H)$; also higher esters, etc.; ROR'OH, HOROH, ROR'OR'', ROR'COOR'' (some specific)
	33 47 (also 34, 35, 45)	61	75 89	103		$C_nH_{2n+1}S$	Alkyl thiols, sulfides (some specific)
				103	117	$C_nH_{2n-3}S_2$	Dithiacyclopentadienyl
				103	117	$C_nH_{2n+3}Si$	$(CH_3)_3SiOCHR\!\!+\!\!R'$

[a]See footnote a, Table A.5; "X" signifies a series separated by other than CH_2 units.
[b]See footnote b, Table A.5.
[c]Might also be formed by H_2 or CH_4 loss from abundant series two mass units higher.

Table A.7. *Common odd-electron fragment ions.*[a]

Δ[b]	m/z	Formula	Compound type
−7	76 90 104 118 132[c]	C_nH_{2n-8}	R-phenyl/HY*, R-phenyl/YY', R-phenyl—C_nH_{2n-1}/HY*
−6	77 91 105 119 133[c]	$C_nH_{2n-7}N$	R-pyridyl/HY*, R-phenyl—N(R')/YY', etc. (see 76, 90, . . .)
−5	92 106 120	C_nH_{2n-6}	R-phenyl—CR'R''/Z/H
	92 106 120 134[c]	$C_nH_{2n-8}O$	(—O—phenyl—R)/HY*, (—O—phenyl—R)/YY', R-phenyl—COCHR'/Z/H
−3	94 108 122	$C_nH_{2n+1}C_6H_4OH$	R-phenyl—O/Z/H, HO-phenyl(R)/Z/H
	66 80 94	$C_nH_{2n+2}S_2$	RSS/Z/H, H/Z/SS/Z/H
−1	66 80 94	C_nH_{2n-4}	Cycloalkene/YY', C_nH_{2n-4}/YY', cycloalkanone/YY'
	68 82 96 110	C_nH_{2n-2}	Cycloalkyl/HY, cycloalkyl/YY', C_nH_{2n-2}/YY', loss of $C_nH_{2n} + H_2O$ from ketones, etc.
0	41 55 69 83 97 111 125[c]	$C_nH_{2n-1}N$	Alkyl cyanides (not specific)
+1	126[c]	C_nH_{2n-14}	R-naphthyl/HY*, R-naphthyl—C_nH_{2n-1}/HY*
	42 56 70 84 98 112 126	C_nH_{2n}	CHR=CR'CHR''/Z/H, cycloalkanes, C_nH_{2n}/RY, H—C_nH_{2n}/Y*
	42 56 70 84 98 112 126	$C_nH_{2n-2}O$	2-Alkylcycloalkanone, cyclobutanones (specific); m/z 98 from Y(CH$_2$)$_{n>5}$—CO—Y
+2	98 112 126[c]	$C_nH_{2n-4}S$	R-thienyl—CHR'/Z/H
+3	85 99 113	$C_nH_{2n-1}NO$	Alkyl isocyanates (not specific)
	44 58 72 86 100	$C_nH_{2n}O$	RCOCHR'/Z/H, O=C(CHR/Z/H)$_2$, CHR=CR'O/Z/H, $C_nH_{2n}O$/HY, cycloalkanols
+4	59 101 115	$C_nH_{2n-1}NS$	Alkyl isothiocyanates
	59 73 87 101 115	$C_nH_{2n+1}NO$	RR'NCOCHR''/Z/H, HON=CRCHR'/Z/H
+5	60 74	$C_nH_{2n}S$	Cyclic sulfides, $C_nH_{2n}S$/HY
	46 60 74 88 102	$C_nH_{2n}O_2$	ROCOCHR'/Z/H, H/Z/OCOCHR/Z'/H, $C_nH_{2n}O_2$/RY

[a] Abbreviations: R, H or alkyl group; X, any halogen atom; Y, a functional group; Y*, an electron-withdrawing functional group—for example, —X, —NO$_2$, —COOR, —CN; Z, a group from which a hydrogen atom is rearranged—for example, RCOCH$_2$/Z/H could be RCOCH$_2$CH$_2$CH$_3^+$ → RC(OH)CH$_2$ + C$_2$H$_4$. Common Z groups are —CH$_2$CH$_2$—, —CH$_2$O—, —OCH$_2$—, —COCH$_2$—, —COOR, and analogs which are substituted or contain other heteroatoms.

[b] See footnote a, Table A.5.

[c] Series can continue to much higher masses.

288

Table A.8. *Common elemental compositions of molecular ions.*[a]

m/z	Composition
16	CH_4
17	H_3N
18	H_2O
26	C_2H_2
27	CHN
28	C_2H_4, CO, N_2
30	C_2H_6, CH_2O, NO
31	CH_5N
32	CH_4O, H_4N_2, H_4Si, O_2
34	CH_3F, H_3P, H_2S
36	HCl
40	C_3H_4
41	C_2H_3N
42	C_3H_6, C_2H_2O, CH_2N_2
43	C_2H_5N, HN_3
44	C_2H_4O, C_3H_8, C_2HF, CO_2, N_2O
45	C_2H_7N, CH_3NO
46	C_2H_6O, C_2H_3F, CH_6Si, CH_2O_2, NO_2
48	C_2H_5F, CH_4S, CH_5P
50	C_4H_2, CH_3Cl
52	C_4H_4, CH_2F_2
53	C_3H_3N, HF_2N
54	C_4H_6, F_2O
55	C_3H_5N
56	C_4H_8, C_3H_4O, $C_2H_4N_2$
57	C_3H_7N, C_2H_3NO
58	C_3H_6O, C_4H_{10}, $C_2H_2O_2$, $C_2H_6N_2$
59	C_3H_9N, C_2H_5NO, CH_5N_3
60	C_3H_8O, C_3H_5F, $C_2H_8N_2$, $C_2H_4O_2$, C_2H_8Si, C_2H_4S, C_2HCl, CH_4N_2O, COS
61	C_2H_7NO, CH_3NO_2, CClN
62	C_3H_7F, C_2H_7P, $C_2H_6O_2$, C_2H_6S, C_2H_3Cl
64	$C_2H_2F_2$, C_2H_5FO, C_2H_5Cl, SO_2
66	C_5H_6, $C_2H_4F_2$, CF_2O, F_2N_2
67	C_4H_5N, CH_3F_2N, ClO_2
68	C_5H_8, C_4H_4O, $C_3H_4N_2$, C_3O_2, CH_2ClF
69	C_4H_7N, C_3H_3NO, $C_2H_3N_3$
70	C_5H_{10}, C_4H_6O, $C_3H_6N_2$, CH_2N_4, CHF_3
71	C_4H_9N, C_3H_5NO, F_3N
72	C_4H_8O, C_5H_{12}, $C_3H_4O_2$, C_4H_5F
73	$C_4H_{11}N$, C_3H_7NO, C_2H_3NS, $C_2H_7N_3$

Table A.8. *Common elemental compositions of molecular ions* [a] (*continued*).

m/z	Composition
74	$C_4H_{10}O$, $C_3H_6O_2$, $C_3H_{10}N_2$, C_3H_6S, $C_2H_6N_2O$, $C_3H_{10}Si$, C_3H_3Cl, $C_2H_6N_2O$, $C_2H_2O_3$, CH_6N_4
75	C_3H_9NO, $C_2H_5NO_2$, C_2H_2ClN
76	$C_3H_8O_2$, C_3H_8S, C_3H_5Cl, C_4H_9F, C_4N_2, C_3H_9P, C_3H_5FO, $C_2H_4O_3$, C_2H_4OS, CH_8Si_2, CH_4N_2S, CS_2
77	CH_3NO_3
78	C_6H_6, C_3H_7Cl, $C_4H_2N_2$, C_2H_6OS, C_2H_3ClO, $C_2H_3FO_2$, CF_2N_2, $CH_6N_2O_2$
79	C_5H_5N
80	C_6H_8, $C_4H_4N_2$, $C_3H_6F_2$, C_2H_5ClO, C_2H_2ClF, CH_4O_2S, HBr
81	C_5H_7N, $C_3H_3N_3$, $C_2H_5F_2N$
82	C_6H_{10}, $C_4H_6N_2$, C_5H_6O, C_2H_4ClF, $C_2H_2N_4$, C_2HF_3, $CClFO$
83	C_5H_9N, $C_3H_5N_3$, C_4H_5NO
84	C_6H_{12}, C_5H_8O, $C_4H_8N_2$, $C_4H_4O_2$, $C_2H_4N_4$, C_4H_4S, $C_2H_3F_3$, CH_2Cl_2
85	$C_5H_{11}N$, C_4H_7NO, C_3H_3NS, CH_3N_5
86	$C_5H_{10}O$, $C_4H_6O_2$, C_6H_{14}, $C_4H_{10}N_2$, C_4H_6S, C_4H_3Cl, $C_3H_6N_2O$, $C_2H_2N_2S$, $CHClF_2$, HF_2O, Cl_2O, F_2OS
87	$C_5H_{13}N$, C_4H_9NO, $C_3H_9N_3$, C_3H_5NS, C_3H_2ClN, ClF_2N, F_3NO
88	$C_5H_{12}O$, $C_4H_8O_2$, $C_4H_{12}N_2$, C_4H_8S, $C_3H_8N_2O$, $C_3H_4O_3$, $C_4H_{12}Si$, C_4H_5Cl, CF_4
89	$C_4H_{11}NO$, $C_3H_7NO_2$, C_3H_4ClN
90	$C_4H_{10}O_2$, $C_4H_{10}S$, C_4H_7Cl, $C_3H_6O_3$, $C_3H_{10}OSi$, C_3H_6OS, C_3H_3ClO, $C_2H_6N_2O_2$, $C_2H_6N_2S$, $C_2H_2O_4$, $C_4H_{11}P$
91	$C_2H_5NO_3$, CH_5N_3S, C_3H_6ClN
92	C_7H_8, C_4H_9Cl, C_3H_5ClO, $C_2H_4O_2S$, $C_3H_8O_3$, C_3H_9FSi, $C_5H_4N_2$, C_6H_4O
93	C_6H_7N, C_5H_3NO, $C_4H_3N_3$
94	$C_5H_6N_2$, C_7H_{10}, C_6H_6O, C_3H_7ClO, $C_2H_2S_2$, $C_2H_6O_2S$, C_3HF_3, $C_2H_3ClO_2$, C_2Cl_2, CH_3Br
95	C_5H_5NO, C_6H_9N, $C_4H_5N_3$, C_2F_3N
96	C_7H_{12}, C_6H_8O, $C_5H_8N_2$, $C_5H_4O_2$, $C_4H_4N_2O$, $C_2H_2Cl_2$, $C_3H_3F_3$, $C_2H_6F_2Si$
97	C_5H_7NO, $C_6H_{11}N$
98	C_7H_{14}, $C_6H_{10}O$, $C_5H_6O_2$, $C_4H_6N_2O$, $C_3H_6N_4$, $C_5H_{16}N_2$, C_5H_6S, $C_4H_2O_3$, $C_2H_4Cl_2$, C_2HClF_2, CCl_2O
99	$C_6H_{13}N$, C_5H_9NO, C_4H_5NS, $C_4H_5NO_2$, CH_3F_2NS
100	$C_6H_{12}O$, $C_5H_8O_2$, C_7H_{16}, $C_4H_4O_3$, $C_3H_4N_2S$, $C_5H_{12}N_2$, C_6H_9F, $C_5H_{12}Si$, C_4H_4OS, $C_3H_4N_2O_2$, $C_2H_3F_3O$, C_2F_4
101	$C_6H_{15}N$, $C_5H_{11}NO$, $C_4H_7NO_2$, C_4H_7NS, $C_2H_3N_3S$
102	$C_6H_{14}O$, $C_5H_{10}O_2$, $C_5H_{10}S$, $C_4H_{10}N_2O$, $C_4H_6O_3$, $C_2H_2F_4$, $C_3H_6N_2S$, C_8H_6, $CHCl_2F$, CHF_3S, HF_2PS, Cl_2S
103	$C_4H_9NO_2$, C_7H_5N, $C_5H_{13}NO$, $C_4H_{13}N_3$, C_4H_6ClN, $C_2H_2ClN_3$
104	$C_5H_{12}O_2$, $C_5H_{12}S$, C_5H_9Cl, $C_4H_8O_3$, C_4H_8OS, C_8H_8, $C_6H_{13}F$, $C_6H_4N_2$, $C_4H_{12}N_2O$, $C_4H_{12}OSi$, C_4H_5ClO, $C_3H_8N_2S$, $CClF_3$, SiF_4
105	C_7H_7N, $C_3H_7NO_3$, $C_4H_{11}NO_2$, C_4H_8ClN, $CBrN$
106	C_8H_{10}, $C_5H_{11}Cl$, $C_6H_6N_2$, C_7H_6O, $C_4H_{10}O_3$, C_4H_7ClO, $C_4H_{10}OS$, $C_3H_6O_2S$, C_2H_3Br
107	C_7H_9N, C_6H_5NO, $C_2H_5NO_4$

Table A.8. *Common elemental compositions of molecular ions* [a] *(continued).*

m/z	Composition
108	$C_6H_8N_2$, C_8H_{12}, C_7H_8O, C_4H_9ClO, C_3H_9ClSi, $C_3H_3ClO_2$, $C_6H_4O_2$, $C_3H_8S_2$, C_2H_5Br, $C_2H_4O_3S$, SF_4
109	C_6H_7NO, $C_7H_{11}N$, $C_2H_4ClNO_2$
110	C_8H_{14}, $C_7H_{10}O$, $C_5H_6N_2O$, $C_3H_4Cl_2$, C_7H_7F, $C_6H_6O_2$, $C_4H_6N_4$, $C_2H_6O_3S$, $C_2H_7O_3P$, $C_3H_7ClO_2$, $C_6H_{10}N_2$
111	$C_7H_{13}N$, C_6H_9NO, $C_5H_5NO_2$, $C_4H_5N_3O$, C_2ClF_2N, C_6H_6FN, $C_5H_9N_3$, C_5H_5NS
112	C_8H_{16}, $C_7H_{12}O$, $C_6H_8O_2$, $C_6H_{12}N_2$, C_6H_8S, $C_5H_8N_2O$, $C_5H_4O_3$, $C_4H_4N_2O_2$, $C_3H_6Cl_2$, C_6H_5FO, C_6H_5Cl, C_5H_4OS, C_3F_4, $C_3H_3F_3O$, CH_2BrF
113	$C_7H_{15}N$, $C_6H_{11}NO$, $C_5H_7NO_2$, C_5H_7NS, C_5H_4ClN, $C_3H_3NO_2$, $C_2H_6F_2NP$
114	$C_7H_{14}O$, $C_6H_{10}O_2$, C_8H_{18}, $C_6H_{14}N_2$, $C_4H_6N_2O_2$, C_5H_6OS, $C_6H_4F_2$, $C_5H_2O_3$, C_6H_7Cl, $C_4H_6N_2S$, $C_2H_4Cl_2O$, C_2HCl_2F, C_2HClF_2O, $C_2HF_3O_2$
115	$C_6H_{13}NO$, $C_7H_{17}N$, $C_5H_9NO_2$, C_5H_9NS, C_4H_5NOS, $C_3H_5N_3S$, $C_5H_{13}N_3$,
116	$C_6H_{12}O_2$, $C_7H_{16}O$, $C_6H_{12}S$, C_9H_8, $C_6H_{16}N_2$, $C_6H_{16}Si$, C_6H_9Cl, $C_5H_{12}N_2O$, $C_5H_8O_3$, $C_4H_8N_2O_2$, $C_4H_4O_4$, $C_4H_8N_2S$, $C_4H_4S_2$, $C_2H_3Cl_2F$, C_2ClF_3
117	C_8H_7N, $C_6H_{15}NO$, $C_5H_{11}NO_2$, C_5H_8ClN, $C_4H_7NO_3$, $C_3H_4ClN_3$
118	$C_6H_{14}O_2$, $C_6H_{14}S$, $C_5H_{10}O_3$, C_9H_{10}, $C_7H_{16}N_2$, $C_5H_{10}OS$, $C_6H_{15}P$, $C_7H_{15}F$, $C_6H_{11}Cl$, $C_5H_{14}OSi$, $C_4H_{10}N_2O_2$, $C_4H_6O_4$, C_4H_3ClS, $C_4H_{10}N_2S$, $C_4H_{14}Si_2$, C_3H_3Br, $C_2H_2ClF_3$, $CHCl_3$
119	C_7H_5NO, $C_6H_5N_3$, C_8H_9N, $C_4H_9NO_3$, C_4H_9NOS
120	C_9H_{12}, C_8H_8O, $C_4H_8O_2S$, $C_4H_8S_2$, $C_7H_8N_2$, $C_6H_{13}Cl$, $C_5H_{12}O_3$, $C_6H_4N_2O$, $C_5H_4N_4$, $C_4H_8O_4$, C_5H_9OCl, $C_4H_{12}O_2Si$, C_3H_5Br, CCl_2F_2
121	$C_8H_{11}N$, C_7H_7NO, $C_6H_7N_3$, $C_4H_9N_5$, C_7H_4FN
122	$C_8H_{10}O$, $C_7H_{10}N_2$, $C_7H_6O_2$, $C_4H_7ClO_2$, $C_4H_{10}S_2$, C_9H_{14}, C_3H_7Br, C_8H_7F, $C_6H_6N_2O$, $C_4H_{10}O_2S$, $C_4H_4Cl_2$, $C_3H_6O_3S$, $C_2H_6N_2S_2$, C_2H_3BrO
123	C_7H_9NO, $C_6H_5NO_2$, $C_8H_{13}N$, $C_6H_9N_3$, $C_4H_4F_3N$, $C_3H_9NO_2S$
124	C_9H_{16}, $C_8H_{12}O$, $C_7H_8O_2$, C_7H_8S, $C_6H_8N_2O$, $C_5H_8N_4$, $C_4H_9ClO_2$, $C_4H_6Cl_2$, $C_3H_8O_3S$, C_8H_9F, $C_7H_{12}N_2$, C_7H_5FO, $C_5H_4N_2O_2$, $C_4H_6F_2O_2$, $C_3H_5ClO_3$, C_2H_5BrO, $C_2H_4S_3$
125	$C_8H_{15}N$, $C_6H_{11}N_3$, C_6H_7NS, $C_7H_{11}NO$, $C_6H_7NO_2$, $C_2H_8NO_3P$, $C_2H_4ClNO_3$
126	C_9H_{18}, $C_8H_{14}O$, $C_4H_8Cl_2$, $C_7H_{10}O_2$, $C_5H_6N_2O_2$, C_7H_7Cl, $C_6H_{10}N_2O$, $C_6H_6O_3$, $C_7H_{10}S$, $C_3H_4Cl_2O$, $C_2H_6S_3$, $C_7H_{14}N_2$, C_7H_7FO, C_6H_6OS, $C_4H_6N_4O$, $C_4H_5F_3O$, $C_3H_2N_6$, $C_2Cl_2O_2$
127	$C_7H_{13}NO$, $C_8H_{17}N$, C_6H_2ClN, $C_5H_9N_3O$, $C_5H_5NO_3$, $C_6H_9NO_2$, $C_4H_5N_3S$, C_2Cl_2FN
128	$C_8H_{16}O$, C_9H_{20}, $C_7H_{12}O_2$, $C_7H_{12}S$, $C_8H_4N_2$, $C_6H_8O_3$, C_6H_8OS, C_6H_5ClO, $C_4H_4N_2OS$, $C_3H_6Cl_2O$, $C_{10}H_8$, $C_2H_6Cl_2Si$, $C_2H_2Cl_2O_2$, CH_2BrCl, $C_2H_5ClO_2S$, $C_8H_{13}F$, HI
129	$C_8H_{19}N$, $C_7H_{15}NO$, $C_6H_{11}NO_2$, C_9H_7N, $C_7H_3N_3$, $C_6H_{15}N_3$, $C_5H_7NO_3$, C_5H_7NOS, $C_4H_7N_3S$, $C_4H_4ClN_3$, $C_4H_3NO_2S$
130	$C_7H_{14}O_2$, $C_8H_{18}O$, $C_6H_{10}O_3$, $C_8H_6N_2$, $C_{10}H_{10}$, C_9H_6O, $C_7H_{14}S$, $C_6H_{14}N_2O$, C_6H_4ClF, $C_5H_6O_4$, $C_5H_6S_2$, $C_3H_5Cl_2F$, $C_5H_{10}N_2O_2$, $C_3H_6N_4S$, $C_3H_2ClF_3$, $C_3H_2N_2O_2S$, C_2HCl_3, $CHBrF_2$
131	C_9H_9N, $C_7H_{17}NO$, $C_5H_9NO_3$, $C_7H_5N_3$, $C_6H_{17}N_3$, $C_6H_{13}NO_2$, $C_4H_9N_3S$, CF_3NOS
132	$C_7H_{16}O_2$, $C_6H_{12}O_3$, $C_{10}H_{12}$, C_9H_8O, $C_7H_{16}S$, $C_8H_8N_2$, $C_6H_{16}OSi$, $C_5H_8O_4$, $C_6H_{12}OS$, C_6H_9ClO, $C_5H_{12}N_2O_2$, $C_5H_{12}N_2S$, $C_5H_8O_2S$, $C_5H_5ClO_2$, $C_4H_4OS_2$, $C_3H_{12}Si_3$, $C_3H_4N_2S_2$, $C_2H_3Cl_3$, $C_2Cl_2F_2$, $C_2F_4O_2$
133	$C_9H_{11}N$, C_8H_7NO, $C_7H_7N_3$, $C_5H_{11}NO_3$, $C_4H_7NO_2S$, C_3H_4BrN, $C_2H_3ClF_3N$
134	$C_9H_{10}O$, $C_{10}H_{14}$, $C_6H_{14}O_3$, $C_6H_6N_4$, $C_5H_{10}O_2S$, $C_8H_6O_2$, $C_8H_{10}N_2$, C_8H_6S, $C_7H_{15}Cl$, $C_7H_6N_2O$, $C_6H_{14}OS$, $C_6H_{11}ClO$, $C_5H_{11}ClSi$, $C_5H_{10}S_2$, $C_5H_7ClO_2$, $C_3H_3F_5$, $C_3Cl_2N_2$, $C_2H_2Cl_2F_2$

Table A.8. *Common elemental compositions of molecular ions* [a] *(continued)*.

m/z	Composition
135	$C_9H_{13}N$, C_8H_9NO, C_7H_5NS, $C_5H_5N_5$, $C_7H_5NO_2$, $C_6H_5N_3O$, $C_4H_9NO_2S$, C_3H_6BrN, $C_3F_3N_3$
136	$C_9H_{12}O$, $C_{10}H_{16}$, $C_8H_8O_2$, C_4H_9Br, $C_8H_{12}N_2$, $C_8H_{12}Si$, C_8H_8S, C_8H_5Cl, $C_7H_8N_2O$, $C_7H_4O_3$, $C_5H_{12}S_2$, $C_6H_4N_2O_2$, $C_6H_4N_2S$, $C_5H_{12}O_4$, $C_5H_{12}OS$, $C_5H_9ClO_2$, $C_5H_4N_4O$, C_2HClF_4, CCl_3F
137	$C_8H_{11}NO$, $C_9H_{15}N$, $C_7H_7NO_2$, C_7H_7NS, C_7H_4ClN, $C_6H_7N_3O$, $C_5H_6F_3N$, $C_5H_3N_3S$, $C_3H_7NO_5$
138	$C_{10}H_{18}$, $C_9H_{14}O$, $C_8H_{10}O_2$, $C_8H_{10}S$, $C_7H_6O_3$, $C_6H_6N_2O_2$, C_8H_7Cl, $C_7H_{10}N_2O$, C_7H_6OS, $C_6H_{10}N_4$, $C_5H_6N_4O$, $C_4H_{11}O_3P$, $C_4H_{10}O_3S$, C_3H_7BrO, $C_3H_6S_3$, $C_2H_3BrO_2$, C_2F_6
139	C_7H_9NS, $C_9H_{17}N$, $C_7H_{13}N_3$, $C_7H_9NO_2$, $C_6H_5NO_3$, $C_5H_5N_3O_2$, C_6H_5NOS
140	$C_9H_{16}O$, $C_{10}H_{20}$, C_8H_9Cl, $C_8H_{12}O_2$, $C_8H_{12}S$, $C_7H_8O_3$, $C_5H_{10}Cl_2$, C_7H_8OS, C_8H_9FO, $C_8H_{16}N_2$, $C_8H_6F_2$, C_7H_5ClO, $C_6H_8N_2O_2$, $C_6H_8N_2S$, $C_6H_4O_4$, $C_6H_4S_2$, $C_4H_6Cl_2O$, C_2H_2BrCl
141	$C_8H_{15}NO$, $C_7H_{11}NO_2$, $C_6H_{11}N_3O$, $C_6H_7NO_3$, $C_9H_{19}N$, $C_7H_{15}N_3$, $C_5H_7N_3S$, $C_4H_6F_3NO$, $C_4H_3F_4N$
142	$C_8H_{14}O_2$, $C_9H_{18}O$, $C_{10}H_{22}$, $C_6H_6O_4$, $C_8H_{14}S$, $C_3H_4Cl_2O_2$, C_7H_7ClO, $C_{11}H_{10}$, $C_9H_{15}F$, $C_8H_{18}N_2$, $C_7H_{14}N_2O$, $C_7H_{10}O_3$, $C_6H_{10}N_2O_2$, $C_6H_6O_2S$, $C_5H_6N_2OS$, $C_4H_8Cl_2O$, $C_4H_5ClF_2O$, C_2H_4BrCl, C_2HBrF_2, CH_3I
143	$C_{10}H_9N$, $C_8H_{17}NO$, $C_7H_{13}NO_2$, $C_9H_{21}N$, $C_7H_{10}ClN$, $C_6H_{13}N_3O$, $C_6H_9NO_3$, C_4H_9NOS, $C_5H_9N_3S$, C_2Cl_3N
144	$C_8H_{16}O_2$, $C_9H_{20}O$, $C_9H_8N_2$, $C_7H_{12}O_3$, $C_6H_8O_4$, $C_{10}H_8O$, $C_{11}H_{12}$, $C_8H_{16}S$, $C_8H_{13}Cl$, $C_7H_{16}N_2O$, $C_7H_{12}OS$, $C_6H_9ClN_2$, $C_5H_8N_2O_3$, $C_3H_3Cl_3$
145	$C_{10}H_{11}N$, C_9H_7NO, $C_8H_7N_3$, $C_7H_{15}NO_2$, $C_6H_{11}NO_3$, $C_5H_{11}N_3S$, $C_3H_3N_3O_2S$
146	$C_8H_{18}O_2$, $C_7H_{14}O_3$, $C_8H_{18}S$, $C_6H_{10}O_4$, $C_6H_{10}O_2S$, $C_{11}H_{14}$, $C_{10}H_{10}O$, $C_9H_{10}N_2$, $C_{10}H_7F$, $C_8H_6N_2O$, $C_7H_{18}OSi$, $C_7H_{14}OS$, $C_7H_6N_4$, $C_6H_{14}N_2O_2$, $C_6H_4Cl_2$, $C_5H_6OS_2$, $C_3H_5Cl_3$, $C_3H_3BrN_2$, C_2HCl_3O, $CHBrClF$
147	$C_{10}H_{13}N$, C_9H_9NO, $C_8H_9N_3$, $C_8H_5NO_2$, $C_7H_5N_3O$, $C_6H_{13}NO_3$, $C_5H_9NO_4$, $C_2H_2BrN_3$
148	$C_{11}H_{16}$, $C_{10}H_{12}O$, $C_9H_8O_2$, $C_7H_8N_4$, $C_9H_{12}N_2$, $C_6H_{12}O_2S$, $C_8H_{17}Cl$, $C_8H_8N_2O$, $C_8H_4O_3$, C_9H_8S, $C_7H_{16}O_3$, $C_7H_{16}OS$, $C_7H_4N_2O_2$, $C_6H_{12}O_4$, $C_6H_4N_4O$, $C_5H_8O_3S$, C_3HF_5O, $C_2H_3Cl_3O$, C_2Cl_3F, $CBrF_3$
149	$C_{10}H_{15}N$, $C_9H_{11}NO$, $C_7H_7N_3O$, $C_8H_{11}N_3$, $C_8H_7NO_2$
150	$C_{10}H_{14}O$, $C_9H_{10}O_2$, $C_6H_{14}S_2$, $C_8H_{10}N_2O$, $C_6H_{11}ClO_2$, $C_5H_{11}Br$, $C_{11}H_{18}$, $C_8H_{14}N_2$, $C_9H_{10}S$, $C_8H_6O_3$, $C_7H_6N_2S$, $C_6H_{14}O_4$, $C_6H_{14}O_2S$, $C_6H_6N_4O$, $C_6H_2F_4$, C_3F_6, $C_2H_2Cl_3F$
151[b]	(see m/z 137) $C_7H_5NO_3$
152	(see m/z 138) C_8H_5ClO, $C_6H_{10}Cl_2$, CCl_4
153	(see m/z 139) $C_9H_{15}NO$, C_7H_5ClNO
154	(see m/z 140; $C_{10}H_{18}O$ has highest occurrence of data base) C_8H_7ClO, C_2ClO, C_2ClF_5, $C_8H_{10}OS$
155	(see m/z 141) $C_8H_{13}NS$
156	(see m/z 142) $C_7H_5ClO_2$, C_6H_5Br, $C_6H_4O_5$
157	(see m/z 143) C_5H_4BrN
158	(see m/z 144) CF_6S
159	(see m/z 145) $C_6H_9NS_2$, $C_5H_6ClN_3O$
160	(see m/z 146) $C_7H_{20}Si_2$, C_6H_9Br, $C_5H_{16}Si_3$
161	(see m/z 147) $C_6H_5Cl_2N$

Table A.8. *Common elemental compositions of molecular ions* [a] *(continued).*

m/z	Composition
162	(see m/z 148) $C_{10}H_7Cl$, $C_8H_6N_2S$, $C_6H_{11}Br$, C_4F_6, $C_6H_4Cl_2O$, $CHBrCl_2$
163	(see m/z 149) $C_7H_9N_5$, C_3H_2BrNS, CCl_3NO_2
164	(see m/z 150) C_9H_8OS, $C_8H_8N_2O_2$, $C_8H_5ClN_2$, $C_7H_{16}O_4$, $C_6H_4N_4O_2$, C_5H_9BrO, $C_5H_8O_2S_2$, C_2Cl_4, $CBrClF_2$
165	(see m/z 137, 151) $C_6H_7N_5O$
166	(see m/z 138, 152) $C_{13}H_{10}$, $C_8H_6O_4$, $C_8H_6O_2S$, $C_8H_6S_2$, $C_6H_6N_4O_2$, $C_6H_6N_4S$, $C_3H_6N_2O_6$, C_3ClF_5, C_3F_6O
167	(see m/z 139, 153) $C_7H_{13}N_5$, $C_7H_5NO_4$, $C_7H_5NS_2$
168	(see m/z 140, 154) $C_{13}H_{12}$, $C_9H_{16}N_2O$, $C_8H_8O_4$, $C_8H_8O_2S$, $C_8H_8S_2$, $C_6H_4N_2O_4$, C_6HF_5, C_3H_5I, $C_2HCl_3F_2$, Cl_4Si
169	(see m/z 141, 155) C_8H_8ClNO, $C_7H_7NO_2S$
170	(see m/z 142, 156) $C_9H_{16}NO_2$, $C_8H_{10}S_2$, $C_3H_6S_4$, $C_2Cl_2F_4$, C_2F_6S
171	(see m/z 143, 157) $C_7H_9NO_2S$, $C_6H_9N_3O_3$, $C_5H_2ClN_3S$
172	(see m/z 144, 158) $C_{12}H_9F$, $C_8H_6Cl_2$, $C_8H_9ClO_2$, $C_7H_8O_3S$, $C_6H_8N_2O_2S$, C_6H_5BrO, $C_6H_4O_6$, CH_2Br_2
•173	(see m/z 145, 159) C_6H_4ClNOS, $C_5H_8ClN_5$
174	(see m/z 146, 160) $C_{11}H_{10}S$, $C_{10}H_6O_3$, $C_{10}H_6OS$, $C_9H_6N_2O_2$, C_6H_4BrF, C_5F_6
175	(see m/z 147, 161) $C_8H_5N_3O_2$, $C_5H_9N_3S_2$, $C_7H_{10}ClNO_2$

[a]Compositions are listed by m/z value, ranked in decreasing order of occurrence probability for compounds in the *Registry of Mass Spectral Data* (Stenhagen *et al.* 1974). Only the more probable combinations of the elements H, C, N, O, F, Si, P, S, Cl, Br, and I are included. Note that these are *odd-electron* ion compositions; many common even-electron fragment ions have compositions differing by ±1 hydrogen atom, and can therefore be found ±1 mass unit from those listed. Tables A.5, A.6, and A.7 can also be used to suggest possible elemental compositions of fragment ions, and a much more complete list with possible structural assignments and occurrence probabilities is given in McLafferty and Venkataraghavan (1982).

[b]As can be seen above, the most probable molecular ion compositions tend to fall in homologous series. For masses above 150 the only compositions included are those for which a corresponding composition differing by having CH_2 less units is not included in the list at the next lower homologous mass (14 mass units less).

INDEX